Lecture Notes in Networks and Systems 658

The series "Lecture Notes in Networks and Systems" publishes the latest developments in Networks and Systems—quickly, informally and with high quality. Original research reported in proceedings and post-proceedings represents the core of LNNS.

Volumes published in LNNS embrace all aspects and subfields of, as well as new challenges in, Networks and Systems.

The series contains proceedings and edited volumes in systems and networks, spanning the areas of Cyber-Physical Systems, Autonomous Systems, Sensor Networks, Control Systems, Energy Systems, Automotive Systems, Biological Systems, Vehicular Networking and Connected Vehicles, Aerospace Systems, Automation, Manufacturing, Smart Grids, Nonlinear Systems, Power Systems, Robotics, Social Systems, Economic Systems and other. Of particular value to both the contributors and the readership are the short publication timeframe and the world-wide distribution and exposure which enable both a wide and rapid dissemination of research output.

The series covers the theory, applications, and perspectives on the state of the art and future developments relevant to systems and networks, decision making, control, complex processes and related areas, as embedded in the fields of interdisciplinary and applied sciences, engineering, computer science, physics, economics, social, and life sciences, as well as the paradigms and methodologies behind them.

Indexed by SCOPUS, INSPEC, WTI Frankfurt eG, zbMATH, SCImago.

All books published in the series are submitted for consideration in Web of Science.

For proposals from Asia please contact Aninda Bose (aninda.bose@springer.com).

Sotir Sotirov · Tania Pencheva ·
Janusz Kacprzyk · Krassimir T. Atanassov ·
Evdokia Sotirova · Simeon Ribagin
Editors

Recent Contributions to Bioinformatics and Biomedical Sciences and Engineering

Springer

Editors
Sotir Sotirov
Faculty of Technical Sciences
"Prof. Assen Zlatarov" University
Bourgas, Bulgaria

Janusz Kacprzyk
Polish Academy of Sciences
Systems Research Institute
Warsaw, Poland

Evdokia Sotirova
Faculty of Technical Sciences
"Prof. Assen Zlatarov" University
Bourgas, Bulgaria

Tania Pencheva
Institute of Biophysics and Biomedical Engineering
Bulgarian Academy of Sciences
Sofia, Bulgaria

Krassimir T. Atanassov
Institute of Biophysics and Biomedical Engineering
Bulgarian Academy of Sciences
Sofia, Bulgaria

Simeon Ribagin
Institute of Biophysics and Biomedical Engineering
Bulgarian Academy of Sciences
Sofia, Bulgaria

ISSN 2367-3370 ISSN 2367-3389 (electronic)
Lecture Notes in Networks and Systems
ISBN 978-3-031-31068-3 ISBN 978-3-031-31069-0 (eBook)
https://doi.org/10.1007/978-3-031-31069-0

This Springer imprint is published by the registered company Springer Nature Switzerland AG
The registered company address is: Gewerbestrasse 11, 6330 Cham, Switzerland

Preface

The present book collects selected and peer-reviewed papers from the Second *International Symposium on Bioinformatics and Biomedicine* (*BioInfoMed'2022*) held in Burgas, Bulgaria, on October 5–7, 2022. More than 50 abstracts papers were submitted for the symposium scientific forum, yet only those which were topically relevant and scientifically sound were considered and accepted.

Six invited lecturers, highly reputed scientists from different countries around the world—Australia, New Zealand, Mexico, the UK, France, and Belgium—took part online, while the invited speaker from the UK succeeded to present her lecture in presence. More than 60 participants, including a large number of young scientists, made 39 oral or poster presentations in presence and gave new ideas about the future of bioinformatics and biomedical engineering. The undeniable success of *BioInfoMed'2022* was mainly due to the high-quality investigations in the fields of biomedicine and human medicine, health care, bioinformatics, artificial intelligence, and mathematical modeling, which were submitted for presentation by the Bulgarian and the international scientific community.

Some of the presentations at the *BioInfoMed'2022* symposium have been distinguished with diplomas, and the authors have been especially invited for full-length paper submissions. After a double-blind peer-reviewed process, the current book comprises 29 selected papers as the most representative of the topics and the scientific level of the symposium.

The book is organized in accordance with the topics of the symposium, namely:

- Data Mining in Biomedicine and Health care
- Decision making in biomedicine and health care
- Biomedical approaches and applications
- Mathematical modelling in biomedicine and health care

At the end of *BioInfoMed'2022,* all members of the program and organizing committees of the Second *International Symposium on Bioinformatics and Biomedicine* expressed a strong desire and determination to make the needed efforts for this successful event to be the starting point of a biennial forum for exchanging novel knowledge in the extremely fast developing topics in the fields of bioinformatics and biomedicine.

The Editors

Contents

Data Mining in Biomedicine and Healthcare

Rhythm Analysis During Cardio-Pulmonary Resuscitation with Convolutional and Recurrent Neural Networks Using ECG and Optional Impedance Input

Vessela Krasteva and Irena Jekova(✉)

Institute of Biophysics and Biomedical Engineering, Bulgarian Academy of Sciences, Acad. G, Bonchev str. bl. 105, 1113 Sofia, Bulgaria
{vessika,irena}@biomed.bas.bg

Abstract. Chest compressions (CC) during cardiopulmonary resuscitation (CPR) produce strong artifacts in the electrocardiogram (ECG) via defi-pads. Heart rhythm can hardly be determined visually, but also by the shock-advisory algorithms in automated external defibrillators (AED). This study aims to investigate the potential of deep neural networks (DNN) as a powerful unsupervised feature extraction and classification algorithm that can give a shock advisory decision during CPR, regardless the CC fraction in the analysis interval. Our research objective is focused on detecting whether the rhythm is shockable or non-shockable from the primary raw ECG input, but also to verify the hypothesis that the secondary impedance (IMP) channel, which is generally modulated by the thorax movements and correlated to CC artifacts may contribute to performance. We designed 7 DNN architectures for processing of one (ECG) or two (ECG, IMP) input channels, involving fully-convolutional or convolutional-recurrent layers (LSTM or BiLSTM). In 30:2 compression-to-ventilation CPR during out-of-hospital cardiac arrest, the start of the CC period preceding the regular AED rhythm analysis was used as a time anchor to extract 62987 CPR strips (ECG, IMP) at 5 offsets (–10 s, –5 s, 0 s, +5 s, +10 s), thus representative of different CC durations in the analysis interval (10 s). They are divided patient-wise to training/test datasets: 1797/1747 ventricular fibrillations (VF), 730/768 normal sinus rhythms (NSR), 8583/8226 other non-shockable rhythms (ONR), 21609/19527 asystoles. Comparative study rejects the hypothesis that the impedance contributes to efficiency of rhythm analysis during CPR, considering that DNNs with two (ECG, IMP) inputs have specificity drop up to 3% points for non-shockable rhythms compared to one (ECG) input. The use of a recurrent layer after fully-convolutional architecture adds about 1% improvement in VF sensitivity (93.8%), keeping compatible specificity for Asystole (95.6%), NSR (99.2%), ONR (96.8%). The applied deep learning strategy justifies that convolutional-recurrent DNN architectures with a single ECG input are able to satisfy AHA recommendations for rhythm analysis with an arbitrary CC fraction distribution during CPR.

Keywords: Deep learning · ECG · impedance · AED · ventricular fibrillation · chest compressions · CPR · CNN · LSTM · BiLSTM

S. Sotirov et al. (Eds.): BioInfoMed 2022, LNNS 658, pp. 3–15, 2023.
https://doi.org/10.1007/978-3-031-31069-0_1

1 Introduction

Chest compressions (CC) during cardiopulmonary resuscitation (CPR) produce strong artifacts in the electrocardiogram (ECG) acquired through the defi-pads [1]. Although CC are provided at a periodic rate (100–120 min^{-1}) [2], their behavior in the ECG channel is quite nondeterministic in respect to waveform patterns and frequency content, mostly depending on the personal rescuer abilities to apply and to continuously manage cardiac massage. ECG rhythm overlapping with CC artifacts can hardly be determined visually by experts, but also automatically by the shock-advisory algorithms implemented in automated external defibrillators (AED). Existing approaches for CC artefact suppression mostly rely on adaptive filters that use external reference signals from sensors for impedance or compression depth [3–11]; compression force [12]; compression acceleration [13]; arterial blood pressure [14], etc. Several approaches seeking a simpler AED implementation use only one ECG input, where periodic CC artifacts are suppressed by pattern matching algorithms [15], coherent line removal [16], Kalman filter [17], condition-based filtering algorithm based on the ECG spectrum [18], etc. Whether with or without reference signals, the aforementioned methods have common disabilities to output filtered ECG signals either with insufficiently suppressed CC artefact components or distorted ECG waves. These warped ECG signals limit the accuracy of the automated shock advisory algorithms whether they are conventionally trained for non-artefacted ECG signals [6, 8–10, 12, 13, 18] or apply specially optimized decision rules for detection of ventricular fibrillation (VF) during CPR [15, 19].

Presently, technologies for rhythm analysis during CPR follow two strategies:

1. Two-stage algorithms are implemented in the real-time AED analysis process during out-of-hospital cardiac arrest (OHCA) interventions, applying the first stage during uninterrupted CC (analysis duration 11–30 s), eventually followed by a second reconfirmation stage on clean ECG (5–9 s). Delayed shock decision with reconfirmation analysis is required in 26–100% of OHCA interventions analyzed by several commercial AED algorithms [20–23]. Such two-stage schemes demand synchronization with additional algorithms for detection of the start and stop of CC in a standard CPR protocol with compression-to-ventilation ratios of 30:2, 15:2, 15:1 [24–26].
2. Single-stage algorithms based on deep neural networks (DNN) are run in PC workstations with OHCA databases during CPR. The DNN input feature maps and architectures depend on study-specific processing concepts, e.g. supplying unfiltered raw ECG signals with continuous CC artifacts to the input of fully-convolutional neural networks (CNN) [27], or pre-filtered raw ECG signals to CNN [28, 29], or a hybrid DNN architecture, including a combination of convolutional layers, residual blocks and bidirectional Long short-term memory (LSTM) layers [30].

Neither the two-stage algorithms nor DNN models trained on ECG during uninterrupted CC can benefit from analyzing the rhythm during the short insufflation periods, even though they are the unique time-slots with clean ECG samples that are commonly used by experts for visual determination of the rhythm during CPR. Although there are shock advisory technologies with reliable performance for short analysis intervals

(3–10 s) on clean ECG [26, 31, 32], there are certain limitations that restrict the real-life application of this technology during insufflations in OHCA. These limitations are mainly related to uncontrollable factors, concerning the ECG signal quantity (indefinite or very short duration of insufflations, typically < 3 s) and quality (presence of movement artefacts, unreliable localization using ECG and/or impedance signal). Also considering that the rhythm may spontaneously convert from non-shockable to shockable (refibrillation) or vice-versa (return of spontaneous circulation or conversion of a shockable rhythm to asystole) at any time during the OHCA resuscitation procedure [33], continuous rhythm monitoring during CPR may be of particular benefit to patients' outcome by early treatment of refibrillation (stop of CC and minimal pre-shock pauses) or by maintaining uninterrupted CPR for asystole and organized rhythms (minimum hands-off time, maximum CC fraction).

This study aims to investigate the DNN as a state-of-the-art powerful unsupervised feature extraction and classification algorithm that can give a shock advisory decision during CPR, regardless the CC fraction in the analysis interval. Our research objective is focused on detecting whether the rhythm is shockable or non-shockable from the primary raw ECG input in AED, but also to verify the hypothesis that the secondary impedance (IMP) channel, which is generally modulated by the thorax movements and correlated to CC artifacts may contribute to performance. To confirm or reject this hypothesis, different DNN architectures with one input (ECG) or two parallel inputs (ECG, IMP) are compared in terms of their shock advisory performance during CPR.

2 Database

The database with CPR contaminated ECG and IMP signals were collected with commercial AEDs (DEFIGARD TOUCH 7, Schiller Medical SA, France) used during OHCA interventions by the fire-fighters of Paris in the period January–December 2017. The reanimation protocol applied CPR with 30:2 compression to ventilation ratio and CC rate of 100–120 min^{-1}, paused every 2 min for regular AED rhythm analysis, as described in the European Research Council (ERC) Adult Basic Life Support guidelines [34]. Both ECG and IMP signals were acquired from the defi-pads in antero-apical position with a resolution: ECG (500 Hz, 4.9 uV/LSB, 1–30 Hz), IMP (125 Hz, 12.2 mΩ/LSB). This study considers two AED events recorded during the OHCA intervention: start of regular AED analysis (*StartAnAED*) and start of CC period (*StartCC*).

StartAnAED events were used to identify ECG rhythms during regular AED analysis, which were visually reviewed and annotated by consensus of three experts according to the AHA classification scheme [35]: VF (coarse ventricular fibrillation with amplitude $> 200\ \mu$V), NSR (normal sinus rhythm with visible P-QRS-T waves and heart rate 40–100 bpm), ONR (other non-shockable rhythm, including atrial fibrillation/flutter, sinus bradycardia, supraventricular tachycardia, premature ventricular contractions, heart blocks, etc.), Asystole (low-amplitude ECG with peak-to-peak signal deflection $\leq 100\ \mu$V during more than 4 s). All cases with lack of annotation consensus or noise during regular AED analysis were not used in this study.

The shock advisory decision (shockable for VF, non-shockable for NSR, ONR and Asystole) was verified to be consistent over the 30 s interval preceding the *StartAnAED* event. The nearest *StartCC* event preceding *StartAnAED* event was searched in this scope interval, and if found, then five signal strips (ECG, IMP) of 10 s duration were extracted at offsets (*StartCC, StartCC* ± *5 s and StartCC* ± *10 s)*, where each offset was valid if it was within the specified scope interval of 40 s around *StartAnAED* [–30 s; + 10 s]. As illustrated in the three examples of VF, ONR and Asystole (Fig. 1), CC periods have arbitrary duration and location within the extracted 10 s signal strips, thereby creating a CPR database with a random CC fraction distribution.

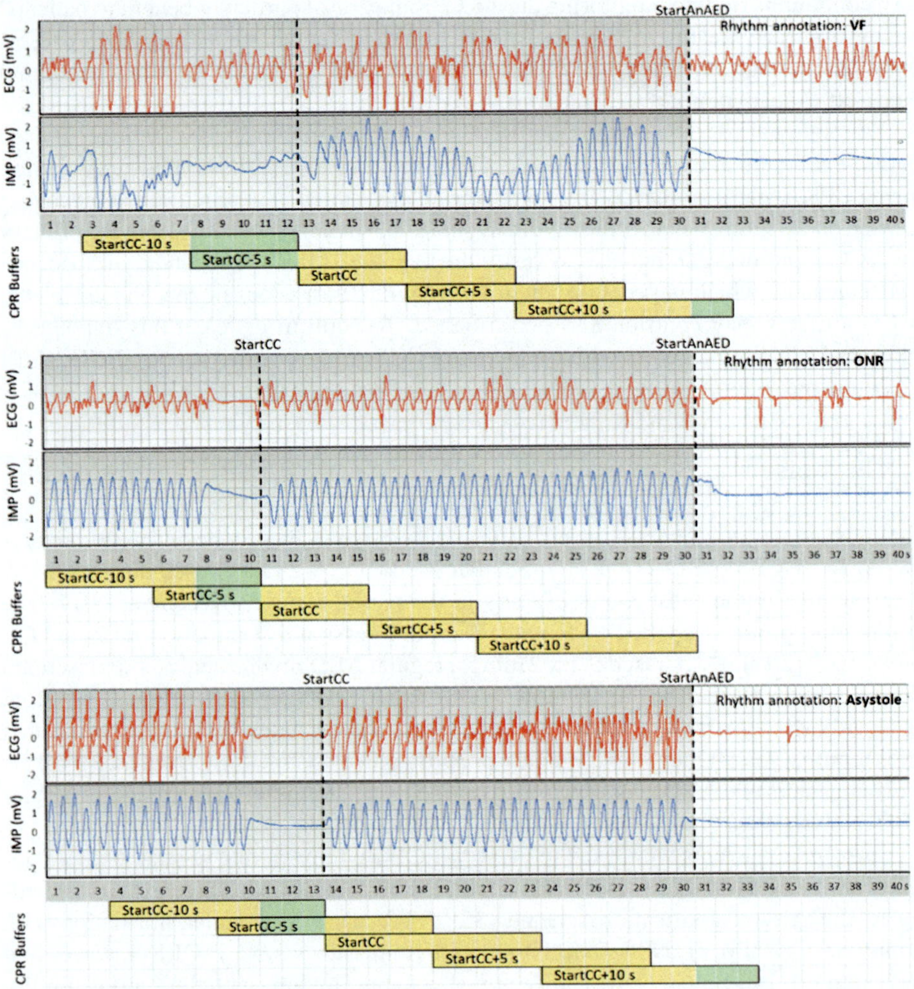

Fig. 1. Illustration of the concept for CPR buffer extraction (ECG – red trace, IMP – blue trace) from OHCA interventions in the scope interval *StartAnAED* [–30 s; + 10 s] with five offsets (*StartCC*, *StartCC* ± 5 s, *StartCC* ± 10 s). CC-fraction in CPR buffers is indicated by different colors (yellow during CC, green during no CC). ECG rhythm (VF, ONR, Asystole) is annotated by observation of 10 s clean ECG part during AED analysis (after the event *StartAnAED*).

Using the defined above scheme, a total number of 62987 CPR strips with 10s buffers (ECG, IMP) were extracted, including 3544 (VF), 1498 (NSR), 16809 (ONR), 41136 (Asystoles). As detailed in Table 1, the database was patient-wise split in two relatively equal parts for independent learning and test. The learning database was additionally partitioned to two training/validation subsets in a ratio 70/30%.

Table 1. CPR database with 10 s (ECG, DZ) divided to training, validation and test subsets.

Dataset		VF	NSR	ONR	Asystole
Learning		**1797**	**730**	**8583**	**21609**
	Training set (70%)	1228	478	5889	14921
	Validation set (30%)	569	252	2694	6688
Test		**1747**	**768**	**8226**	**19527**

3 Methods

3.1 DNN Model Design

The DNN input is with size of 1250 × 1 (one channel ECG) or 1250 × 2 (two channel ECG and IMP), considering 10 s analysis interval and 125 Hz sampling rate. Both signals are used without pre-filtering in the original AED acquisition bandwidth. Z-score normalization is applied by removing the mean and scaling to unit variance:

$$ECG = \frac{ECG - mean(ECG)}{ECGstd.dev.},\ IMP = \frac{IMP - mean(IMP)}{IMPstd.dev.} \tag{1}$$

where the denominator is the standard deviation computed for all training samples.

The basic DNN design uses a fully-convolutional architecture, which hyperparameters have been optimized in a previous study for VF detection during uninterrupted CC [27]. This basic model, denoted as *Seq3CONV (ECG)* in Fig. 2 has one ECG input fed to a sequence of three convolutional blocks – $CONV_1$(5, 10, 2, 0.3), $CONV_2$(25, 20, 2, 0.3), $CONV_3$(50, 20, 1, 0.3), where $CONV_i$(number of filters, kernel size, max pooling size, dropout rate) denotes the layout of each convolutional block, including convolutional, activation (ReLU), max-pooling and dropout layers. The last layer includes one dense neuron with a binary probability output [0–1] for prediction of shockable rhythm.

In order to study the effect of impedance in addition to ECG signal, we design three modifications of the basic architecture with two input channels (ECG, IMP), including: (1) A sequential model with joint input, denoted as *Seq3CONV (ECG, IMP)* in Fig. 2; (2) Two models with parallel branches for ECG and IMP, which are concatenated in different layers, denoted in Fig. 3 as *Par1CONV + Seq2CONV (ECG, IMP)* (with concatenation after the first convolutional block) and *Par3CONV (ECG, IMP)* (with concatenation after the third (last) convolutional block).

In order to study the effect or recurrent layers, we include LSTM and bidirectional BiLSTM layers after the third convolutional block in three architectures, denoted as

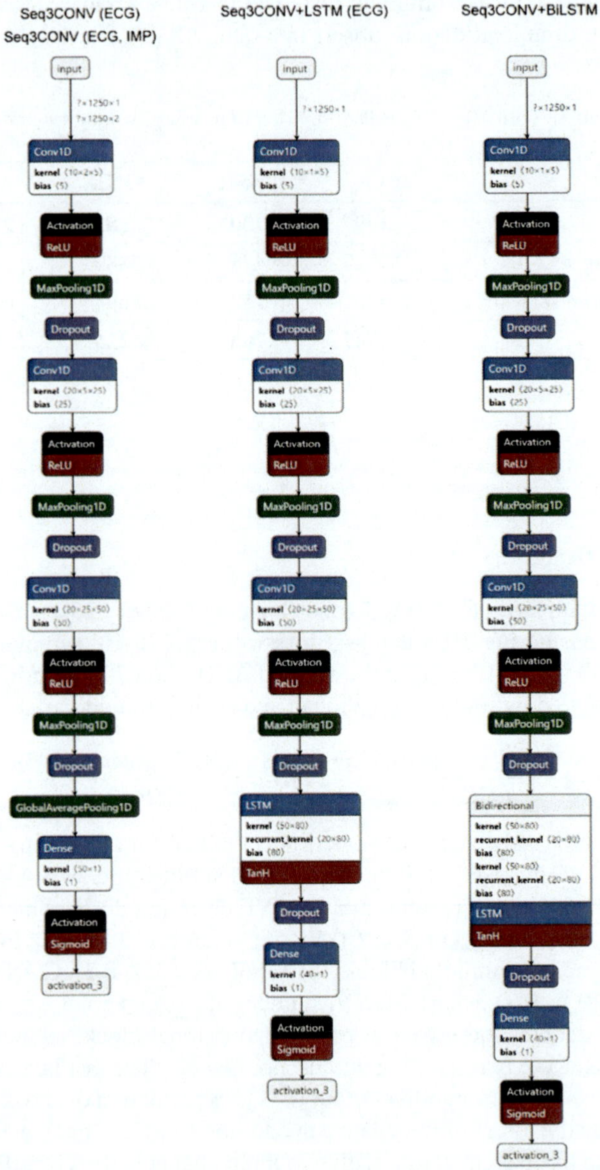

Fig. 2. Visualization of three sequential DNN architectures for VF detection during CPR. The input size is denoted as (? × 1250 × 1), corresponding to batch size (?), number of time samples (1250 in 10 s), number of channels (1) for single ECG input. The most left DNN architecture is also trained with (ECG, IMP) input, denoted as input size (? × 1250 × 2).

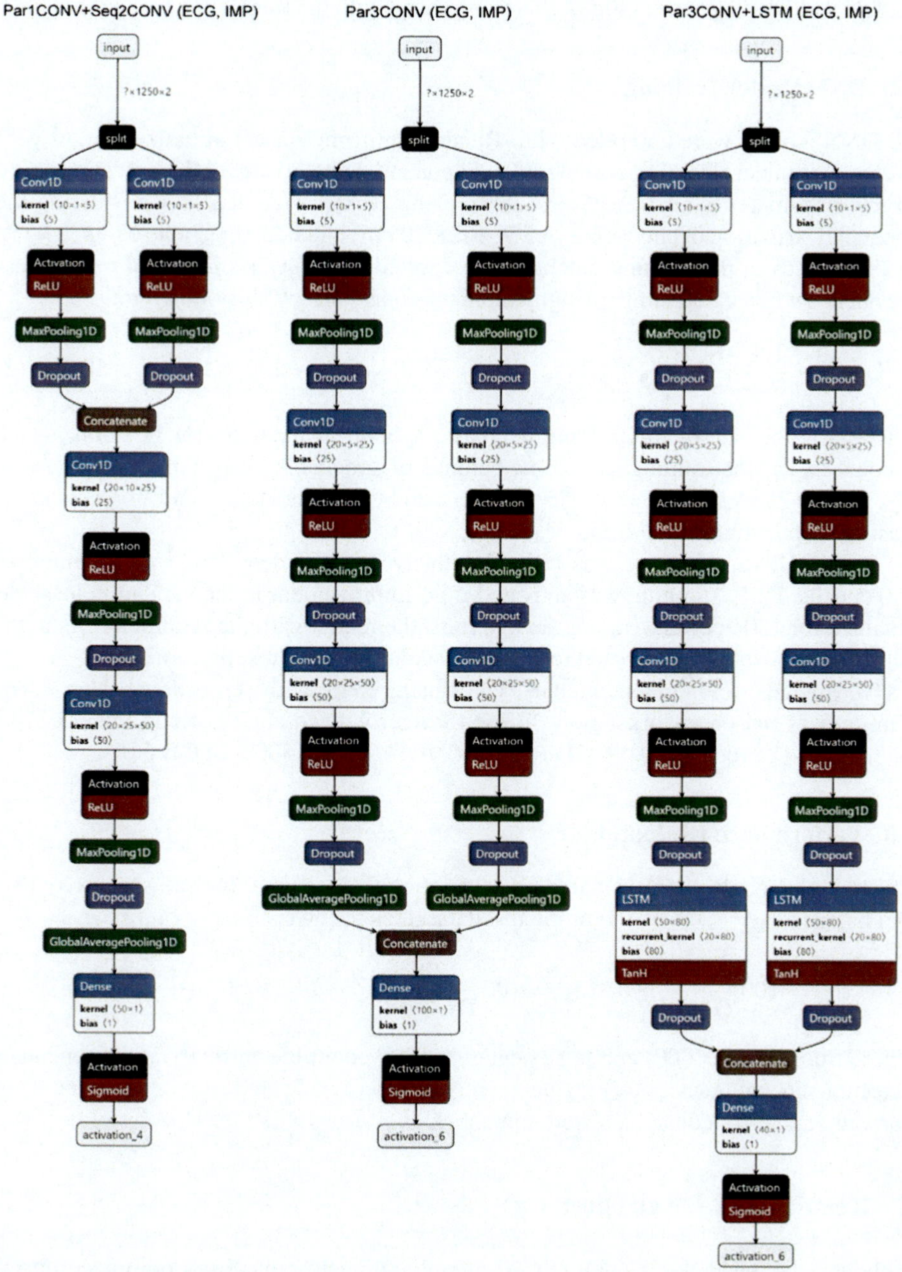

Fig. 3. Visualization of three parallel DNN architectures for VF detection during CPR. The input size is denoted as (? × 1250 × 2), corresponding to batch size (?), number of time samples (1250 in 10 s), number of channels (2) for ECG and IMP signals.

Seq3CONV + LSTM (ECG), *Seq3CONV + BiLSTM (ECG)* in Fig. 2, *Par3CONV + LSTM (ECG, IMP)* in Fig. 3.

DNN block diagrams in Fig. 2, 3 are visualized with the free online tool Netron [36].

3.2 DNN Model Training

All DNN models were compiled with 'Random uniform' kernel initializer and fit with 'Adam' optimizer (learning rate of 0.001, exponential decay rate for the first and second moment estimates $\beta 1 = 0.9$, $\beta 2 = 0.999$), using a batch size of 128. Considering the unequal distribution of shockable (5.5%, 1228/22516) and non-shockable cases (94.5%, 21288/22516) in the training database, we applied a penalty proportional to the class prevalence when calculating 'Weighted binary cross-entropy' loss-function:

$$\text{Loss} = -\frac{1}{N}\sum\nolimits_{n}^{N} \delta_n w_{Sh} \log(P(x_n \in Sh)) + (1 - \delta_n) w_{NSh} \log(1 - P(x_n \in Sh)) \quad (2)$$

where N is the size of the training database; δ_n is a binary indicator function, which is equal to 1 if the training sample x_n belongs of shockable class, otherwise $\delta_n = 0$; $w_{Sh} = 0.945$ and $w_{NSh} = 0.055$ are the weights for shockable and non-shockable classes, respecting the condition $w_{Sh} + w_{NSh} = 1$.

Each DNN architecture was trained with five independent runs for maximum of 750 epochs. Early stopping was activated if no improvement in the validation loss was obtained for 150 epochs. Among the five runs, the model with the minimum loss in the validation dataset was subjected to further evaluation of the test performance.

The models were implemented in Python using Keras with Tensorflow backend. The training was run on a workstation PERSY Stinger with Intel CPU Xeon Silver 4214R @ 2.4 GHz (2 processors), 96 GB RAM, NVIDIA RTX A5000-24 GB GPU.

3.3 Performance Evaluation

The trained models were evaluated by calculating their sensitivity (Se), specificity (Sp) and balanced accuracy (BAC) in the test dataset, as follows:

$$Se = 100\frac{TP}{TP + FN}(\%)\ Sp = 100\frac{TN}{TN + FP}(\%)\ BAC = 100\frac{Se + Sp}{2}(\%) \quad (3)$$

where true positives (TP) and false negatives (FN) count the correctly and erroneously detected shockable cases (VF); true negatives (TN) and false positives (FP) count the correctly and erroneously detected non-shockable cases (NSR, ONR, Asystole).

4 Results and Discussion

Although the basic *Seq3CONV (ECG)* model architecture has been optimized during continuous CC in [27], this study applied retraining of the basic model and the other six augmented DNN architectures using the extended set of CPR samples with a random CC fraction, as defined in Table 1. Additional optimization was applied for:

- The number of filters in the first $CONV_1$ layer, which concatenates ECG and IMP channels in the architecture *Seq3CONV (ECG, IMP)*. Two settings were tested: (1) 5 filters corresponding to the original setting in [27] with one input channel (ECG); (2) 10 filters (5 × 2) – doubled for 2 input channels (ECG, IMP). The test performance in Fig. 4(a) justifies that the model with 5 filters is more balanced in respect of Se and Sp, having the largest BAC = 93.1%, which outperforms by 0.7% points the wider model with 10 input filters. Therefore, *Seq3CONV (ECG, IMP)* retained the same layout as the basic architecture *Seq3CONV (ECG)*.
- The recurrent kernel (RK) size in the LSTM layer of the recurrent model *Seq3CONV + LSTM (ECG)*. Three settings were tested (RK = 10, 20, 40 neurons). The test performance in Fig. 4(b) proves that RK has less impact on Sp (96.3–96.7%) but more on Se (92.1–93.4%). The maximal performance is considered for the LSTM setting with RK = 20 units. The same optimal RK is used in all architectures, including LSTM and BiLSTM layers.

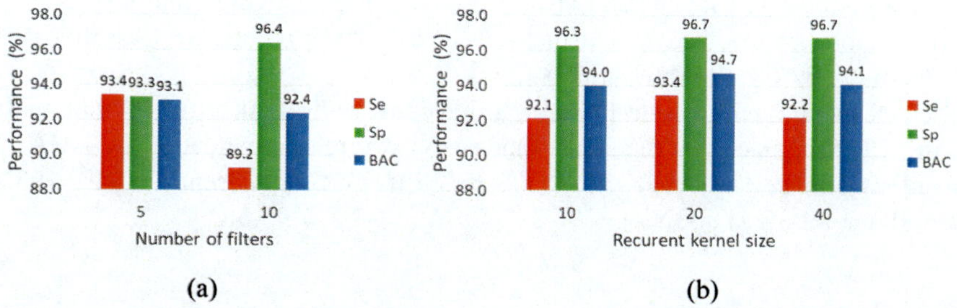

Fig. 4. Test performance used for optimal choice of: (a) number of filters in CONV1 layer in the architecture *Seq3CONV (ECG, IMP)*; (b) RK size in the LSTM layer of the recurrent model *Seq3CONV + LSTM (ECG)*.

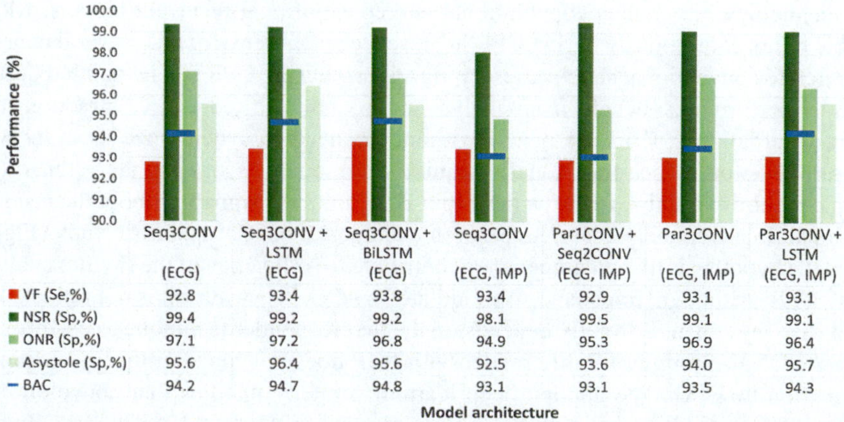

Fig. 5. Test performance of the designed seven DNN architectures for VF detection during CPR.

Comparative study of the seven DNN model architectures for VF detection during CPR with random CC fraction distribution is presented in Fig. 5. Detailed analysis of categorized graphs for rhythms (VF, NSR, ONR, Asystole) and global BAC shows that:

- The three DNN models with one input channel (ECG) are better than any of the four concatenating the two input channels (ECG, IMP), noting that:
  - The fully convolutional models are most vulnerable to the additional IMP input, presenting remarkable Sp drop by up to 3% points (Asystole), 2% points (ONR), 1.3% points (NSR). This results in BAC drop by 0.7–1.1% points: 94.2% (ECG) vs. 93.1–93.5% (ECG, IMP).
  - The convolution-recurrent model with LSTM layer is better managed to use the additional IMP input by improving Sp, although it is still drop by up to 0.8% points (ONR and Asystole) compared to the model with a single ECG input.
  - The additional use of IMP has a slight impact on Se(VF) $<$ 0.5% points.
- The use of recurrent layer (LSTM or BiLSTM) after convolutional blocks gives a slight performance improvement of Se (VF) up to 1% points, Sp (Asystole) up to 0.7% points, BAC up to 0.6% points.
- All DNN models with single ECG input have test performance for rhythm analysis during CPR that satisfies the AED shock advisory recommendation by AHA [35], considering the best model *Seq3CONV + BiLSTM (ECG)* to present: Se(VF) = 93.8%, Sp(NSR) = 99.2%, Sp(ONR) = 96.8%, Sp(Asystole) = 95.6%.

5 Conclusions

In a setting with large number of cardiac arrest rhythms during CPR, this study presents a deep learning methodology for unsupervised rhythm analysis of one (ECG) and two (ECG, IMP) channels with a random distribution of the CC fraction in analysis interval. This technology has trained convolutional and convolutional-recurrent models, which are able to run continuously during CPR in presence and absence of CC. Several benefits compared to conventional shock advisory algorithms during CPR can be pointed out: (i) Potential performance benefit from analysis of signals with reduced CC fraction (e.g. during insufflations); (ii) Ease of use without the need to synchronize analysis with CC start and stop, which traditionally requires additional sensors and algorithms; (iii) Internal feature extraction from the raw input ECG and IMP signals without the need for additional time-frequency feature measurement algorithms. Comparative study (Fig. 5) rejects the hypothesis that the impedance contributes to efficiency of the rhythm analysis during CPR. Although, impedance is modulated by CC, its waveform is not reflected by the cardiac rhythm and strongly depends on the first responder technique to perform the massage. DNN training of either model with IMP does not find significant additional information to ECG. The applied deep learning strategy justifies that convolutional-recurrent DNN architectures with ECG input are able to satisfy AHA goals for rhythm analysis with arbitrary CC fraction distribution during CPR.

Acknowledgement. This work was supported by the Bulgarian National Science Fund, grant number KП-06-H42/3 "*Computer aided diagnosis of cardiac arrhythmias based on machine learning and deep neural networks*".

References

1. Fitzgibbon, E., Berger, R., Tsitlik, J., Halperin, H.: Determination of the noise source in the electrocardiogram during cardiopulmonary resuscitation. Crit. Care Med. **30**, S148–S153 (2002)
2. Olasveengen, T., Semeraro, F., Ristagno, G., et al.: European resuscitation council guidelines 2021: basic life support. Resus **161**, 98–114 (2021)
3. Isasi, I., Irusta, U., Rad, A., et al.: Automatic cardiac rhythm classification with concurrent manual chest compressions. IEEE Access **7**, 115147–115159 (2019)
4. Isasi, I., Irusta, U., Elola, A., et al: A robust machine learning architecture for a reliable ECG rhythm analysis during CPR. In: 41st Annual International Conference of the IEEE Engineering in Medicine and Biology Society, pp. 1903–1907, Berlin, Germany (2019)
5. Isasi, I., Irusta, U., Aramendi, E., Idris, A., Sörnmo, L.: Restoration of the electrocardiogram during mechanical cardiopulmonary resuscitation. Physiol. Meas. **41**, 105006 (2020)
6. Irusta, U., Ruiz, J., de Gauna, S., Eftestøl, T., Kramer-Johansen, J.: A least mean-square filter for the estimation of the cardiopulmonary resuscitation artifact based on the frequency of the compressions. IEEE Trans. Biomed. Eng. **56**, 1052–1062 (2009)
7. Ayala, U., Irusta, U., Ruiz, J., et al.: A reliable method for rhythm analysis during cardiopulmonary resuscitation. Biomed. Res. Int. **2014**, 872470 (2014)
8. Ruiz, J., Irusta, U., De Gauna, S., Eftestøl, T.: Cardiopulmonary resuscitation artefact suppression using a Kalman filter and the frequency of chest compressions as the reference signal. Resus **81**, 1087–1094 (2010)
9. Aramendi, E., Ayala, U., Irusta, U., Alonso, E., Eftestøl, T., Kramer-Johansen, J.: Suppression of the cardiopulmonary resuscitation artefacts using the instantaneous chest compression rate extracted from the thoracic impedance. Resus **83**, 692–698 (2012)
10. Babaeizadeh, S., Firoozabadi, R., Han, C., Helfenbein, E.: Analyzing cardiac rhythm in the presence of chest compression artifact for automated shock advisory. J. Electrocardiol. **47**, 798–803 (2014)
11. Gong, Y., Gao, P., Wei, L., Dai, C., Zhang, L., Li, Y.: An enhanced adaptive filtering method for suppressing cardiopulmonary resuscitation artifact. IEEE Trans. Biomed. Eng. **64**, 471–478 (2016)
12. Berger, R., Palazzolo, J., Halperin, H.: Rhythm discrimination during uninterrupted CPR using motion artifact reduction system. Resus **75**, 145–152 (2007)
13. Tan, Q., Freeman, G., Geheb, F., Bisera, J.: Electrocardiographic analysis during uninterrupted cardiopulmonary resuscitation. Crit. Care Med. **36**, S409–S412 (2008)
14. Rheinberger, K., Steinberger, T., Unterkofler, K., Baubin, M., Klotz, A., Amann, A.: Removal of CPR artifacts from the ventricular fibrillation ECG by adaptive regression on lagged reference signals. IEEE Trans. Biomed. Eng. **55**, 130–137 (2007)
15. Krasteva, V., Jekova, I., Dotsinsky, I., Didon, J.P.: Shock advisory system for heart rhythm analysis during cardiopulmonary resuscitation using a single ECG input of automated external defibrillators. Annals Biomed. Eng. **38**, 1326–1336 (2010)
16. Amann, A., Klotz, A., Niederklapfer, T., et al.: Reduction of CPR artifacts in the ventricular fibrillation ECG by coherent line removal. Biomed. Eng. Online **9**, 2 (2010)

17. De Gauna, S., Ruiz, J., Irusta, U., Aramendi, E., Eftestøl, T., Kramer-Johansen, J.: A method to remove CPR artefacts from human ECG using only the recorded ECG. Resus **76**, 271–278 (2008)
18. Hajeb-Mohammadalipour, S., Cascella, A., Valentine, M., Chon, K.H.: Automated condition-based suppression of the CPR artifact in ECG data to make a reliable shock decision for AEDs during CPR. Sensors **21**, 8210 (2021)
19. Li, Y., Bisera, J., Geheb, F., Tang, W., Weil, M.: Identifying potentially shockable rhythms without interrupting cardiopulmonary resuscitation. Crit. Care Med. **36**, 198–203 (2008)
20. Fumagalli, F., Silver, A., Tan, Q., Zaidi, N., Ristagno, G.: Cardiac rhythm analysis during ongoing cardiopulmonary resuscitation using the analysis during compressions with fast reconfirmation technology. Heart Rhythm **15**, 248–255 (2018)
21. Hu, Y., Tang, H., Liu, C., et al.: The performance of a new shock advisory algorithm to reduce interruptions during CPR. Resus **143**, 1–9 (2019)
22. de Graaf, C., Beesems, S., Oud, S., et al.: Analyzing the heart rhythm during chest compressions: performance and clinical value of a new AED algorithm. Resus **162**, 320–328 (2021)
23. Didon, J.P., Menetre, S., Jekova, I., Stoyanov, T., Krasteva, V.: Analyze whilst compressing algorithm for detection of ventricular fibrillation during CPR: a comparative performance evaluation for automated external defibrillators. Resus **160**, 94–102 (2021)
24. Ayala, U., Irusta, U., Kramer-Johansen, J., et al.: Automatic detection of chest compression pauses for rhythm analysis during 30:2 CPR in an ALS scenario. Resus **85S**, S9 (2014)
25. González-Otero, D., Ruiz de Gauna, S., Ruiz, J., Ayala, U., Alonso, E.: Automatic detection of chest compression pauses using the transthoracic impedance signal. Comput. Cardiol. **39**, 21–24 (2012)
26. Didon, J.P., Krasteva, V., Ménétré, S., Stoyanov, T., Jekova, I.: Shock advisory system with minimal delay triggering after end of chest compressions: Accuracy and gained hands-off time. Resus **82S**, S8–S15 (2011)
27. Jekova, I., Krasteva, V.: Optimization of end-to-end convolutional neural networks for analysis of out-of-hospital cardiac arrest rhythms during cardiopulmonary resuscitation. Sensors **21**, 4105 (2021)
28. Isasi, I., Irusta, U., Aramendi, E., Eftestøl, T., Kramer-Johansen, J., Wik, L.: Rhythm analysis during cardiopulmonary resuscitation using convolutional neural networks. Entropy **22**, 595 (2020)
29. Isasi, I., Irusta, U., Aramendi, E., Olsen, J.-Å., Wik, L.: Detection of shockable rhythms using convolutional neural networks during chest compressions provided by a load distributing band. Comput. Cardiol. **47**, 1–4 (2020). https://doi.org/10.22489/CinC.2020.045
30. Hajeb, M., Cascella, A., Valentine, M., Chon, K.: Deep neural network approach for continuous ECG-based automated external defibrillator shock advisory system during cardiopulmonary resuscitation. J. Am. Heart Assoc. **10**, e019065 (2021)
31. Irusta, U., Ruiz, J., Aramendi, E., Ruiz de Gauna, S., Ayala, U., Alonso, E.: A high-temporal resolution algorithm to discriminate shockable from nonshockable rhythms in adults and children. Resus **83**, 1090–1097 (2012)
32. Krasteva, V., Ménétré, S., Didon, J.-P., Jekova, I.: Fully convolutional deep neural networks with optimized hyperparameters for detection of shockable and non-shockable rhythms. Sensors **20**, 2875 (2020)
33. Skjeflo, G.W., Nordseth, T., Loennechen, J.P., Bergum, D., Skogvoll, E.: ECG changes during resuscitation of patients with initial pulseless electrical activity are associated with return of spontaneous circulation. Resus **127**, 31–36 (2018)
34. Koster, R., et al.: European resuscitation council guidelines for resuscitation 2010 section 2. Adult basic life support and use of automated external defibrillators. Resus **81**, 1277–1292 (2010)

35. Automatic External Defibrillation: Recommendations for specifying and reporting arrhythmia analysis algorithm performance, incorporating new waveforms, and enhancing safety. Circulation **95**, 1677–1682 (1997)
36. Roeder, L.: Netron, Visualizer for neural network, deep learning, and machine learning models. Comput. Softw. (2017). https://doi.org/10.5281/zenodo.5854962

Parallel Technique on Bidirectional Associative Memory Cohen-Grossberg Neural Network

Gani Stamov[1], Stanislav Simeonov[2], Ivan Torlakov[2(✉)], and Marina Yaneva[3]

[1] Technical University of Sofia, 8800 Sliven, Bulgaria
gstamov@abv.bg
[2] Department of Computer Systems and Technologies,
Burgas "Prof. Dr. Assen Zlatarov" University, 8010 Burgas, Bulgaria
stanislav_simeonov@btu.bg, itorlakov@gmail.com
[3] Department of Social health and Health care,
Burgas "Prof. Dr. Assen Zlatarov" University, 8010 Burgas, Bulgaria
marinaianeva@abv.bg

Abstract. The present paper is devoted to software analysis on applied method used on parallel technology, in particular `CUDA` and `OpenMPI`, to find stable areas of single mathematical model of Bidirectional Associative Memory (BAM) Cohen-Grossberg neural network with time-varying delays. The given type of neural networks give opportunity of modelling and study of biological problems.

Keywords: neural network · Cohen-Grossberg · parallelism · CUDA · OpenMPI

1 Introduction

First introduction of a neural network of type Cohen-Grossberg was made by M. A. Cohen and S. Grossberg in 1983 [1]. Extensive research has been made in the area due to the opportunity of using various models in scientific and engineering fields [11,12,16–18].

The dynamic behaviour of such neural networks using the model are greatly effected by the synaptic transmission delays. More commonly the name "time delays" is used in neural networks. The limited speed of the signal transmissions determines the existence of such delays. Numerous researches have published their results based on those time delays in neural networks [5–10,13,14].

Real-world problems solutions using neural networks are better suited with time-varying delays than fixed delays since they provide a more realistic description.

The current study is based on paper [2], where it is reviewed the global exponential stability of solutions with respect to a manifold defined by a function for neural network of type Bidirectional Associative Memory Cohen-Grossberg with time-varying delays and variable impulsive perturbations.

S. Sotirov et al. (Eds.): BioInfoMed 2022, LNNS 658, pp. 16–20, 2023.
https://doi.org/10.1007/978-3-031-31069-0_2

The aim on this paper is to present a single result of usage of High Performance Computing on calculating the stability requirements of a single neural network of type BAM Cohen-Grossberg.

In this paper the focus is to investigate the different scalability using two technologies - `OpenMPI` [3] and `CUDA` [15], while searching for applied method used on parallel technology to find stable areas of the given mathematical model. With the help of High Performance Computing we calculate and validate the required conditions for stability of the respected given model of Bidirectional Associative Memory Cohen-Grossberg-type neural network with time-varying delays. Such technologies and methodology can be used on other types of neural networks to achieve similar results.

2 Mathematical Model

Consider the following impulsive Bidirectional Associative Memory Cohen-Grossberg neural networks with time-varying delays (System 1 in [2]):

$$\begin{cases} \dot{x}_i(t) = -a_i(x_i(t)) \Big[b_i(x_i(t)) - \sum_{j=1}^{2} c_{ji} f_j(y_j(t)) \\ \qquad - \sum_{j=1}^{2} d_{ji} g_j(y_j(t - \sigma_j(t))) - I_i \Big], \quad t \neq \tau_k(x(t), y(t)), \\ \dot{y}_j(t) = -\hat{a}_j(y_j(t)) \Big[\hat{b}_j(y_j(t)) \sum_{j=1}^{2} p_{ij} \hat{f}_i(x_i(t)) \\ \qquad - \sum_{i=1}^{2} q_{ij} \hat{g}_j(x_i(t - \hat{\sigma}_i(t))) - J_j \Big], \quad t \neq \tau_k(x(t), y(t)), \end{cases} \tag{1}$$

with impulsive perturbations of the type:

$$\begin{aligned} x(t^+) - x(t) &= \begin{pmatrix} -1 + \frac{1}{2k} & 0 \\ 0 & -1 + \frac{1}{2k} \end{pmatrix} x(t), \quad t = \tau_k(x(t), y(t)), \quad k = 1, 2, ..., \\ y(t^+) - y(t) &= \begin{pmatrix} -1 + \frac{1}{3k} & 0 \\ 0 & -1 + \frac{1}{3k} \end{pmatrix} y(t), \quad t = \tau_k(x(t), y(t)), \quad k = 1, 2, ..., \end{aligned} \tag{2}$$

Using the mathematical analysis in [2] of a system 1 as generalization of existing models of impulsive Bidirectional Associative Memory Cohen-Grossberg-type neural networks with time-varying delays, a conclusion can be made for system stability requirements. Solving the practical problems is essential to find the intervals at which the requirements for stability are satisfied. Given the different available WAYS of solving the task we use only parallel technique. Searching the stability points of neural networks according to Theorem 1 in [2] is achieved using two parallel techniques. First implementation was created using `NVIDIA CUDA` ™technology, second results were obtained using software realization available `OpenMPI` libraries.

2.1 CUDA

GPU computation power can be utilized via `CUDA` parallel platform and API model such as multiplying matrices or other algebra operations. `CUDA` is an extension of C programming language created and developed by NVIDIA ™. GPUs focus on execution throughput of massively-parallel programs.

The algorithm of main function aims to connect the device CPU code with the GPU executable source. Another function it takes is to manage memory allocation and cleaning and external variable validation. The algorithm can be presented with the following pseudo-code:

```
int main(int argc, char** argv) {
    create local variables;
    validate input values and set to local variables;
    allocate memory for results on host via malloc();
    allocate memory for results on device via cudaMalloc();
    send values from host to device via cudaMemcpy);
    receive results from device to host via cudaMemcpy();
    save results to file via fputs();
    free memory on device via cudaFree();
    free memory on host via free();
}
```

The algorithm of the `__device__` function for validating the μ values:

```
__device__ void alg_calc(double* min, double* max, double* step) {
    shared memory variables included;
    __syncthreads();
    create local variables;
    calculate array index element based on blockIdx, threadIdx,
    gridDim and blockDim;
    double result = μ formula from Theorem 1, point 2 in [2];
    return result;
}
```

2.2 OpenMPI

`OpenMPI` is an open source Message Passing Interface implementation. The first release of the standard, known as `MPI-1.0` is published in 1993 [4].

The approach under `OpenMPI` used in this paper is similar to the software implementation with the `CUDA` technique. Main difference comes from the different approach used in both technologies - `CUDA` uses single machine GPU while `OpenMPI` uses cluster of machines using CPUs.

Main algorithm handles the initialization of the `MPI` required functions and variables. Handles the local variable initialization, external parameter validation and ranking distribution. Can be seen as pseudo-code as follow:

```
int main(int argc, char **argv) {
    MPI_Init(&argc, &argv);
    int rank, size;
    MPI_Comm_rank(MPI_COMM_WORLD, &rank);
    MPI_Comm_size(MPI_COMM_WORLD, &size);
    create local variables;
    validate input values and set to local variables;
```

```
    if (rank == 0) {
        iterate all available settings;
        send information via to next available process MPI_Send();
    } else {
        double result = μ formula from Theorem 1, point 2 in [2];
        return and save result to file;
    }
    MPI_Finalize();
    exit program with success;
}
```

3 Results

Both software implementation generate data and are validated according to Theorem 1 from [2]. Validation rules are supplied to verify if Hypotheses 1 through 5 are satisfied. Given the results a positive number μ is checked for existence. Having zero or negative value to the according data gives non-stable neural network. Functions P_{ik} and Q_{jk} as abrupt changes of the state at the impulsive moment are validated. According to Theorem 1, point 4 in [2] a function $h(t, z)$ is search if it exists in the given context. If any of those validation rules are not met given in Theorem 1 we can conclude that the given input parameters used for the neural network lead to non-stable results which are not satisfying the given goal of searching stable models.

Given all validation rules according to the methodology in Theorem 1, a large amount of data have been generated with both software implementations amounting to over a milliard and half situations. Only limited results are being reviewed and summarized with respect to Theorem 1.

Each of the two used technologies have it's pros and cons. Using `CUDA` we have limited amount of parallel code and fixed memory that can be handled. Given the shared memory paradigm making writing code more easily compared to `OpenMPI`. The GPUs run only parallel code, while using host base CPU for non-parallel source.

Using the `OpenMPI` technology we can achieve great scalability since it runs in distributed clusters. Given that all synchronizations are explicit leads to more complex source code development. Even that since all code is run in parallel with `OpenMPI` it can easily be achieved great speed up.

Acknowledgements. This research was funded in part by the European Regional Development Fund through the Operational Program "Science and Education for Smart Growth" under contract UNITe No BG05M2OP001–1.001–0004 (2018–2023).

References

1. Cohen, M.A., Grossberg, S.: Absolute stability of global pattern formation and parallel memory storage by competitive neural networks. IEEE Trans. Syst. Man Cybern. SMC **13**(5), 815–826 (1983)

2. Stamov, G., Stamova, I., Simeonov, S., Torlakov, I.: On the stability with respect to H-manifolds for Cohen-Grossberg-type bidirectional associative memory neural networks with variable impulsive perturbations and time-varying delays. Mathematics **8**(3), 335 (2020)
3. Gabriel, E., et al.: Open MPI: goals, concept, and design of a next generation MPI implementation. In: Proceedings, 11th European PVM/MPI Users' Group Meeting, pp. 97–104 (2004)
4. The MPI Forum: MPI: a message passing interface. In: Proceedings of the 1993 ACM/IEEE Conference on Supercomputing, pp. 878–883. Association for Computing Machinery (1993)
5. Aouiti, C., Assali, E.A.: Nonlinear Lipschitz measure and adaptive control for stability and synchronization in delayed inertial Cohen-Grossberg-type neural networks. Int. J. Adapt. Control Signal Process. **33**(10), 1457–1477 (2019)
6. Gan, Q.: Adaptive synchronization of Cohen-Grossberg neural networks with unknown parameters and mixed time-varying delays. Commun. Nonlinear Sci. Numer. Simul. **17**(7), 3040–3049 (2012)
7. Li, Y., Zhao, L., Zhang, T.: Global exponential stability and existence of periodic solution of impulsive Cohen-Grossberg neural networks with distributed delays on time scales. Neural Process. Lett. **33**(1), 61–81 (2011)
8. Song, Q., Cao, J.: Stability analysis of Cohen-Grossberg neural network with both time-varying and continuously distributed delays. J. Comput. Appl. Math. **197**(1), 188–203 (2006)
9. Yuan, K., Cao, J., Li, H.X.: Robust stability of switched Cohen-Grossberg neural networks with mixed time-varying delays. IEEE Trans. Syst. Man Cybern. Part B (Cybern.) **36**(6), 1356–1363 (2006)
10. Sader, M., Wang, F., Liu, Z., Chen, Z.: Projective synchronization analysis for BAM neural networks with time-varying delay via novel control. Nonlinear Anal. Model. Control **26**, 41–56 (2021)
11. Guo, S., Lihong, H.: Stability analysis of Cohen-Grossberg neural networks. IEEE Trans. Neural Netw. **17**, 106–17 (2006)
12. Hongtao, L.: Global exponential stability analysis of Cohen-Grossberg neural networks. IEEE Trans. Circuits Syst. II Express Briefs **52**(8), 476–479 (2005)
13. Stamova, I., Stamov, G.: On the stability of sets for reaction-diffusion Cohen-Grossberg delayed neural networks. Discrete Contin. Dyn. Syst.-Ser. S **14**(4), 1429–1446 (2021)
14. Stamov, G., Stamova, I., Venkov, G., Stamov, T., Spirova, C.: Global stability of integral manifolds for reaction-diffusion delayed neural networks of Cohen-Grossberg-type under variable impulsive perturbations. Mathematics **8**(7), 1082 (2020)
15. NVIDIA Corporation: CUDA Zone — NVIDIA Developer (2022)
16. Song, Q., Zhang, J.: Global exponential stability of impulsive Cohen-Grossberg neural network with time-varying delays. Nonlinear Anal. Real World Appl. **9**(2), 500–510 (2008)
17. Yang, F., Zhang, C., Dongqing, W.: Global stability analysis of impulsive BAM type Cohen-Grossberg neural networks with delays. Appl. Math. Comput. **186**(1), 932–940 (2007)
18. Lisena, B.: Dynamical behavior of impulsive and periodic Cohen-Grossberg neural networks. Nonlinear Anal. Theory Methods Appl. **74**(13), 4511–4519 (2011)

Prediction of the Granulometric Composition of the Silt Loading on Transport Arteries in the City of Bourgas Based on Artificial Neural Networks

Dimitrinka Ivanova, Yordanka Tasheva, Evdokia Sotirova, Aleksandar Dimitrov(✉), and Sotir Sotirov

University "Prof. Dr. Assen Zlatarov", 8010 Burgas, Bulgaria
divanova@btu.bg, al_dim_2000@mail.bg

Abstract. This study integrates data from laboratory road silt analyses and predicts these values using artificial neural networks.

It's known that the emissions from vehicular traffic are a significant contributor to total particulate matter (PM) concentrations in urban areas and have detrimental effects on human health.

The characteristics of these emissions, from different traffic-related sources, are essential to study their impact on human health.

The deposition of particulate matter (road silt) on road surfaces and its suspension from roadways is the main mechanism by which road transport causes secondary PM pollution.

The analysis of road silt was performed by sieve and laser diffraction analyses.

The obtained data were used to predict the grain size composition of the fractions below 75 μm from the road silt samples, by neural networks, after sampling along the main and secondary traffic arteries of the city of Burgas, and analyzing the fractions up to 75 μm.

Keywords: Neural Network · Silt loading · Granulometric composition · Transport arteries

1 Introduction

With each passing year, the problem of air pollution and air cleanliness becomes more and more urgent. Pollution inevitably affects human health and especially cancerous growths in the human body [1].

In this paper, the authors proposed that in measuring the concentrations of total particulate matter (PM), the data should be analyzed using artificial intelligence methods and some of them should be predicted.

When accumulating data obtained from different types of instrumentation, different measurement principles and methods are very often used. Some of these methods are very complex, have different measurement error and have different data acquisition times. It

S. Sotirov et al. (Eds.): BioInfoMed 2022, LNNS 658, pp. 21–31, 2023.
https://doi.org/10.1007/978-3-031-31069-0_3

is only natural that different instruments also need different consumables. When more than one instrument is used to complement the different data, questions arise about the efficiency, speed, accuracy and cost of the measurements made. In this case, one can make some of the measurements and predict the other (which in reality will be slower, expensive and inaccurate) with the proposed method.

It is well known that emissions due to vehicular traffic are a major contributor to total particulate matter (PM) concentrations in urban areas and that exposure to particulate matter emissions from vehicles has detrimental effects on human health [2–10]. Hence, a better understanding of the characteristics of emissions from different traffic-related sources is essential to study their impact on human health [11]. Particulate matter emissions from vehicles include exhaust emissions and emissions caused by wear of vehicle parts such as brakes, tires, and clutch, as well as dust resuspension (non-exhaust emissions). Tailpipe emissions contribute mainly to particulate matter fractions in the range of 2.5 to 10 μm, while exhaust emissions contribute mainly to fine particulate matter with aerodynamic diameter <2.5 μm [12–15].

In determining suspended particulate matter emissions from motor vehicles, either the European or the US approach, of the respective Environmental Protection Agencies, is applied. In view of the intermittent maintenance of the transport network in R. Bulgaria, the American approach is more appropriate. This approach is based on the presence of road debris on the roadways.

The aim of this paper is to predict the grain size composition of the fractions below 75 μm from road silt samples, by neural network analysis of the fractions up to 75 μm.

Road silt - this is the total amount of unbound solid particles (usually soil, sand, etc.) that have fallen on the roadway in all possible ways. It is measured in g/m^2 of road surface and is an average. Only particles with an aerodynamic diameter of up to 75 μm are considered as silt (by pre-screening, larger particles are separated). The silt is distributed unevenly over the road surface. It is least around the centreline of the road and increases towards the road shoulder or kerb. In urban conditions, the kerb plays a holding role, so that the density of silt there can reach very high values.

In real conditions, the road silt is a constantly changing quantity. Its values can vary over a very wide range - 0.02 ÷ 400 g/m^2 [16].

The main causes of particulate pollution of roadways can be classified as natural (natural) and anthropogenic (caused by various types of human activity). The latter will be discussed in more detail:

- ***Particulate matter from exhaust gases***
  Motor vehicles are an important source of carbonaceous aerosols, especially in the fine particle size range (aerodynamic diameter <2.5 μm) [17]. Particulate emissions from vehicles depend on the engine type, age, maintenance and organic and elemental carbon content with trace amounts of other metals and ions [3, 5, 10, 18, 19].
- ***Particulate matter from tyre and brake wear***
  The wear particles of automobile tires are generated either by friction between the tire tread and the road surface or by evaporation [15, 19–21]. The amount, size, and chemical composition of particles released have been shown to be influenced by multiple factors, including tire characteristics, pavement design, vehicle characteristics, and vehicle condition.

- ***Particulate matter from road wear***

  Pavement type is another important factor in road dust emissions. The purpose of road surfaces is to absorb and distribute loads on the ground, to create the necessary traction and flatness for vehicle movement. They consist of a drainage layer, base course and road surface. The bases of asphalt concrete pavements, depending on the class of the road and the traffic volume, are crushed stone, crushed stone improved, ballasted, concrete, pavement, base cobble and others [22].

  There is little information in the literature on the exact chemical composition of road pavements and the PM generated by their wear.
- ***Suspension of dust from roadways***

  Road surfaces are a place for the deposition of particulate matter. This deposited material is called road dust. Suspension of dust from roadways is the primary mechanism by which vehicular traffic causes secondary PM pollution [23, 24].

2 Materials and Methods

- ***Sieve analysis***

  We used sieve analysis to determine the particle size composition (greater than 75 μm) of the road pavement samples collected (total dust load) and the silt content of the samples.

  Sieve analysis is a method of estimating the particle size distribution of granular material by allowing the material to pass through a series of progressively smaller mesh size sieves and weighing the amount of material that is stopped by each sieve as a fraction of the total mass.

  A CISA BA200N vibration machine was used (Picture 1). A set of sieves were used with sizes 10 mm, 2 mm, 1 mm, 850 μm, 500 μm, 150 μm, 106 μm, 75 μm.

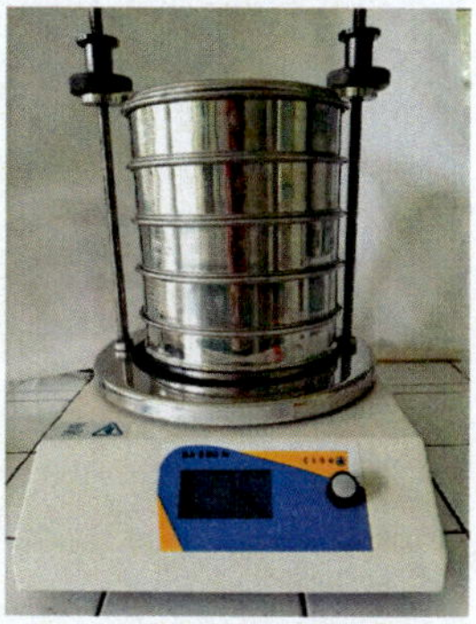

Picture 1. Vibrating machine CISA BA200N with sieves.

- ***Laser diffraction analysis***

  We used laser diffraction analysis and a standard [25] to determine the grain size composition of the road silt (below 75 μm). Laser diffraction is a widely used technique for analyzing the fineness and particle size distribution of various powdery materials. The apparatus used is presented in Picture 2.

Picture 2. "Analysette 22" laser diffraction analyzer.

- ***Artificial neural networks***

  When solving various tasks in computer science, the so-called algorithmic approach. In this approach, an abstract model of the task under consideration is made, after an algorithm (a precise sequence of steps) is developed for the solution and, finally, the algorithm is implemented with a computer program. The use of neural networks is based on learning by examples, with which an attempt is made to model their biological element - the brain [26].

  In principle, any computable function can be calculated with neural networks, i.e. with neural networks everything can be done with standard digital computers. Neural networks work on a black-box basis, using information from the process or phenomenon under consideration for training purposes while adjusting the network's weights. Therefore, they are a convenient tool for modeling processes and phenomena for which a lot of input-output information (from measurement, etc.) can be collected, but which can hardly be described analytically [27].

  Neural networks are effective in solving classification and approximation tasks, when there is a lot of information (measurement data, for example) for training, when it is difficult to find sufficiently simple analytical relationships to describe the relevant objects and processes. Given enough data and computing resources for training, almost any image between two vector spaces can be approximated with desired accuracy by means of a multi-layer one-way feedforward neural network (neural networks are the most commonly used in practice) [28].

  In this case, the available information is perceived, which is perceived by the neural network. This can serve to predict values based on the knowledge already acquired by the neural network [29].

3 Results and Discussion

The object of our study are 12 road silt samples, collected from 12 road segments (streets or parts of them) from the territory of the city of Burgas presented in Fig. 1 [30] The sampling was carried out before the winter sanding, in the period October - November.

After collecting the samples, sieve analysis was carried out to determine their particle size distribution and to determine the amount of road silt representing particles with sizes less than or equal to 75 μm. Nine fractions were obtained from each analysed sample, as presented in Tables 1.

Fig. 1. Location of sampling points

Table 1. Results of sieve analysis of the samples from the studied streets in the territory of the city of Sofia. The results of the samples from the streets of Burgas.

No of road segment	Fractions, %								
	over 10 mm	10 ÷ 2 mm	2 ÷ 1 mm	1 mm ÷ 850 μm	850 ÷ 500 μm	500 ÷ 150 μm	150 ÷ 106 μm	106 ÷ 75μm	under 75 μm
1	0,28	11,67	11,7	4,17	18,04	39,86	6,59	4,05	3,64
2	3,8	16,97	12,48	3,36	13,63	33,5	6,97	4,77	4,52
3	2,74	32,44	15,68	3,5	11,98	22,37	5,11	2,95	3,23
4	36,1	36,21	11,31	2	5,52	7,47	0,68	0,33	0,38
5	2,67	14,62	16,11	3,75	13,19	34,64	7,34	4,41	3,27
6	1,34	29,93	15,96	3,32	11,42	24,36	4,82	3,98	4,87
7	3,07	23,74	11,07	2,25	8,92	30,37	7,31	4,19	9,08
8	0,74	17,04	16,03	3,53	9,31	30,75	9,26	6,55	6,79
9	1,47	18,14	13,26	3,28	13,34	36,14	6,28	4,1	3,99
10	1,21	23,06	17,99	3,24	10,79	25,3	5,15	4,59	8,67
11	0,73	14,49	14,68	1,93	7,56	35,9	8,53	9,12	7,06
12	1,21	17,77	11,05	2,06	8,39	33,03	8,74	7,38	10,37

The results showed that particles with sizes above 150 μm accounted for more than 73.5% of the sample, and particles below 150 μm accounted for less than 26.5% or an average of 16.59%. The particle size fractions below 75 μm ranged from 3.23% to 10.37% with the exception of street #4 where the amount was 0.38% of the sample.

The fraction below 75 μm, called road silt, is of most interest. As a result of laser diffraction analysis, the percentage of particles 75 ÷ 30 μm, 30 ÷ 15 μm, 15 ÷ 10 μm,

$10 \div 2.5$ μm and smaller than 2.5 μm was determined. The results obtained are presented in Table 2.

Table 2. Result of the laser diffraction analysis of the samples from the studied streets in the territory of the city of Burgas.

No of road segment	Fractions, %				
	75 ÷ 30 μm	30 ÷ 15 μm	15 ÷ 10 μm	10 ÷ 2,5 μm	2,5 ÷ 0 μm
1	53,9515	21,2625	10,449	7,4115	6,9255
2	54,4485	24,3485	5,825	10,6015	4,7765
3	53,9918	26,0676	5,9042	9,8032	4,2332
4	56,6589	23,5135	6,2279	9,7867	3,813
5	59,0416	22,4296	4,876	9,1425	4,5103
6	52,356	27,4784	5,9832	10,0828	4,0996
7	47,1008	28,5092	7,3712	12,3576	4,6612
8	60,0733	23,0208	4,5562	8,0333	4,3164
9	66,19	22,05	3,43	5,39	2,94
10	56,1208	25,7184	5,3016	8,7984	4,0608
11	47,8423	30,0951	7,1757	10,8171	4,0698
12	47,954	31,694	6,996	9,54	3,816

The main idea behind using the neural network here is to be able to predict the amounts of fine dust particles by using only a sieve method to separate the individual dust sizes from the roads.

For the purpose of the inputs of the neural network, we provide the parameters from a sieve analysis of the samples from the investigated streets in the territory of the city of Burgas. The goal is to obtain the results of the laser diffraction analysis of road sediment at the output of the neural network. For this purpose, we use a 4-layer neural network that uses deep learning. The structure of the neural network is 9:15:5:5:5. Deep learning was used to train the neural network. The root mean square error was used to evaluate the performance. In Fig. 2 the structure of the neural network is given.

The network was trained for 33 iterations, achieving a root mean square error of 5.93.

A graph of the learning process can be seen in Fig. 3 and 4. As can be seen from it, the total number of vectors is divided into three parts. 70% of the vectors are for training, 20% are for testing and the last 10% are for verification.

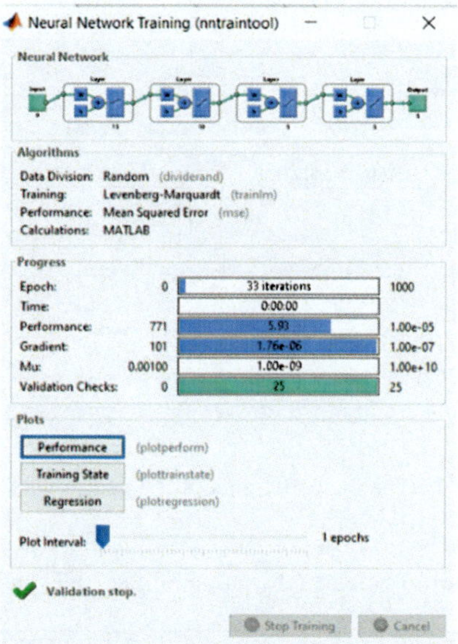

Fig. 2. Structure of the neural network

The best error from verification is at iteration 8 and is 4.7419.

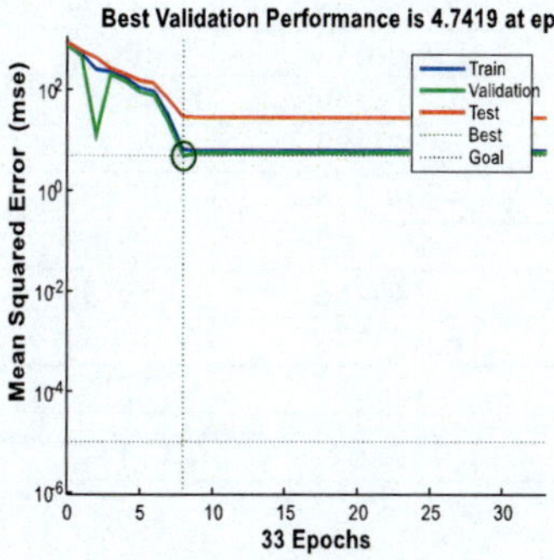

Fig. 3. Learning process of the neural network

At the output of the neural network, values are obtained that must be verified. For this purpose, in table. At the Table 3 are shown the error between the output of the ANN and the real data.

For individual fractions, the sum of the individual errors and the arithmetic mean error are different.

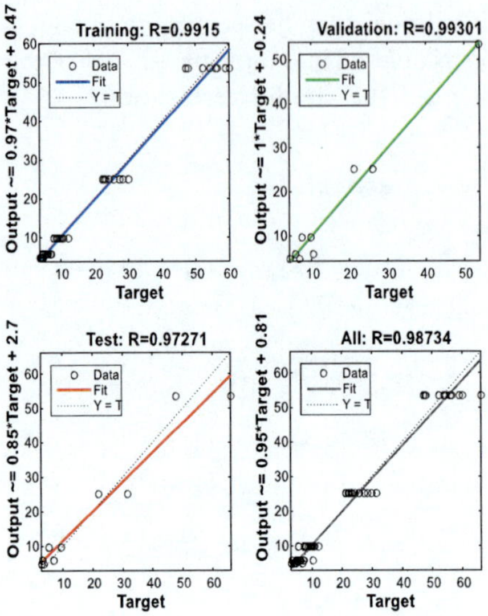

Fig. 4. The R coefficients of the neural network

Table 3. The error between the output of the ANN and the real data

№ of the segment	Error between the output of the ANN and the real data				
	75 ÷ 30 µm	30 ÷ 15 µm	15 ÷ 10 µm	10 ÷ 2,5 µm	2,5 ÷ 0 µm
1	-1.13774	-3.75213	4.147055	-1.64016	2.382961
2	-0.64048	-0.66638	-0.47696	1.549965	0.233841
3	1.00179	-0.95435	-0.49685	1.734138	-1.2848
4	0.149258	-0.14291	-0.00699	0.070186	-0.06951
5	3.952363	-2.58503	-1.42594	0.090844	-0.03224
6	-0.8653	0.677614	-0.40693	1.905476	-1.31092
7	-6.17315	1.758759	0.983552	4.155632	-0.72486
8	4.984063	-1.99383	-1.74574	-1.01836	-0.22614
9	11.10133	-2.96518	-2.87197	-3.66139	-1.6028
10	1.037528	0.698062	-1.00063	-0.25046	-0.48451
11	-7.24694	5.080466	0.873755	1.765444	-0.47274
12	-7.13521	6.679337	0.694054	0.488359	-0.72655
SUM	0.97248	1.834407	1.73361	5.189682	4.31827
AVR	0.08104	0.152867	0.14447	0.432474	0.35986

For the fraction 75 ÷ 30 µm, the sum is 0.97248, and the average arithmetic error is 0.08104. For the 30 ÷ 15 µm fraction, the sum is 1.834407, and the average arithmetic error is 0.152867. For the 15 ÷ 10 µm fraction, the sum is 1.73361, and the average arithmetic error is 0.14447. For the 10 ÷ 2.5 µm fraction, the sum is 5.189682, and the arithmetic mean error is 0.432474. For the smallest fraction 2.5 ÷ 0 µm the sum is 4.31827 and the average arithmetic error is 0.35986.

The mean square error of the comparison between the results obtained from the ANN and the values from the laser diffraction analysis of the samples from the investigated streets is 2.925912.

All these values define ANN as one of the best tools for information perception and prediction, in the case of fine dust particles. The proposed method will save performing laser diffraction analysis measurements on samples using only sieve analysis data.

4 Conclusions

The moving, cars continuously suspend silt in the air and cause pollution. The high motor vehicle traffic, leads to 'cleaning' of the roadway and, on the other hand, to greater use of the road surface and to deterioration of its condition, which creates conditions for the formation and retention of silt.

For modelling of transport-related PM pollution and assessment of the environmental impact of PM, information on the road litter content along transport arteries is needed. In practice, such a database does not exist in Bulgaria. In this regard, the use of a neural network can facilitate the creation of a database on the amounts of fine particulate matter with sizes less than or equal to 75 μm.

The method is proposed in which more than one instrument is used to complement the various data, the question arises as to the efficiency, speed, accuracy and cost of the measurements made. In this case, one can make part of the measurements and the other (which in the real case will be slower and more expensive) can be predicted with the proposed method.

Acknowledgment. This research has been supported by contract №KP-06-H27/12 of 11.12.2018 "Modeling and elaboration of complex system for selection of transport technology in transport network" funded by the National Science Fund of the Ministry of Education and Science of Bulgaria.

References

1. Zhang, Y., et al.: Fine particle-bound polycyclic aromatic hydrocarbons (PAHs) at an urban site of Wuhan, central China: characteristics, potential sources and cancer risks apportionment. Environ. Pollut. **246**, 319–327 (2019)
2. Buckeridge, D.L., Glazier, R., Harvey, B.J., Escobar, M., Amrhein, C., Frank, J.: Effect of motor vehicle emissions on respiratory health in an urban area. Environ. Health Perspect. **110**(3), 293–300 (2002)
3. Bhaskar, V.S., Sharma, M.: Assessment of fugitive road dust emissions in Kanpur, India. Transp. Res. Part D **13**, 400–403 (2008)
4. Fan, Y., et al.: Acute short-term exposures to PM2.5 generated by vehicular emissions and cardiopulmonary effects in older adults. Epidemiology **17**(6), S213–S214 (2006)
5. Fulper, C.R., et al.: Methods of characterizing the distribution of exhaust emissions from light-duty, gasoline-powered motor vehicles in the U.S. fleet. J. Air Waste Manag. Assoc. **60**, 1376–1387 (2010)

6. HEI Panel on the Health Effects of Traffic-Related Air Pollution, 2010. Traffic-Related Air Pollution: A Critical Review of the Literature on Emissions, Exposure, and Health Effects, HEI Special Report, 17, Health Effects Institute, Boston, MA
7. Mathissen, M., Scheer, V., Vogt, R., Benter, T.: Investigation on the potential generation of ultrafine particles from the tire–road interface. Atmos. Environ. **45**(34), 6172–6179 (2011)
8. Masiol, M., Hofer, A., Squizzato, S., Piazza, R., Rampazzo, G., Pavoni, B.: Carcinogenic and mutagenic risk associated to airborne particle-phase polycyclic aromatic hydrocarbons: a source apportionment. Atmos. Environ. **60**, 375–382 (2012)
9. Mauderly, J.L.: Toxicological and epidemiological evidence for health risks from inhaengine emissions. Environ. Health Perspect. **102**, 165–171 (1994)
10. Robert, M.A., VanBergen, S., Kleeman, M.J., Jakober, C.A.: Size and composition distributions of particulate matter emissions: Part 1 - Light-duty diesel vehicles. J. Air Waste Manag. Assoc. **57**, 1414–1428 (2007)
11. Schauer, J.J., et al.: Characterization of metals emitted from motor vehicles. Health Effects Inst. 88 (2006)
12. Kam, W., Liacos, J.W., Schauer, J.J., Delfino, R.J., Sioutas, C.: Size-segregated composition of particulate matter (PM) in major roadways and surface streets. Atmos. Environ. **55**, 90–97 (2012)
13. Laidlaw, M.A.S., Zahran, S., Mielke, H.W., Taylor, M.P., Filippelli, G.M.: Re-suspension of lead contaminated urban soil as a dominant source of atmospheric lead in Birmingham, Chicago, Detroit and Pittsburgh, USA. Atmos. Environ. **49**, 302–310 (2012)
14. Tervahattu, H., Kupiainen, K.J., Raisanen, M., Makela, T., Hillamo, R.: Generation of urban road dust from anti-skid and asphalt concrete aggregates. J. Harazdous Mater. **132**, 39–46 (2006)
15. Thorpe, A., Harrison, R.M.: Sources and properties of non-exhaust particulate matter from road traffic: a review. Sci. Total Environ. **400**(1–3), 270–282 (2008)
16. Compilation of Air Pollutant Emission Factors. 5th ed. (AP-42), 2011. Vol I: Stationary Point and Area Sources. Section 13.2.1 Paved Roads: Measurement Policy Group Office of Air Quality Planning and Standards, U.S. Environmental Protection Agency (2011)
17. Haddad, E., et al.: Comprehensive primary particulate organic characterization of vehicular exhaust emissions in France. Atmos. Environ. **43**(39), 6190–6198 (2009)
18. Mancilla, Y., Mendoza, A.: A tunnel study to characterize PM2.5 emissions from gasoline-powered vehicles in Monterrey Mexico. Atmos. Environ. **59**, 449–460 (2012)
19. Rogge, W.F., Hildemann, L.M., Mazurek, M.A., Cass, G.R.: Sources of fine organic aerosol. 3. Road dust, tire debris, and organometallic brake lining dust: roads as sources and sinks. Environ. Sci. Technol. **27**, 1892 (1993)
20. Ntziachristos, L., Ning, Z., Geller, M.D., Sheesley, R.J., Schauer, J.J., Sioutas, C.: Fine, ultrafine and nanoparticle trace element compositions near a major freeway with a highheavy-dutydiesel fraction. Atmos. Environ. **41**, 5684–5696 (2007)
21. Peltier, R.E., Cromar, K.R., Ma, Y., Fan, Z.-H., Lippmann, M.: Spatial and seasonal distribution of aerosol chemical components in New York City: (2) Road dust and other tracers of traffic-generated air pollution. J. Exposure Sci. Environ. Epidemiol. **21**(5), 484–494 (2011)
22. Tsonev, M.: Road Construction, Textbook for Technical Schools of Construction. Technika Publishing House, Sofia (1992)
23. Bender, G., Terstriep, M.L.: Effectiveness of street sweeping in urban runoff pollution control. Sci. Total Environ. **33**, 185 (1984)
24. Bogacki, M., Oleniacz, R., Rzeszutek, M., Szulecka, A., Mazur, M.: The impact of street cleaning on particulate matter air concentrations: a case study of a street canyon in Krakow (Poland). In: E3S Web of Conferences 45, 00009, INFRAEKO 2018 (2018)
25. ISO 13320:2020 Particle size analysis - Laser diffraction methods

26. Hagan, M., Demuth, H., Beale, M.: Neural Network Design. PWS Publishing, Boston (1996)
27. Mater, A.C., Coote, M.L.: Deep learning in chemistry. J. Chem. Inf. Model. **59**(6), 2545–2559 (2019)
28. Kelleher, J.D.: Deep Learning. MIT Press (2019)
29. Esteva, A., et al.: A guide to deep learning in healthcare. Nat. Med. **25**(1), 24–29 (2019)
30. Ivanova, D., Nikolaeva, Z.: Pollution with silt loading on major roads in the city of Burgas Bulgaria. J. Environ. Prot. Ecol. **22**(3), 968–975 (2021)

Emotion Recognition Using Convolutional Neural Network

Todor Petkov(✉), Aleks Titanyan, Veselina Bureva, and Stanislav Popov

Assen Zlatarov Univeristy, Burgas, Bulgaria
{todor_petkov,vbureva}@btu.bg

Abstract. Human emotions are significant to a person's mental health. In the field of Affective Computing human emotions are one of the main subjects that play an important role. In order to observe emotions of the human a lot of characteristics can be used such as speech, skin responses, heart rate or facial expressions. In the paper the process of emotion recognition using artificial neural network is described. The training set consists of grayscale images which are 48×48 pixels and are divided into 7 categories Angry, Disgust, Fear, Happy, Sad, Surprise, Neutral. The Convolutional Neural Network is used for the purpose of emotion recognition along with its pooling and convolutional layers. At the end of the process the trained neural network can successfully recognize the human emotion.

Keywords: Emotion Recognition · Artificial Neural Network · Convolutional Neural Network

1 Introduction

1.1 Artificial Neural Networks

Artificial neural networks (ANNs) [1, 2] are inspired by the biological neural network of humans and animals. The main function of the biological neuron is to receive and send bio-chemical information to other connected neurons. The model of the biological neuron is shown in Fig. 1.

In the cell body the incoming signals from the other neurons are summarized and transferred to the other neurons. The synapses are the points of connections of the neuron and other neurons. The dendrites carry electrical signals to the cell body. The axon carries the signals from the cell body out to the other neurons [2, 8].

The model of Artificial neuron is shown in Fig. 2. On the input of the neuron, the input signals denoted by p and representing the data features are multiplied by the weight vector of the neuron denoted by w and the result is summed with the bias value b. The result n enters into a functional block which applies the activation function of the neuron and the actual output of the neuron is denoted by a [2, 8].

A single neuron is not appropriate to solve complex problems. Thus, in such cases a different kind of ANNs architectures is used [3, 4].

S. Sotirov et al. (Eds.): BioInfoMed 2022, LNNS 658, pp. 32–38, 2023.
https://doi.org/10.1007/978-3-031-31069-0_4

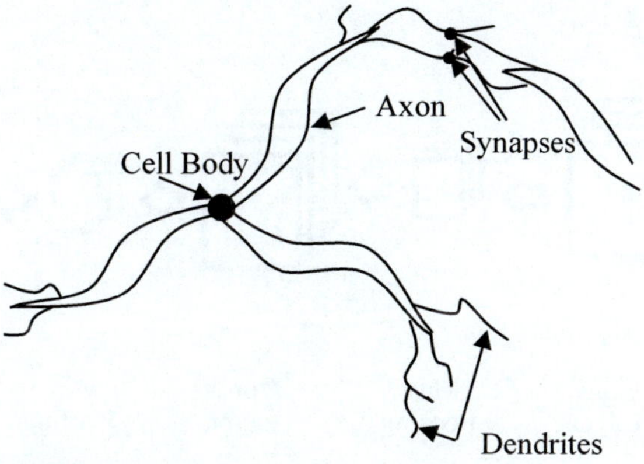

Fig. 1. A figure of the model of biological neuron

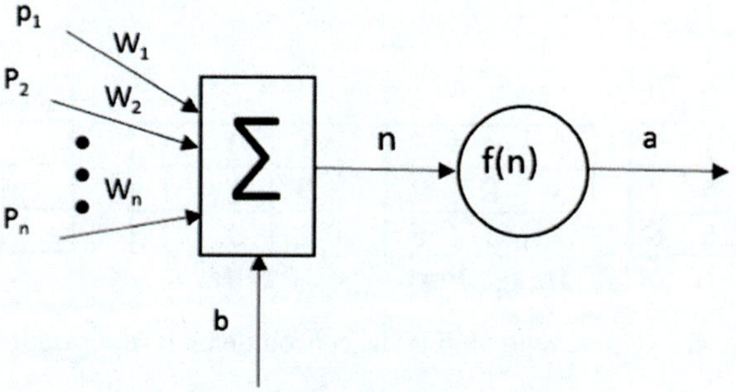

Fig. 2. Model of artificial neuron

1.2 Convolutional Neural Networks

Convolutional Neural Networks (CNNs) take advantage of the fact that the input consists of images and they constrain the architecture in a more sensible way [5, 6, 9, 10]. The layers of a CNNs have neurons arranged in 3 dimensions: width, height and depth. Unlike standard ANNs, the neurons within any given layer will only connect to a small region of the layer preceding it. In Fig. 3 is shown a simple CNN architecture:

The input layer of the architecture is the actual image input of the CNN, it holds the pixel values. The convolutional layer will determine the output of the neurons which are connected to local regions on the input through the calculation of the product between weights and connected input volume [5–7, 9, 11]. The pooling layer simplifies the spatial dimensionality of the given input by reducing the number of parameters within the activation. The Neural Network performs the standard ANNs calculations and tries to produce the actual classes as an output of the ANN. The CNNs are able to transform the original input layer by using convolutional and downsampling techniques to produce class scores for classification. In the Fig. 4 is shown a visual representation of convolutional layer operations.

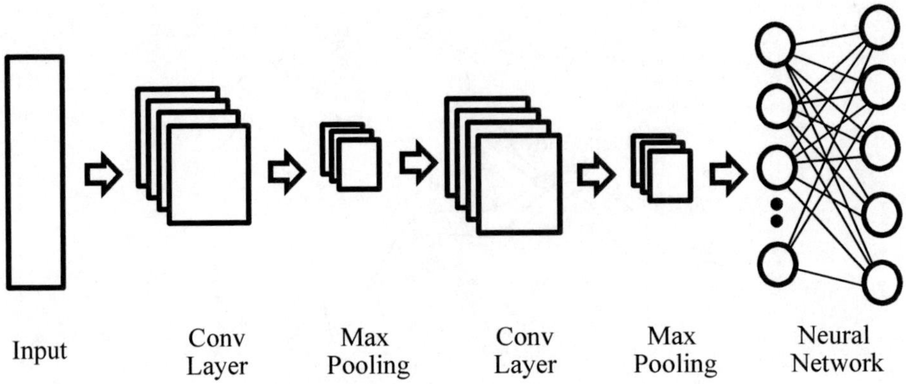

Fig. 3. A simple architecture of CNN

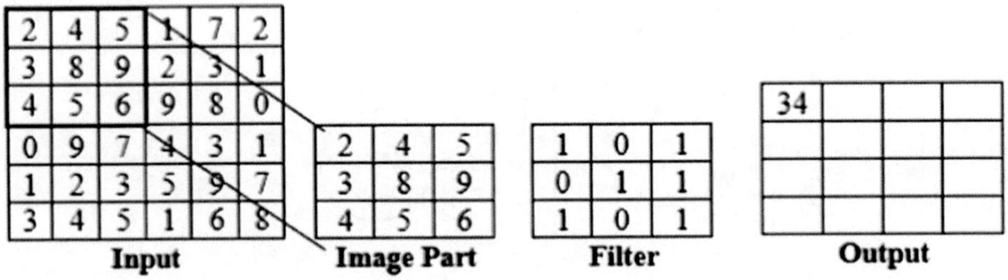

Fig. 4. A figure representing the convolutional layer operations

The input image is divided into a grid of pixels as each part is convolved with a filter and the actual output is calculated.

There are different kind of methods for recognition of human emotions for instance "Four-layer ConvNet to facial emotion recognition with minimal epochs and the significance of data diversity" is the one way to recognize the emotions the method for recognition is based on CNN also. The other way to recognize human emotion is by using facial expressions by using coefficient that describes the elements of facial expressions, the method uses MLP neural network and k-NN classifier.

2 Results and Discussion

The goal of the research is to create a CNN model that is able to recognize human emotions. The input of the network represents the grayscale image (48 × 48 pixels) of the human face emotion. The training set consists of 28709 images as each of them can be in one of the classes shown in Table 1.

Table 1. Classes and corresponding emotion of the image

Class	Emotion
1	Angry
2	Disgust
3	Fear
4	Happy
5	Neutral
6	Sad
7	Surprise

In Table 2 are shown some of the training images that are used to train the CNN.

Table 2. A set of training images

Image	Emotion
	Angry
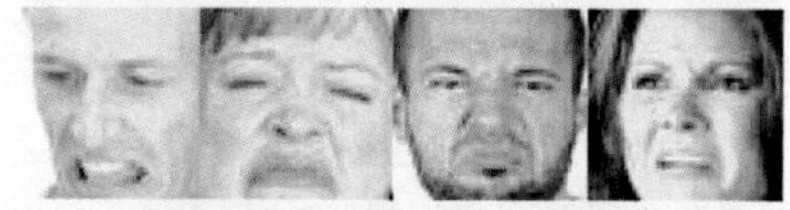	Disgust
	Fear
	Happy
	Neutral
	Sad
	Surprise

The architecture of the CNN is shown in Fig. 5. The input layer is of 48×48 size while the next three layers are convolutional and each of them has different number of filters.

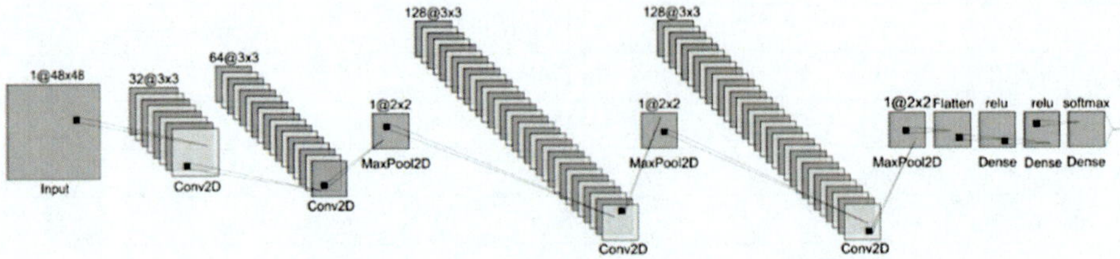

Fig. 5. The architecture of CNN of Emotion Recognition

Each input image enters into the CNN network and after each convolutional layer the network extracts more abstract information for the emotion. At the end of the convolutional layers there is a standard ANN where each layer applies Rectified Linear Unit (ReLU) activation function (see Fig. 5) (Fig. 6).

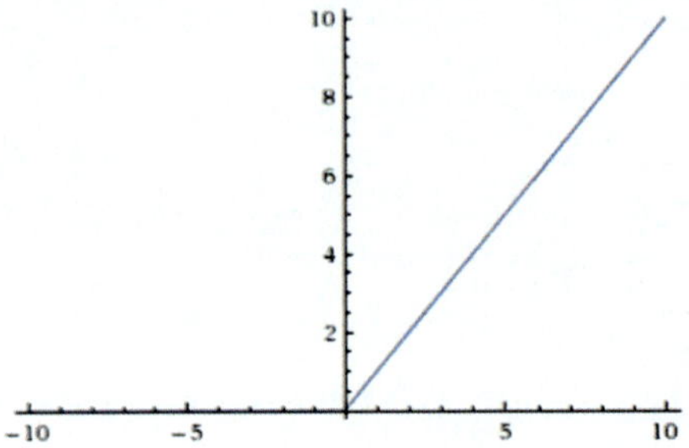

$$f(x) = \max(0, x)$$

Fig. 6. The ReLU activation function

The last layer of the CNN has 7 outputs; each output represents one of the emotions which has to be recognized by the network, it applies softmax activation function:

$$f(x)_i = \frac{e^{x_i}}{\sum_{j=1}^{n} e^{x_j}}$$

In the Fig. 7 is shown the graphic of model loss of each epoch of the CNN. It can be seen that during each epoch the actual loss decreases of training vectors. In Fig. 8 is shown the graphic of accuracy of the process. The accuracy can be between 0 and 1 and it can be seen that with each iteration the algorithm reaches better predictability in comparison with the previous iteration.

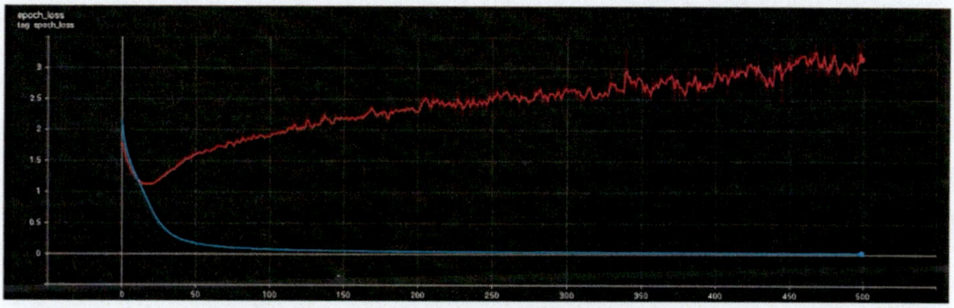

Fig. 7. The loss of each epoch of the algorithm

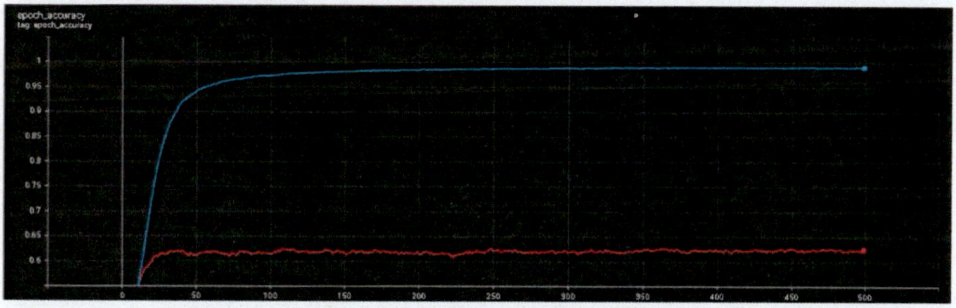

Fig. 8. Accuracy of the algorithm

3 Conclusion

The CNN artificial neural network is used to recognize seven human emotions (angry, disgust, fear, happy, neutral, sad and surprise). The CNN is trained with a set of images for each human emotion. During the learning trial it adjusts its weights and biases for better performance. The CNN was validated with a set of images for each emotion and the result of the testing and training performance was visualized with a graphic. At the end of the process the CNN was trained good enough to predict the human emotion and the model can be successfully integrated in different kinds of applications. In comparison with other methods this method for emotion recognition can be successfully applied for such kind of applications.

Acknowledgements. The authors are thankful for the support provided by Project "Analysis and modelling of artificial intelligence algorithms and their application", №NIH – 462/2021.

References

1. Kukeja, H., Bharath, N., Siddesh, C.S.: An Introduction to Artificial Neural Network, IJARIIE-ISSN (O)-2395-4396, Vol-1 Issue-5 (2016)
2. Hagan, M.: Neural Network Design, Second Edition

3. da Silva, I.N., Hernane Spatti, D., Andrade Flauzino, R., Liboni, L.H.B., dos Reis Alves, S.F.: Artificial neural network architectures and training processes. In: Artificial Neural Networks. Springer, Cham (2017). https://doi.org/10.1007/978-3-319-43162-8_2
4. Suykens, J.A.K., Vandewalle, J.P.L., De Moor, B.L.R.: Artificial neural networks: architectures and learning rules. In: Artificial Neural Networks for Modelling and Control of Non-Linear Systems. Springer, Boston, MA (1996). https://doi.org/10.1007/978-1-4757-2493-6_2
5. Manaswi, N.K.: Convolutional Neural Networks. In: Deep Learning with Applications Using Python. Apress, Berkeley, CA (2018). https://doi.org/10.1007/978-1-4842-3516-4_6
6. Dhillon, A., Verma, G.K.: Convolutional neural network: a review of models, methodologies and applications to object detection. Prog. Artif. Intell. **9**(2), 85–112 (2019). https://doi.org/10.1007/s13748-019-00203-0
7. Li, Z., Liu, F., Yang, W., Peng, S., Zhou, J.: A survey of convolutional neural networks: analysis, applications, and prospects. In: IEEE Transactions on Neural Networks and Learning Systems. https://doi.org/10.1109/TNNLS.2021.3084827
8. Tacchino, F., Macchiavello, C., Gerace, D., et al.: An artificial neuron implemented on an actual quantum processor. NPJ Quantum Inf. **5**, 26 (2019)
9. Sotirov, S., Sotirova, E., Surchev, S., Petkov, T., Georgieva, V.: Generalized net model of the deep convolutional neural network. In: Uncertainty and Imprecision in Decision Making and Decision Support: New Challenges, Solutions and Perspectives. IWIFSGN 2018. AISC, vol. 1081. Springer, Cham (2021). https://doi.org/10.1007/978-3-030-47024-1_14
10. Todorov, M., Avramova-Todorova, G., Sotirov, S.: Generalized network model for career guidance using emotional intelligence, self-assessment and attitudes toward the employer. In: 2021 Big Data, Knowledge and Control Systems Engineering (BdKCSE), 2021, pp. 1–4 (2021)
11. Hristov, S., Baltov, A., Sotirov, S.: Functional outcome prediction of operated proximal humerus fractures by means of artificial neural networks. In: Sotirov, S.S., Pencheva, T., Kacprzyk, J., Atanassov, K.T., Sotirova, E., Staneva, G. (eds.) Contemporary Methods in Bioinformatics and Biomedicine and Their Applications. BioInfoMed 2020. LNNS, vol. 374. Springer, Cham (2022). https://doi.org/10.1007/978-3-030-96638-6_23

Water Safety and Toxicity Assessment Using Real Time Sensor Measurements and Fuzzy Logic Data Processing

Husein Yemendzhiev[1](✉), Plamena Zlateva[2,3], and Valentin Nenov[1]

[1] Prof. Asen Zlatarov University, Prof. Yakimov Blvd. No. 1, 8010 Burgas, Bulgaria
{hyemendzhiev,vnenov}@btu.bg

[2] University of National and World Economy, 8 December Str., 1700 Sofia, Bulgaria
plamzlateva@unwe.bg

[3] Institute of Robotics, Bulgarian Academy of Sciences, Acad. Georgi Bonchev Str., Bl. 2, 1113 Sofia, Bulgaria

Abstract. The water quality monitoring is a crucial part of the efficient water management. However, the conventional chemical analyses could not offer real time toxicity assessment which creates significant risks for human and ecosystems health. Analysing the entire diversity of toxic compounds potentially presented in the water is not an easy, nor economical task and sometimes it is impossible, as well. This group of chemicals varies from heavy metals ions to complex organic molecules and bio-toxins. Finding a general indicator for the toxicity presented in water and aquatic environments could be an important step toward the development of early warning systems to support decision making in water monitoring. In this study, Fuzzy logic approach is used for integration and evaluation of the specific signal patterns obtained from the electro conductivity measurement combined with the data received form a custom build bio-electrochemical sensor. Two different scenarios were studied, being model cases of organic toxic compounds and heavy metals in fresh water.

Keywords: Toxicity evaluation · hazardous materials · water · fuzzy logic

1 Introduction

Various parameters are used to describe the biological, chemical and physical properties of water. However, there is not a general method for assessing the toxicity in water which puts a high level of risk for human and environmental health. In recent years, the fuzzy logic based methods have demonstrated to be appropriated to address uncertainty and subjectivity in environmental issues [1]. Presence of toxic compounds in water could affect some of the water quality parameters which are not directly related to the inhibitory influence towards the living organism. Our recent experiments and results demonstrated good level of correlation between conventionally measured pH and electric conductivity (EC) of water and the presence and concentration of phenol and copper ions in water [2, 3]. Although, these data alone could not be used for toxicity evaluation it could

S. Sotirov et al. (Eds.): BioInfoMed 2022, LNNS 658, pp. 39–46, 2023.
https://doi.org/10.1007/978-3-031-31069-0_5

be a secondary criterion for approximation of the chemical nature of toxic compound presented. The toxic influence itself could be evaluated only by measuring the inhibition of specific biological function of living organisms such as respiration, genes structure and expression or catabolic activity towards specific substrates.

Such a complex evaluation needs a lot of data of different quality which have to be gathered and processed. The available information is usually described by a number of variables characterizing the water quality dynamics. The toxic influence of chemical compounds on the living organisms is non-linear in terms of concentration-chemical structure and toxic effect observed. Furthermore, many of the variables can be estimated only by expensive laboratory analyses. Fuzzy logic seems to be and effective approach to process such data sets [4–6]. The advantage of this approach is the opportunity to use qualitative (expert) knowledge and imprecise (inexact) information about the water status and eventual solving of wrong data interpretations of a complex sets of multidimensional data by using only empirical methods [7]. Nevertheless, there is not enough information about fuzzy modelling of the toxicity as a general parameter [8, 9].

Two major groups of toxic chemicals, namely heavy metals and organic toxic compounds such as aromatic hydrocarbons are the most commonly found in environment. Heavy metals are highly poisonous even at very small concentration, becoming toxic when they are not metabolized by the body and accumulate in the cells and tissues. Copper for instance, has the highest numbers of cases reported worldwide. It is an essential element for plant nutrition but it becomes toxic at high concentrations. It should be mentioned that, the toxic effect of some heavy metals could vary in different type of water. Water quality parameters such as total hardness, alkalinity, pH and dissolved organic carbon can modify the biological response to heavy metals. As an example, the copper toxicity is higher in water with low hardness and low concentration of dissolved organic carbone [10]. Phenolic compounds form the other hand are toxic aromatic hydrocarbons and they are common industrial substances involve in the production of chemicals such as xylenes, oils, plastics and several pharmaceuticals. The industrial effluent containing phenol derivatives are very resistant to biological treatment and could became a great environment and human health concern due to the negative influence of this group of chemicals on the general biochemistry of the cell, nephro- and hepato-toxicity [11].

The aim of the study is to apply Fuzzy logic modelling to process the data obtained from the measurements of a physical parameter of water such as the electrical conductivity and the response of bio-electrochemical sensor (based on living bacterial cells) when coper ions and phenol in different contraptions are presented in freshwater samples. Finding signal profiles specific for these two groups of pollutants could be used for toxicity level evaluation and for initial assessment of the chemical nature of the toxic compounds in water. Chemicals.

2 Experimental

2.1 Bio-electrochemical Sensor Construction and Operation

In this study, Microbial Fuel Cell type micro reactor (MFC) was used as a bio-electrochemical sensor. The MFC was assembled as a cylindrical plexiglass body consists of two identical chambers (for anode and cathode placement) with total volume of 50 dm^3 each. The two compartments were separated by Nafion® 424 perfluorinted proton exchange membrane with 2 cm diameter. The electrodes were made of carbon cloth and stainless steel mesh as a current collector. The bio-anode (which contained electrochemically active bacteria attached to the surface of the carbon cloth) and cathode were with circular shape, diameter of 45 mm and distance between them 20 mm. To close the circuit, the electrodes were connected with external electric circuit loaded by 1 kΩ resistor. The anode compartment of the MFC reactor which reached the operational state was fed by the tested water-toxic compound models. The cathode compartment was loaded 2% ferricyanide as a terminal electron acceptor.

2.2 Microorganisms and Cultivation Conditions

The electrochemically active microorganisms (electrogenes) were isolated from bottom sediments of "Poda" Protected Site located on the outfall of Lake "Uzungeren", south of Burgas and cultivated according to the procedure described by Mersinkova et al. 2020 [12]. The culture was suspended in fresh nutrient medium (LB with acetate) to a concentration of 10^7 CFU/dm^3. It was then loaded in the anode chamber of the MFC for biofilm formation and cell immobilization.

2.3 Toxicity Detection

Toxicity was evaluated in a model system of water with different contrations of phenol and copper ions. The tested concentrations of the pollutants were as follows: 0; 5 and 20 mg/dm^3 which are bellow mentioned as "low", "middle" and "high toxicity" levels respectively. Electro conductivity and biosensor signal for each model systems were profiled. Electrical conductivity (EC) was recorded by a HQ30d flexi conductometer (Hach Lange). The electrogenic bacteria on the anode surface of the MFC and their inhibition after presenting the toxic agents was used for evaluation of the bio-toxicity. As a result, the signal obtained from the MFC (voltage) decreases which indicates toxic influence.

All the measurements are performed in triple and the presented data are the mean values of this measurements.

2.4 Fuzzy Logic Modelling

The main idea of this study is to propose a simple and reliable way to both determine type of the water pollutant and to evaluate the degree of its toxicity/concentration. The proposed fuzzy logic model is based on the assumption that it is not necessary to carry out numerous laboratory analyses to obtain an initial, indicative and rapid assessment of the safety and toxicity of water. The fuzzy logic model is designed as a system with two inputs and one output which provides information for the type and concentration of certain water polluting hazardous materials or suggests that additional testing have to be performed.

The two inputs of the fuzzy system are determined based on expert knowledge and real experiments: *Input 1* "Conductivity" in μS/cm and *Input 2* "MFC Signal" in mV.

The fuzzy logic system output is defined as the linguistic variable "Result". The two inputs are described by four linguistic variables: 1) "Very Low (VL)", 2) "Low (L)", 3) "Middle (M)", and 4) "High (H)". These linguistic variables are assessed in the given intervals using trapezoidal membership functions as shown in Table 1.

Table 1. Numeral values of membership functions and linguistic variables of the *Input 1* and *Input 2*

Input 1 *Conductivity*, μS/cm	*Input 2* *MFC Signal*, mV	Linguistic variable
[0, 0, 2, 3.6]	[0, 0, 0.9, 1.3]	*Very Low*
[2, 3.6, 4.4, 20]	[0.9, 1.3, 2.3, 7.7]	*Low*
[4.4, 20, 26, 65]	[2.3, 7.7, 10.5, 18]	*Middle*
[26, 65, 80, 90]	[10.5, 18, 23, 23]	*High*

The output variable "*Result*" is described by six linguistic variables as follows:

- *R1* "*Clean water*";
- *R2* "*Water with a low concentration of phenol*";
- *R3* "*Water with a high concentration of phenol*";
- *R4* "*Water with a low concentration of copper ions* ";
- *R5* "*Water with a high concentration of copper ions*";
- *R6* "*Uncertainty – more research is needed*".

The linguistic variables, corresponding of the output are defined by six trapezoid membership functions in the interval [0, 30] as shown in Table 2.

The inference rules of the fuzzy logic system are defined through the "IF - THEN"-clause based on expert knowledge and real experiments. Here, the number of rules in the knowledge base (logic matrix) is 16 and their description is presented in Table 3. In particular, some of the inference rules are defined as follows:

IF "*Conductivity*" is "Very Low" and "*MFC SIGNAL*" is "Very Low" THEN "*Result*" is "R3";
IF "*Conductivity*" is "Very Low" and "*MFC SIGNAL*" is "Low" THEN "*Result*" is "R2";
IF "*Conductivity*" is "Low" and "*MFC SIGNAL*" is "High" THEN "*Result*" is "R1".

The fuzzy logic system is designed in MATLAB computer environment using Fuzzy Logic Toolbox. The calculation of the output variable is based on Mamdani's inference machines, max/min operations and center of gravity defuzzification. The crisp value of the output variable belongs to interval [0, 30].

For determination of the type of water pollutant component and its toxicity/concentration the interval [0, 30] is divided into six equal parts as shown in Table 2. Thus, the crisp value of the output variable "*Result*" is interpreted as a classification

Table 2. Numeral values of membership functions and crisp intervals of the variable *Result*

	*R*1	*R*2	*R*3	*R*4	*R*5	*R*6
Fuzzy Output	[0, 0, 4, 6]	[4, 6, 9, 11]	[9, 11, 14, 16]	[14, 16, 19, 21]	[19, 21, 24, 26]	[24, 26, 30, 30]
Crisp Output	[0, 5]	[5, 10)	[10, 15)	[15, 20)	[20, 25)	[25, 30]

Table 3. The inference rules of the fuzzy logic system

Input 1 *Conductivity*, μS/cm	Input 2 *MFC SIGNAL*, mV	System output *Result*
Very Low	Very Low	R3
Very Low	Low	R2
Low	High	R1
Middle	Middle	R4
High	Low	R5
Very Low	Middle or High	R6
Low	Very Low	R6
Low	Low or Middle	R6
Middle	Very Low	R6
Middle	Low or High	R6
High	Very Low	R6
High	Middle or High	R6

criterion. The following rules for the crisp values of the output variable "*Result*" are defined:

IF "*Result*" ∈ [0, 5] THEN *the water is clean* (R1);
IF "*Result*" ∈ [5, 10] THEN *the water has a low concentration of phenol* (R2);
IF "*Result*" ∈ [10, 15] THEN *the water has a high concentration of phenol* (R3);
IF "*Result*" ∈ [15, 20] THEN *the water has a low concentration of copper ions* (R4);
IF "*Result*" ∈ [20, 25] THEN *the water has a low concentration of copper ions* (R5);
IF "*Result*" ∈ [25, 30] THEN *there is uncertainty and more research is needed* (R6).

3 Results and Discussions

For the purposes of this study, a lab scale installation has been prepared including the sensors and electrodes used. The installation was fed with the tested model solutions representing fresh water with two different toxic components in different concentrations. The data recorded about the electrical conductivity of the water samples and bio-electrochemical signal during these set of experiments are presented in Figs. 1 and 2.

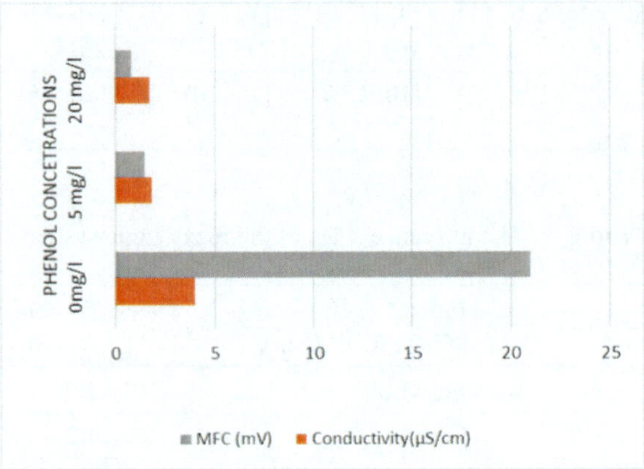

Fig. 1. Effect of different concentrations of phenol in freshwater on the EC and MFC biosensor signal.

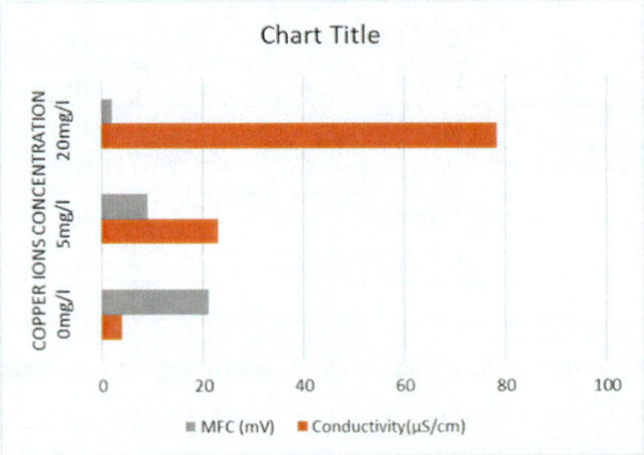

Fig. 2. Effect of different concentrations of Cu^{2+} in freshwater on the EC and MFC biosensor signal.

From this initial data, even in their raw format it is obvious that there is a certain amount of correlation between the status of the tested water and parameter values observed. Even more, in dependence with the type of the toxic compound applied we observed some characteristic profiles of the signals. For instance, both phenol and copper ions in the water influences negatively the bacteria in the bio-electrochemical sensor and the signal drops by the increasing concentration of the toxic agent. This observation is an evidence for the biological response of bacteria to the negative influence of the tested substances. However, the two chemical pollutants differ by their influence on the conductivity and it could be easily used as a criterion for evaluation of the source of the toxic influence – if the conductivity is significantly increased then the toxic compound should be copper ions. Contrary, if there is not a change in this parameter, then the presented pollutant is aromatic hydrocarbon.

Various scenarious have been simulated using the proposed Fuzzy Logic model. Some of the rules of logical inference with specific numerical values of the input variables are listed bellow:

- IF "Conductivity" = 4 and "MFC Signal" = 21 THEN *"Safe water" = R1*;
- IF "Conductivity" = 5 and "MFC Signal" = 23 THEN *"Safe water" = R1*;
- IF "Conductivity" = 1.82 and "MFC Signal" = 1.5 THEN *"Water with a low level of toxicity – phenolic compounds" = R2*;
- IF "Conductivity" = 1.98 and "MFC Signal" = 2.2 THEN *"Water with a low level of toxicity - phenolic compounds" = R2*;
- IF "Conductivity" = 1.72 and "MFC Signal" = 0.8 THEN *"Water with a high level of toxicity - phenolic compounds" = R3*;
- IF "Conductivity" = 1.3 and "MFC Signal" = 1.0 THEN *"Water with a high level of toxicity - phenolic compounds" = R3*;
- IF "Conductivity" = 23.1 and "MFC Signal" = 9 THEN *"Water with a low level of toxicity – heavy metals ions" = R4*;
- IF "Conductivity" = 33 and "MFC Signal" = 12 THEN *"Water with a low level of toxicity – heavy metals ions" = R4*;
- IF "Conductivity" = 78.3 and "MFC Signal" = 2.0 THEN *"Water with a high level of toxicity – heavy metals ions" = R5*;
- IF "Conductivity" = 63 and "MFC Signal" = 1.2 THEN *"Water with a high level of toxicity – heavy metals ions" = R5*;
- IF "Conductivity" = 16 and "MFC Signal" = 21 THEN *"Uncertainty – more data/analyses is needed" = R6*;
- IF "Conductivity" = 3.3 and "MFC Signal" = 9.4 THEN *"Uncertainty – more data/analyses is needed" = R6*;
- IF "Conductivity" = 76 and "MFC Signal" = 10 THEN *"Uncertainty – more data/analyses is needed" = R6*;

The input data and the corresponding output simulation results described above are visualized in Table 4.

Table 4. Input data and corresponding outputs from their Fuzzy Logic processing

	R1	R1	R2	R2	R3	R3	R4	R4	R5	R5	R6	R6	R6
Conductivity	**4**	5	**1.82**	1.98	**1.72**	1.3	**23.1**	33	**78.3**	63	16	3.3	76
MFC Signal	**21**	23	**1.5**	2.2	**0.8**	1.0	**9**	12	**2.0**	1.2	21	9.4	10

Comparing the simulation results with the experimental data confirms the adequacy of the Fuzzy Logic model and its possible application for initial assessment of the water safety and toxicity in the studied two scenarios.

In conclusion, this study reveals the potential contribution of the Fuzzy Logic method in the overall assessment of the water safety in terms of the toxic pollutants and human and environmental health. Application of mathematical modelling and data processing is the first step toward sophistication of the water quality monitoring by building a new generation of systems for integrated data collection and processing with the capability to evaluate the levels of toxicity in water and to make predictions about the source and nature of the toxic compounds presented.

Acknowledgements. The work supported by Operational Program "Science and Education for Smart Growth" under the UNITe BG05M2OP001-1.001-0004-C01 and Clean&Circle BG05M2OP001-1.002-0019 projects.

References

1. Pauzi Abdullah, Md., Waseem, S., Raman Bai, V., Mohsin, I.: Development of new water quality model using fuzzy logic system for Malaysia. Open Environ. Sci. **2**, 101–106 (2008)
2. Yemendzhiev, H., Koleva, R., Nenov, V., Georgieva, V.: Opportunity to detect hazardous materials in water using intercriteria analysis. In: Sotirov, S.S., Pencheva, T., Kacprzyk, J., Atanassov, K.T., Sotirova, E., Staneva, G. (eds.) BioInfoMed 2020. LNCS, vol. 374, pp. 285–295. Springer, Cham (2022). https://doi.org/10.1007/978-3-030-96638-6_31
3. Yemendzhiev, H., Mersinkova, Y., Koleva, R., Peeva, G., Nenov, V.: Bio-electrochemical systems – state of the art: biology, electrochemistry and application in wastewater management. In: Serbula, S. (ed.) Proceedings of the EcoTER22, Beograd, Serbia, vol. 29, pp. 25–32 (2022). ISBN 978-86-6305-123-2
4. Alexieva, Z., Gerginova, M., Zlateva, P., Manasiev, J., Ivanova, D., Dimova, N.: Monitoring of aromatic pollutants biodegradation. Biochem. Eng. J. **40**, 233–240 (2008)
5. Satishkumar, B., Chidambaram, M.: Bioproc. Eng. **20**, 127–132 (1999)
6. Riid, A., Rustern, E.: Fuzzy modelling and control of fed-batch fermentation. In: Computational Intelligence and Applications. Physica Verlag (1999)
7. Driankov, D., Hellendoorn, H., Reinfrank, M.: An Introduction to Fuzzy Control. Springer, New York (1993)
8. Yoo, C., Lee, I.V.: J. Biotechnol. **105**, 135–163 (2003)
9. Annadurai, G., Babu, S.R., Srinivasamoortthy, V.: Bioproc. Eng. **23**, 599–666 (2000)
10. Gao, Y., Feng, J., Zhu, J., Zhu, L.: Predicting copper toxicity in zebrafish larvae under complex water chemistry conditions by using a toxicokinetic–toxicodynamic model. J. Hazardous Mater. **400** (2020)
11. Medical Management Guidelines for Phenol (C6H6O). Agency for Toxic Substances and Disease Registry. U.S. Department of Health and Human Services 21 October 2014. Accessed 8 Aug 2015
12. Mersinkova, Y., Yemendzhiev, H.: Identification and characterization of natural habitats of electrochemically active bacteria. J. Adv. Biol. Biotechnol. **23**(1), 19–25 (2020)

An Intuitionistic Fuzzy Estimation Approach on a Magnetic Resonance Imaging

Sotir Sotirov, Todor Kostadinov(✉), and Stoyan Hristov

Prof. Assen Zlatarov University, Burgas, Bulgaria
ssotirov@btu.bg, kostadinov.todor@yahoo.com, hristov@dr.com

Abstract. This work addresses a novel approach of image estimation using fuzzy logic. The research aims to devise a technique for estimating the quality of tomographic images by measuring the level of uncertainty on an image of interest. This approach can be applied in order to improve the quality of a tomographic image of human bones and thus improve the resulted image, especially in case of fracture or anomaly without compromising it. In order to verify the approach, an experiment on real tomographic images has been carried out.

Keywords: Tomography · Tomographic image estimation · Fuzzy logic

1 Introduction

Since the development of non-invasive imaging techniques, there is a vast improvement in the field of security, materials analysis and also healthcare. There are several directions in the medicine, where the remote imaging methods are widely used and became part of the standard procedures. Some of the directions in medicine, where the remote imaging has the most significant impact are neonatology, internal diseases, neurology and orthopedics. Due to the nature and the structure of the human body, different image techniques are used due to the imaging capabilities of the imaging of the waves used. For instance, in order to observe soft tissues a common echography can be used, as the ultrasonic waves are capable to go through the soft tissues without major attenuation, reflect by the tissue of interest and thus provide a projection of the monitored areas of interest. This technique is viable when applied on soft tissues, as the energy of the emitted waves is not great at a low frequency. On the other hand, when monitoring consistent tissues such as bones, the low frequency ultrasound signals are reflected by the solid surface of the bones and it is impossible to estimate the inner structure of the bone as well as the tissues behind. For that reason, in the medical divisions like neurology and orthopedics a tomography method is applied. The method uses x-ray or high frequency imaging [2] to provide a projection of the soft tissues of the brain, located in the hard bone shell. When a bone structure is observed by this method, an intersection of the bone in the plane of observation is presented. Thus, the method can be used to visualize the outer as well as the inner structure of the bones. Due to the difference in the density of the human flesh, bones and cartilage the energy of the reflected wave is absorbed by

S. Sotirov et al. (Eds.): BioInfoMed 2022, LNNS 658, pp. 47–52, 2023.
https://doi.org/10.1007/978-3-031-31069-0_6

the tissue, which reflects the resulting image, providing a level of uncertainty regarding the boundaries of the bone structure.

The scope of this article is focused on the estimation of the uncertainty [5–7, 10], caused by noises, misfocus or poor resolution, which degrades the quality of the image. In order to test the approach on estimating the bone shape and the quality of the image taken, an experiment on the real images of a shoulder joint have been used. The images have been taken by an "Emotion 16" tomographic device, then by applying fuzzy estimation [8] over the resulted images an assessment will be made in order to decide whether to use the current images or repeat the scan.

2 Image Acquisition

The initial research data consists of series of images in "Digital Imaging and Communication in Medicine" known as DICOM. The images depict the shoulder joint of a female patient in an age of 58 years. In order to make an assessment it is necessary to extract the image raw data from the digital image format – DICOM, which has an object-oriented structure depicted in Fig. 1.

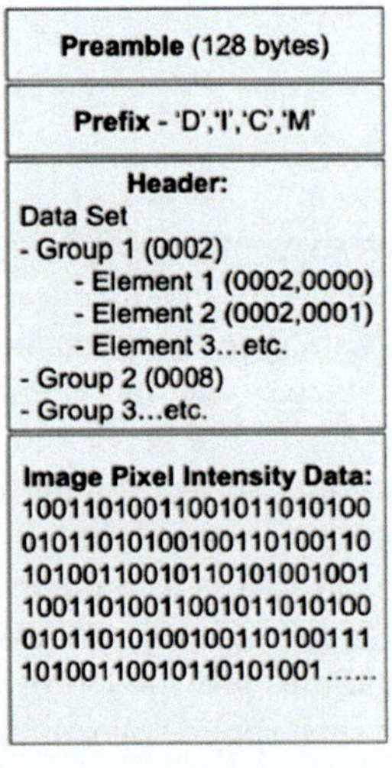

Fig. 1. Structure of the DICOM image format.

The byte stream of the Data Set is placed into the file after the DICOM File Meta Information. The File Meta Information includes identifying information on the encapsulated Data Set. The header consists of a 128 bytes File Preamble, followed by a 4 bytes DICOM prefix, followed by the File Meta Elements [1]. The last data element of a data set is the raw data, which represents the pixels intensities in grayscale. After reading the metainformation file, consisting of encapsulation data as well as patient data and the raw image data, the image of the bone structure can be acquired for later processing and visualized. The dataset consists of 268 images sampled during the scan process of a shoulder joint. The scan is performed while the patient is situated stationary in the center of the tomographic scanner and the scanner samples series of 268 images, increasing the scan angle for each recorded image. The images files, the user information and raw image data are imported in Matlab by a specially devised script for that purpose in order to store, sort and visualize the images for further processing. An example of a tomographic image is presented in Fig. 2.

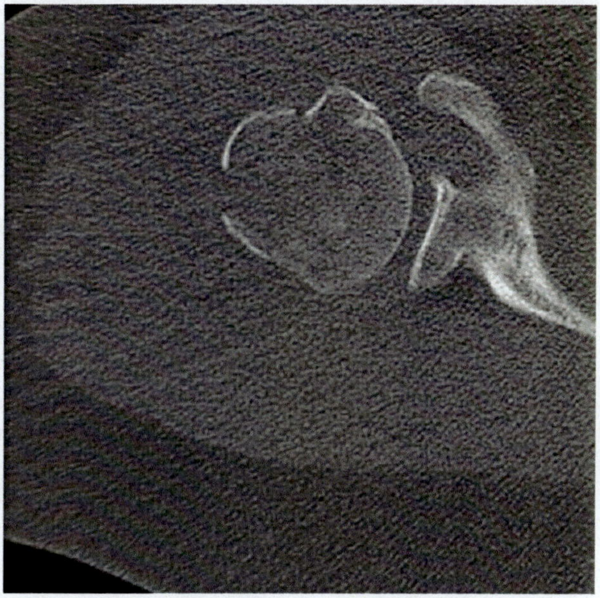

Fig. 2. Arbitrary image of the shoulder joint

As can be seen in Fig. 2, no sharp contour, distinguishing the bones shape can be noted. The image itself is blurred due to the absorption of the energy of the emitted signal by the muscle, skin and tendons.

It can be concluded by Fig. 2 that in case of a low-quality image, the image reading process can be obstructed.

3 Fuzzy Estimation on the Image Quality

As seen from Fig. 2 the tomographic image is depicted in grayscale. It consists of three areas: white, gray and black. The white areas denote the structure of bones, the gray structure denotes flesh and the black area represents the open space. Taking into account a threshold, the shades of gray at a certain level can be interpreted as bone structure and the values above that threshold can be disregarded. In case of noisy picture, the threshold cannot be distinguished, resulting poor image of the bones including uneven surface and lack of bone structure at certain locations. That phenomenon is illustrated in Fig. 3.

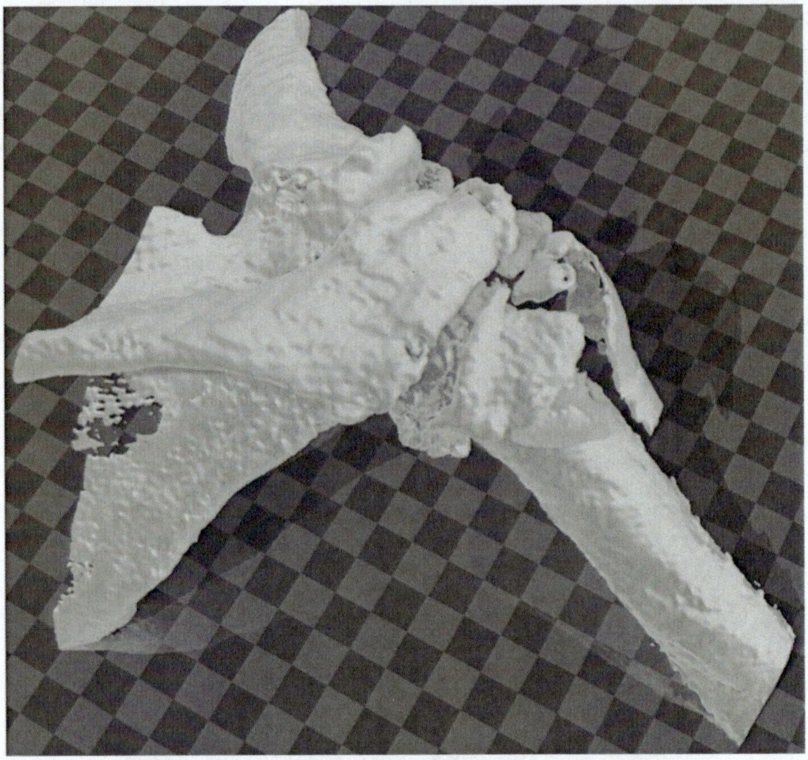

Fig. 3. 3D reconstruction of a scanned shoulder joint

Figure 3 presents the 3D bones and joints reconstruction. Depending on the threshold level there are two possible scenarios – first one is depicted as parts of the bone missing, accompanied by pitted surface of the bones – Fig. 3. The second scenario can be described as excessive bone formations in specific places.

Both of the aforementioned scenarios are caused by improper distinction of flesh structure from the bone structure, resulted by the level of uncertainty for certain boundary pixels.

An alternative way to assess and improve the image approach is the theory of intuitionistic fuzzy sets is presented in [9, 11, 12].

Intuitionistic fuzzy sets see [10] represent an extension of the concept of fuzzy sets, as defined by Zadeh [13], exhibiting function $\mu_A(x)$ defining the membership of an element x to the set A, evaluated in the [0; 1]-interval. The difference between fuzzy sets and intuitionistic fuzzy sets (IFSs) is in the presence of a second function $\nu_A(x)$ defining the non-membership of the element x to the set A, where $\mu_A(x) \in [0; 1]$, $\nu_A(x) \in [0; 1]$, under the condition of $\mu_A(x) + \nu_A(x) \in [0; 1]$.

These intuitionistic fuzzy estimations reflect the degree of each arteries' permeability μ, or arteries' non-permeability ν of the lower limbs, and for them is valid that

$$\mu + \nu \leq 1.$$

In this case

$$\mu = \frac{S_1}{S} \tag{1}$$

where:

S - the total number of pixels.

S_1 - denotes the pixels belonging to the bone structure

$$\nu = \frac{S_2}{S} \tag{2}$$

where:

S_2 – denotes the pixels that do not belong to the bone structure.

In the process of determination whether a pixel belong or not to the bone structure the accuracy of the sampling apparatus must be considered. Usually it has positive or negative deviation in a certain threshold tr or in mathematical terms $\pm\, tr$.

In that case the degree of uncertainty π is where such cases corespond.

$$\pi = \frac{S_3}{S} \tag{3}$$

S_3 - denotes the pixels not belonging to the bone structureпикселите непринадлежащи към костната структура.

The IFS itself is formally denoted by $A = \{\langle x, \mu_A(x), \nu_A(x)\rangle \mid x \in E\}$. Strongly optimistic formula:

$$\langle \mu_1, \nu_1\rangle + \langle \mu_2, \nu_2\rangle + \ldots + \langle \mu_n, \nu_n\rangle = \sum_{i=1}^{n} \langle \mu_i, \nu_i\rangle = \langle 1 - \prod_{i=1}^{n} (1-\mu_i), \prod_{i=1}^{n} \nu_i\rangle; \tag{4}$$

Optimistic formula:

$$\max(\langle \mu_1, \nu_1\rangle, \ldots, \langle \mu_n, \nu_{n)}\rangle) = \langle \max\,(\mu_1, \ldots \mu_n),\ \min\,(\nu_1, \ldots \nu_k)\rangle;$$

Pessimistic formula:

$$\min(\langle \mu_1, \nu_1\rangle, \ldots, \langle \mu_n, \nu_{n)}\rangle) = \langle \min\,(\mu_1, \ldots \mu_n),\ \max\,(\nu_1, \ldots \nu_k)\rangle;$$

Strongly pessimistic formula:

$$\langle \mu_1, \nu_1\rangle . \langle \mu_2, \nu_2\rangle \ldots . \langle \mu_n, \nu_n\rangle = \prod_{i=1}^{n} \langle \mu_i, \nu_i\rangle = \langle \prod_{i=1}^{n} \mu_i, 1 - \prod_{i=1}^{n} (1 - \nu_i)\rangle. \tag{5}$$

It can be stated, that the fuzzy estimation approach [3–5] proposed in this chapter will provide an estimation criterion for the image and if feasible, improve the overall performance of the tomographic scan.

4 Conclusion

An intuitionistic fuzzy approach to improve the quality of the remote imaging technique has been proposed in this paper. The method allows an assessment whether a certain pixel in an image of computer tomography belongs or does not belong to the bone structure and thus allowing to define the level of uncertainty and thus include or exclude the current pixel in the bone structure. This will dramatically improve the visualization of the bone structure and allow better estimation of bone fracture or disease.

Acknowledgment. This work was supported by the European Regional Development Fund through the Operational Programme "Science and Education for Smart Growth" under contract UNITe № BG05M2OP001–1.001–0004 (2018–2023).

References

1. Farahat, Z., Hasni, M., Megdiche, K., Souissi, N., Ngote, N.: Comparative study of DICOM files handling software's: study based on the anatomage table. In: Serrhini, M., Silva, C., Aljahdali, S. (eds.) EMENA-ISTL 2019. LAIS, vol. 7, pp. 390–399. Springer, Cham (2020). https://doi.org/10.1007/978-3-030-36778-7_43
2. Atanassov, K.: On Intuitionistic Fuzzy Sets Theory. Springer, Berlin (2012)
3. Atanassov, K.: Interval-Valued Intuitionistic Fuzzy Sets. Springer, Switzerland (2020)
4. Atanassov, K., Szmidt, E., Kacprzyk, J.: On intuitionistic fuzzy pairs. Notes Intuitionistic Fuzzy Sets **19**(3), 1–13 (2013)
5. Bureva, V., Yovcheva, P., Sotirov, S.: Generalized net model of fingerprint recognition with intuitionistic fuzzy evaluations. In: Kacprzyk, J., Szmidt, E., Zadrożny, S., Atanassov, K.T., Krawczak, M. (eds.) IWIFSGN/EUSFLAT -2017. AISC, vol. 641, pp. 286–294. Springer, Cham (2018). https://doi.org/10.1007/978-3-319-66830-7_26
6. Kostadinov, T., Bureva, V.: Pattern recognition with intuitionistic fuzzy estimations, notes on intuitionistic fuzzy sets, vol. 23, no. 2, pp. 88–94 (2017). Print ISSN: 1310–4926, Online ISSN: 2367–8283
7. Sotirova, E., Shannon A., Atanassova V., Atanassov K., Bureva V.: Interval valued intuitionistic fuzzy evaluations for analysis of students' knowledge. Intuitionistic and Type-2 Fuzzy Logic Enhancements in Neural and Optimization Algorithms (2020)
8. Michalıkova, A.: Intuitionistic fuzzy sets and their use in image classification, notes on intuitionistic fuzzy sets, vol. 25, no. 2, pp. 60–66 (2019)
9. Atanassov, K.: Intuitionistic fuzzy logics as tools for evaluation of data mining processes. Knowl.-Based Syst. **80**, 122–130 (2015)
10. Bozov, H., Bozova, G., Sotirova, E., Shannon, A.: A generalized net model with intuitionistic fuzzy assessments of the process of cardiopulmonary resuscitation. In: Sotirov, S.S., Pencheva, T., Kacprzyk, J., Atanassov, K.T., Sotirova, E., Staneva, G. (eds) Contemporary Methods in Bioinformatics and Biomedicine and Their Applications. BioInfoMed 2020. Lecture Notes in Networks and Systems, vol 374, 100–112. Springer, Cham (2022). https://doi.org/10.1007/978-3-030-96638-6_11
11. Minkov, M., Sotirova, E.: Opportunity for obtaining an intuitionistic fuzzy estimation for health-related quality of life data. In: Atanassov, K.T., et al. (eds.) Advances and New Developments in Fuzzy Logic and Technology: Selected Papers from IWIFSGN'2019 – The Eighteenth International Workshop on Intuitionistic Fuzzy Sets and Generalized Nets held on October 24-25, 2019 in Warsaw, Poland, pp. 151–157. Springer International Publishing, Cham (2021). https://doi.org/10.1007/978-3-030-77716-6_13
12. Zadeh, L.A.: Fuzzy Sets. Inf. Control **8**, 333–353 (1965)

Bioinformatics and Biostatistical Models for Analysis and Prognosis of Antimicrobial Resistance

Maya Zhelyazkova[1(✉)], Roumyana Yordanova[2], Iliyan Mihaylov[1], Stefan Tsonev[3], and Dimitar Vassilev[1]

[1] Faculty of Mathematics and Informatics, Sofia University St. Kliment Ohridski, 5 James Bourchier Blvd, 1164 Sofia, Bulgaria
zhelyazkova@fmi.uni-sofia.bg

[2] Bulgarian Academy of Sciences, Institute of Mathematics and Informatics, Acad. Georgi Bonchev Str., Block 8, Sofia, Bulgaria

[3] AgroBioInstitute, 8 Dragan Tsankov Blvd., 1164 Sofia, Bulgaria

Abstract. The relationship between bacteriophages and antimicrobial resistance could be regraded in the scope of potential impact of microbiology on contemporary medicine and pharmaceuticals. Till now research projects have some ambiguous outcomes by examining and confirming the influence of bacteriophages on variation of antimicrobial resistance genes. The major goal of our study is to acquire a new knowledge how the viruses, their hosts and antimicrobial resistance genes are related and how these relations can be clarified in the context of antimicrobial resistance dissemination through phages. The presented work is oriented towards zooming in the relationship and possible dependencies between bacteriophages and antimicrobial resistance. The data has been collected from different city environments all over the world. Our analyses consist of different bioinformatics and biostatistical methods for assessment of the differential abundance of phages, their diversity across samples, the impact on antimicrobial resistance categories and associations with antimicrobial resistant genes.

Keywords: antimicrobial resistance · bacteriophages diversity indexes · relative risk · compositional data analysis · spatial analysis

1 Introduction

Antimicrobial resistance forms important set of medical problems to be solved in the light of recent spread of microbial infections. One obvious problem is the relationship between the developing of such resistance and the overuse of antibiotic agents in both human and agriculture environments. Another cause of such resistance is the extensive use of antibacterial agents in agriculture. Selective pressures due to presence of antibacterial agents leads to acquiring new genetic traits either by spontaneous mutations or by horizontal gene transfer. In both cases the survival rate of the bacteria will increase.

S. Sotirov et al. (Eds.): BioInfoMed 2022, LNNS 658, pp. 53–62, 2023.
https://doi.org/10.1007/978-3-031-31069-0_7

Horizontal gene transfer may have great epidemiological implications due to transfer of antibiotic resistance genes from non-pathogenic bacteria to human or animal pathogens. Bacterial conjugation is considered the main mechanism for horizontal gene transfer among bacteria [2]. The process is accomplished through physical binding of two bacterial cells and transfer of mobile genetic elements, such as plasmids and integrative conjugative elements. Other mechanisms for horizontal gene-transfers are bacterial transformation and transduction. The former is accomplished by DNA fragments intake from the environment by a bacterial cell without physical contact with another cell and the latter is a gene transfer mediated by phages.

Bacteriophages (or phages) are viruses that infect and replicate within bacteria and since their discovery in the early 1900s became a focus of intensive research on possible causes for antimicrobial resistance. Their effect on the horizontal gene transfer via transduction and transformation can contribute to the development of antimicrobial resistance (AMR) [3]. However, their role in the dissemination of antimicrobial resistance genes (ARGs) is not certain and some studies confirm [1], or question [2] that phages may not necessarily encode ARGs genes. Therefore, understanding of the extent to which phages contribute to the propagation of antibacterial resistance is an important task.

Metagenomics studies using high throughput sequencing which generate a large amount of data in short read fragments sampled from microbial communities. Many of the existing tools for analyzing metagenomics sequences rely on the so-called k-mer approaches to extract microbial taxa from such data. Some recent methods have been developed based on deep learning approaches to find potential mutated or viral sequences [10–12].

The study was based on a set of originally developed, various statistical and machine learning approaches which to a great extent provide answers to the emerged problems. The study was initialized by the CAMDA project (www.camda.info) set of tasks related to metagenomics challenges exploring the coevolution of the bacteria and their corresponding phages and how potentially is possible to predict antimicrobial resistance events from samples collected in monitored areas. CAMDA provides also the data of this sudy. More specifically, there are four tasks for analyses and prediction: (1) discovery of phages and prophages in metagenomic samples with relevance to antimicrobial resistance (AMR); (2) establishing species that may be infected by detected phages, horizontal AMR gene transfers etc.; (3) applications and assessments of relation mining for the occurrence of phages and bacteria in the context of AMR; (4) advancement of algorithms for identification and discovery of phages and prophages from bacterial genomes. In this work we will focus on analyzing the data and complete the first three tasks.

2 Data and Methods of Analysis

Data: CAMDA project challenge provides a set of 124 samples from subways and urban public transport environments across all continents divided into two groups based on their ARGs profiles (62 high and 62 low AMR samples. Operational Taxonomic Units (OTUs) counts for phages taxa normalized, as reads per million (RPM), were extracted of the samples shotgun sequences using *kaiju* metagenomics classifier related to viruses from NCBI database [6]. This classifier uses protein level classification and is proved to attain high sensitivity [5]. The total number of non-zero abundant viruses is 6618 while the total number of non-zero abundant phages is 2687. We also extracted bacteria taxa using the ProGenomes database [6] and we summarized the counts by species. As for the AMR abundances we used the provided from CAMDA and Pangea ARGs which included normalized data of both ARGs and taxa classes.

Methods. The methodological basis for analysis and prognostication in the study comprises several different approaches to explore the relationship between phages and ARGs (AMR) distributions. We first use compositional data analysis to find the specific phages that separate samples, based on their ARGs profiles. We then calculate different alpha diversity measures, as well as beta diversity measures. We use penalized *Lasso* regression to find the best predictor models for AMR diversity based on phage abundances. Finally, we use a Bayesian spatial model to estimate relative risk of phages taking into account their host abundances.

Compositional Data Analysis. Data are considered compositional if they contain multiple parts of nonnegative numbers whose sum is either 1 or any constant-sum constraint. More specifically, this defines the sample space as a hyperplane called Aitchison Simplex [7].

$$S^D = \left\{ x = (x_1, x_2, ..., x_D)' \in R_+^D \mid \forall k > 0 \; \exists! \; \lambda > 0 : x = \lambda C_k(x) \right\},$$

where C_k is rescaling the compositional vector to a constant sum (1 or 100). Metagenomics data (e.g. OTU count data) are compositional data since sum of all component values depend on the sampling procedure and the data are proportional. Taking into account this structure we perform differential abundance analysis using compositional methods. We compare the phages abundance for high and low AMR samples to find for which phages there is a difference between the two groups. We use the R-package ALDEx2 [8] which was developed using Bayesian methods for detecting ANOVA type of differential expressions specifically for compositional data. We also perform Fisher exact test to find possible overrepresentation of families of phages taxa.

Alpha and Beta Diversity Indices. Alpha diversity indices measure the variation of microbes in a single sample while beta diversity indices measure the variation across samples. In this work we use both measures to characterize our data. In particular we use three indexes such as Chao1 diversity index [9], next we used Shannon-Wiener diversity index [10] which is based on information theory. Both Chao1 and Shannon indexes give more importance to less common species. Finally we use Simpson index [11] which presents the species heterogeneity. In contrary to the Shannon's index, Simpson's diversity index gives more importance to more common species.

The Beta diversity Bray-Curtis dissimilarity index [12] is a statistical measure used to quantify the compositional dissimilarity between two different samples, based on counts at each sample. We use the above methods for estimation of species richness, namely Chao 1, Shannon-Wiener and Simpson diversity indices as implemented in the R package *vegan*. We perform Kruskal-Wallis test to compare the diversity measures across AMR status and geographical location. We apply Constrained Analysis of Principal Coordinates with Bray-Curtis dissimilarity (*capscale* from R package *vegan*).

Penalized Lasso Regression. In this study we applied several types of regression models provided in R package *caret* [13] and we focused on *Lasso* (Least Absolute Shrinkage and Selection) penalized regression which is a regularization technique that achieves accurate prediction by avoiding the over-fitting of the data. The *Lasso* model is applied to the phages data to predict the AMR variation within the samples and subsequently to account for the two major classes of AMR - high AMR and low AMR. We split the data into training and testing set (60% to 40% respectively) and use cross validation as provided in the *Lasso* implementation of package *caret*.

Spatial Model to Estimate Relative Risk of Bacteriophages. Further the examination of the relationship between phages and AMR is based on taking into account the geographical location and host bacteria information [14]. There are two main biological hypotheses with respect to the role of the phages in the dissemination of ARG. Namely: 1) the bacteriophages transmit the antimicrobial resistance genes through processes such as bacterial transduction or 2) the ARG transmission is accomplished by another biological process such as bacterial conjugation with no involvement of bacteriophages. In the second case, the phages abundance is expected to be proportional to the host abundance. In order to test the two hypotheses, we use a Bayesian spatial model which incorporates the spatial information of the samples and includes the corresponding host bacteria abundances to estimate relative risk of phages.

Spatial autocorrelation is very common when observations that are close in space have similar values. A proportion of this spatial autocorrelation is usually modelled by known covariate risk factors in a regression model. However, it is common for spatial structure to remain in the residuals after accounting for these covariate effects. Spatial models such as Bayesian hierarchical models are then used to augment the linear predictor with a set of spatially autocorrelated random effects depending on the neighborhood structure of geographic areas. These random effects are typically represented with a conditional autoregressive (CAR) prior, which induces spatial autocorrelation through the adjacent structure of the areal units.

Since the samples in the CAMDA challenge have spatial information such as latitude and longitude and previous studies indicated that metagenomics taxa profiles exhibit spatial correlations we use a Bayesian hierarchical model, in particular Besag-York-Mollie (BYM) [15], which is a convolution model with a CAR prior. The model uses a spatial neighborhood matrix which measures the distance between the samples based on the available geographic information such as latitude and longitude. The model will estimate the phage's risk which is a posterior estimate of the Standardised Incidence Ratio (SIR=Observed/Expected) of a phage across the samples where the expected abundances are proportional to the corresponding host bacteria taxa counts. The model includes the AMR category (low vs high) as a potential factor. If the phage's relative risk is higher than 1 and the AMR category credible set does not include zero, we can conclude that the phage's abundance is higher than what we would expect based on the host bacteria counts and this higher risk can be partially explained by the antimicrobial category status. For example, one AMR group (e.g. high) has an increased risk while the other group (e.g. low) has a decreased risk relative to the expected numbers which are proportional to the phage's host bacteria. The corresponding host bacteria for each phage were queried from the following database https://www.genome.jp/virushostdb/. In case where phages infect more than one bacteria, we add the additional host bacteria abundance in the model. In stricter mathematical terms the model is described in [15].

$$O_i|E_i, \theta_i \sim Poisson(E_i\theta_i), i = 1, \ldots, n$$
$$ln(\theta_i) = x_i^T\beta + \nu_i + u_i,$$

where ν are spatially unstructured random effects and u are random effects that capture the spatial autocorrelation between the samples. The response is assumed to follow Poisson distribution and it accounts for overdispersion $Var(O) > E(O)$ and this is an advantage over the pure Poisson model. We use the Bayesian setting implementation in R 3.6.3 package CARBayes [15], where inference is based on Markov chain Monte Carlo simulation. Finally, we calculate the correlations between phages (on the level of species and families) and antimicrobial resistance genes.

3 Results

From the applied compositional data analysis, we obtained 136 phages which differentiate the two categories: high AMR and low AMR levels as it was set up by the sum of relative ARGs counts (Pangea.org within the CAMDA challenge). 114 of them have significantly higher abundance (by false discovery rate - FDR) in the high AMR group, while 22 have significantly lower abundance (by FDR) in the same high AMR group. The false discovery rate (FDR), is defined as expected proportion of discoveries which are falsely rejected. The two heatmaps on Fig. 1 show the distribution of those phages across the continents. The abundances were pooled by phages families and the Myoviridae family is overrepresented in the group of differentially abundant phages.

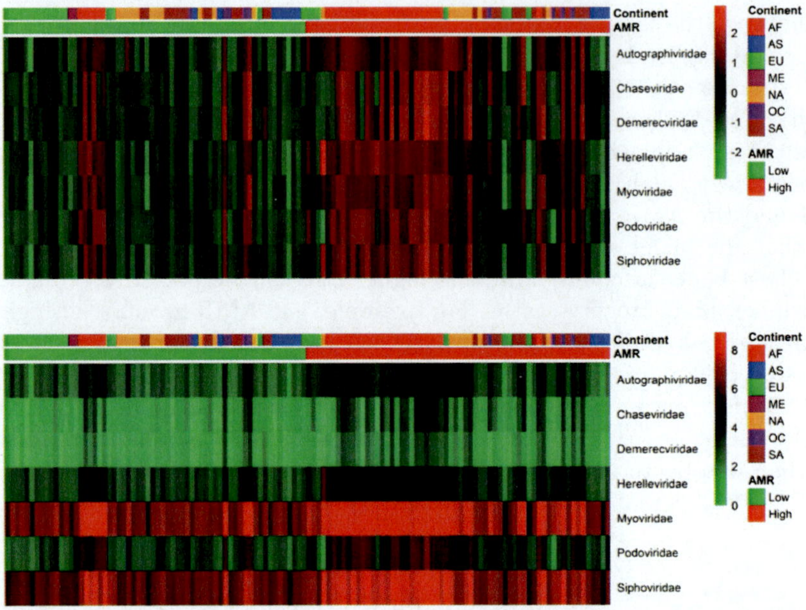

Fig. 1. Scaled and not scaled log transformed RPM (reads per million) top differentially abundant phages (FDR, Welch t test with Benjamini-Hochberg $p < 0.05$) pooled by the corresponding families. The phages in the family Myoviridae are overrepresented ($p < 0.01$) in the set that differentiated high and low AMR samples. AF Africa, AS Asia, EU Europe, NA North America, ME Middle East, OC Oceania, SA South America

There are two major bi-clusters of strongly positive correlations (>0.8). Among the top correlated genes in the larger bi-cluster are arcB, parE, rpoB and tufAB which are correlated mostly with the Myoviridae phage family. The other cluster involves genes such as blaZ, ermX, mphC, tetK, vga and the corresponding correlated phages, mostly from the Siphoviridae family. The phages in one of the clusters are negatively correlated with the ARGs from the other clusters.

Significant differences adjusted for multiple comparisons are found between AF(943) and EU(1209), AF(943) and SA(1356), EU(1209) and OC(685), NA(1165) and OC(685) and between OC(685) and SA(1356) for Chao1 index; AF(5.24) and NA(5.6), AF(5.24) and SA(5.8) and between OC(5.19) and SA(5.8) for Shannon index; AF(0.987) and SA(0.99) and EU(0.988) and SA(0.99) for Simpson index. SA has the highest diversity measures while OC has the lowest Chao1 and Shannon indexes.

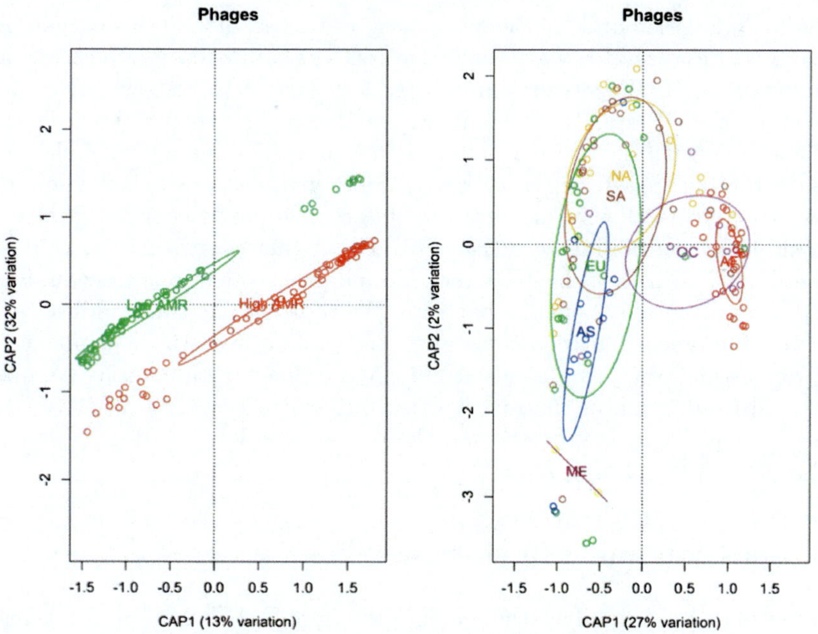

Fig. 2. a. Distributions of phages across Category (high and low AMR) and b. across Continents. AF Africa, AS Asia, EU Europe, NA North America, ME Middle East, OC Oceania, SA South America.

Constrained analysis of Principal Coordinates using Bray-Curtis Dissimilarity index for phages (Fig. 2) showed clear differentiation between the high and low AMR samples (Fig. 2a) and also within the continents (Fig. 2b).

Our next analysis step used *Lasso* regression in the context of machine learning (R package *caret*) to select phages that can differentiate the two AMR categories and can be used as predictors for the AMR status. Sensitivity is 1-p (type I error) while specificity is the power of the test. The model achieves high sensitivity (0.9) and specificity (0.95). These results show very good model fit. The top features with non-zero coefficients are mainly from the two most frequent family classes Myoviridae (Escherichia phage RCS47, Synechococcus phage ACG-2014d, Cyanophage Syn30, Synechococcus phage S-SM2, Psychrobacter phage pOW20-A, Acinetobacter phage Ac42) and Siphoviridae (Skunavirus, Bacillus virus SPbeta, Enterobacteria phage YYZ-2008, Acinetobacter phage Bphi-B1251, Sitaravirus, Enterobacteria phage cdtI) as well as Autographiviridae (Pseudoalteromonas phage RIO-1) and Podoviridae (Escherichia phage TL-2011b).

Bayesian spatial analysis is used in the study mainly to estimate the relative risk of phage abundance compared to their host abundance. We used as examples two bacteria (Salmonella enterica and Staphylococcus Aureus) which are recently studied as part of generalized transduction process and their corresponding phages found in the virus-host database https://www.genome.jp/virushostdb/. Figure 3 shows the relative risk (i.e. adjusted for the host bacteria abundance). AMR category is used as a covariate to test the effect on the phage abundance. Staphylococcus Aureus related phages show differences among AMR category with the credible set not containing zero and average relative risk higher in the high AMR group (Fig. 3a). Salmonella enterica show similar profiles across continents that also are not different among the AMR groups with the credible set containing zero (Fig. 3c). The higher observed relative risk can also be due to additional host bacteria abundance, which is not accounted for, since phages often affect similar bacteria. We took into account additional host bacteria abundance due to phages affecting similar bacteria, e.g. phages that infect Staphylococcus Aureus, also infect Acinetobacter baumanni. We observe on (Fig. 3b) reduction of the relative risk but still the average relative risk for the high AMR group is greater than the one for the low AMR group with the credible set not containing zero.

4 Discussion and Conclusions

Recent studies [1,16] analyze the global and regional distribution of phages and their impact on the transmission of ARGs between bacteria. A significant correlation between phages and ARGs was observed, indicating that phages may play a role in ARG dissemination. Due to a low abundance of the identified phages, further studies are needed to establish the relationship between phages and ARGs. Some of the findings indicate that phages in sewage environment are widely distributed as free-living phages not always associated with bacteria and persist longer than their host. In our work we confirm those findings by observing strong correlations between phages and ARGs as well as increased relative risk (higher than expected phage presence in some samples) which is correlated to the AMR category.

The goal of our study was to examine the relationship between bacteriophages and antimicrobial resistance. We developed a set of different and interlinked approaches to answer this question. Based on the compositional differential abundance analysis we found that phages which differentiate the AMR category and are highly correlated with the ARGs are overrepresented by the Myoviridae family. Beta diversity measures differentiate between the AMR categories and the continents. Samples from South America have the highest alpha diversity indices and Oceania is among the continents with the lowest alpha diversity indices. It is possible to make very good predictions of the AMR categories based on phages abundance using *Lasso* regression.

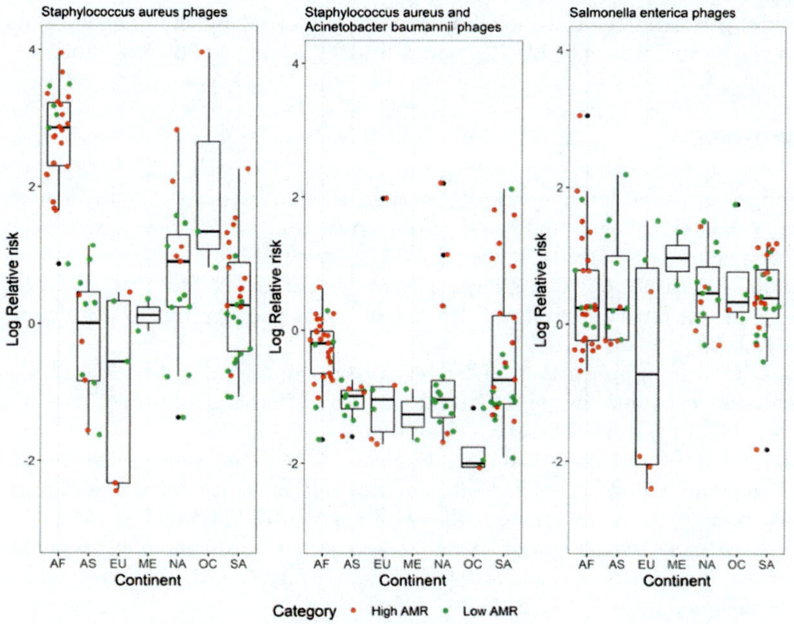

Fig. 3. Relative risk (log scale) plots based on Bayesian hierarchical model for the pulled abundance for all phages that are associated with a. Staphylococcus aureus, b. Staphylococcus aureus and Acinetobacter baumannii, and c. Salmonella enterica respectively. Here the SIR ratio is observed phage abundance while expected is proportional to their host abundance. AF Africa, AS Asia, EU Europe, NA North America, ME Middle East, OC Oceania, SA South America.

The novel part of our analysis is the use of a Bayesian hierarchical model to take into account the geographical location of the samples and estimate the relative risk of the phages by incorporating information about their host bacteria. The bacteriophages can transmit the antimicrobial resistance genes through bacterial transduction or the ARG transmission can be achieved through bacterial conjugation without the involvement of bacteriophages. In the second case, the phages abundance is expected to be proportional to the host abundance. Bayesian spatial analysis help us to test the hypotheses about the role of the phages in the dissemination of ARG by examining how the phages evolved across spatially correlated samples using both phages and hosts abundances. This model will be very useful when we have closely related municipalities and cities so we can track how the phages evolve and if the relative risk changes across borders. This is very similar to the disease mapping model [15] but in this setting the virus is a bacteriophage and the human host is the corresponding bacteria host. For improving the accuracy of the model, we can further use several additional factors as potential covariates: climate conditions and antibiotic usage.

Acknowledgments. This work was partially supported by the financial funds allocated to the Sofia University St. Kliment Ohridski, grant No 80-10-94/2022.

References

1. Strange, J., Leekitcharoenphon, P., Moller, F., Aarestrup, F.: Metagenomics analysis of bacteriophages and antimicrobial resistance from urban sewage. Nature (2021). https://doi.org/10.1038/s41598-021-90880-6
2. Enault, F., Briet, A., Bouteille, L., Roux, S., Sullivan, M., Petit, M.A.: Phages rarely encode antimicrobial resistance genes: a cautionary tale for virome analysis. ISME J. **11**, 237–247 (2017)
3. Balcázar, J.L.: Implications of bacteriophages on the acquisition and spread of antibiotic resistance in the environment. Int. Microbiol. **23**(4), 475–479 (2020). https://doi.org/10.1007/s10123-020-00121-5
4. Subirats, J., Sánchez-Melsió, A., Borrego, C.M., Balcázar, J.L., Simonet, P.: Metagenomic analysis reveals that bacteriophages are reservoirs of antibiotic resistance genes. Int. J. Antimicrob. Agents **48**, 163–167 (2016)
5. Menzel, P., Ng, K.L., Krogh, A.: Fast and sensitive taxonomic classification for metagenomics with Kaiju. Nat. Commun. **7**, 11257 (2016). https://doi.org/10.1038/ncomms
6. Mende, D.R., Letunic, I., Huerta-Cepas, J., Li, S.S., Forslund, K., Sunagawa, S., et al.: proGenomes: a resource for functional and taxonomic annotations of prokaryotic genomes. Nucleic Acids Res. **45**, D529–D534 (2017). https://doi.org/10.1093/nar/gkw989
7. Aitchison, J.: The Statistical Analysis of Compositional Data. Chap-man and Hall Ltd., London, Reprinted in 2003 with additional material by The Blackburn Press (1986)
8. Fernandes, A.D., Macklaim, J.M., Linn, T.G., Reid, G., Gloor, G.B.: ANOVA-like differential gene expression analysis of single-organism and meta-RNA-seq. PLoS One **8**(7), e67019 (2013)
9. Chao, A.: Nonparametric estimation of the number of classes in a population. Scandinavian **11**, 265–270 (1984)
10. Shannon, C.E., Weaver, W.: The Mathematical Theory of Communication. University of Illinois Press, Urbana (1949)
11. Simpson, E.H.: Measurement of diversity. Nature **163**, 688 (1949)
12. Bray, J.R., Curtis, J.T.: An ordination of upland forest communities of southern Wisconsin. Ecol. Monogr. **27**, 325–349 (1957)
13. Tibshirani, R.: Regression Shrinkage and Selection via Lasso. J. R. Statist. Soc. Ser. B **58**(1), 267–288 (1996)
14. Besag, J., York, J., Mollie, A.: Bayesian image restoration with two applications in spatial statistics. Ann. Inst. Stat. Math. **43**, 1–20 (1991). https://doi.org/10.1007/BF00116466
15. Lawson, A.B.: Bayesian Disease Mapping: Hierarchical Modeling in Spatial Epidemiology, 3rd edn. Chapman and Hall/CRC, Boca Raton (2018)
16. Strange, J., Leekitcharoenphon, P., Mooler, F., Aarestrup, F.: Metagenomics analysis of bacteriophages and antimicrobial resistance from global urban sewage. Sci. Rep. **11**(1), 1–11 (2021). https://doi.org/10.1038/s41598-021-80990-6

Decision Making in Biomedicine and Healthcare

InterCriteria Analysis of Data Obtained from Patients with Hypercholesterolemia Treated with Linoprixol

Valentin Vassilev[1], Hristo Hlebarov[1], Simeon Ribagin[2,3](✉), and Krassimir Atanassov[2]

[1] University Hospital Burgas, Burgas, Bulgaria
vailvasilev@mail.bg

[2] Department of Bioinformatics and Mathematical Modelling Institute of Biophysics and Biomedical Engineering, Bulgarian Academy of Sciences, Sofia, Bulgaria
simribagin@gmail.com

[3] Department of Health and Pharmaceutical Care, Medical College, University "Prof. D-r Asen Zlatarov", Burgas, Bulgaria

Abstract. Hypercholesterolemia is one of the main factors contributing to number of diseases, leading to permanent disability and/or worldwide. Nowadays the standard treatment of high cholesterol levels is carried out by drugs known as statins. Despite their main effect they also have a number of ADRs. The purpose of the present paper is to analyze the effects of the alternative treatment of patients with hypercholesterolemia with Linoprixol.using the InterCriteria Analysis method. The method of InterCriteria Analysis was applied to study the dependencies between the total cholesterol, LDL and triglycerides levels obtained in a clinical trial conducted in the Vascular Surgery Clinic of University Hospital Burgas.

Keywords: InterCriteria Analysis · Hypercholesterolemia · Linoprixol

1 Introduction

Lipoprotein disorders are clinically important due to the role of the lipoproteins in the process of atherosclerotic plaque formation, and the associated risk of atherosclerotic cardiovascular diseases. Lipoproteins comprise lipids and protein and can be transported in plasma as such, for delivery of cholesterol, triglycerides, and fat-soluble vitamins to the respective organs as needed [10]. Hypercholesterolemia results from alterations in lipoprotein metabolism that lead to high total cholesterol, LDL-C or triglycerides, and/or low HDL-C [9]. Cholesterol is the normal compound of all cell membranes in the human body. It is necessary for the synthesis of adrenal gland hormones – cortisone, corticosteroids and aldosterone, male sex hormones – testosterone and androsterone, female sex hormones – estrone, estriol, progesterone, vitamin D and bile acids. The metabolism of cholesterol is carried out by the liver where it is bind to the specialized transporting

S. Sotirov et al. (Eds.): BioInfoMed 2022, LNNS 658, pp. 65–71, 2023.
https://doi.org/10.1007/978-3-031-31069-0_8

proteins in the lipoprotein complexes – LDL (known as "bad" cholesterol), HDL (known as "good" cholesterol) and VLDL. Except the cholesterol and its derivatives, there are another compounds known as triglycerides which are main types of lipids taking part in the human metabolism. The reference value for the total cholesterol regarding to the recommendations of the European Atherosclerosis Society [11] are in the range of 3.5 to 5.2 mmol/l (125–200 mg/dL). The reference value for LDL cholesterol is less than 2.59 mmol/l (less than 100 mg/dL), HDL cholesterol – higher than 1.45 mmol/l (higher than 40 mg/dL) for men and higher than 1.68 mmol/l (50 mg/dL) for women, triglycerides less than 1.7 mmol/l (less than 150 mg/dL). Every value (except HDL) higher than the upper limit is accepted as a condition of hypercholesterolemia.

Nowadays the high blood levels of cholesterol and triglycerides are treated with medications known as statins. In their essence, they are inhibitors of HMG Co-A reductase – the main enzyme needed for the formation of cholesterol in the liver. Statins' purpose is to reduce the levels of total cholesterol, "bad" cholesterol and triglycerides and to increase the levels of "good" cholesterol with which actually slows down the formation of atherosclerosis plaques in the blood vessel wall. According to number of studies, statins also have additional effects – they improve endothelial dysfunction [1], increase NO levels in the blood which leads to vasodilatation, have antioxidative effect and stabilize the antisclerotic plaques [8], reducing the risk of detachment. Such as other medications, statins have adverse drug reactions (ADR) which are numerous and should not be underestimated by both patients and physicians.

The main ADR include:

1. Increasing levels of liver transaminases – in the most cases these changes develop within the first three months of treatment and can be observed at around 3% of the patients,
2. Myopathy – the main symptom limiting the usage of statins. It is characterized by pain and weakness in the main muscle groups. Myopathy is manifested at around 10-15% of patients subjected to statin therapy. Proceeding with the treatment it is known to develop rhabdomyolysis,
3. Kidney failure – it is presented with acute tubule necrosis, proteinuria and decreasing tubule reabsorbtion of normally filtrated proteins,
4. Gastrointestinal dysfunction – nausea, dyspepsia, abdominal pain, flatulence, diarrhea and constipation. These reactions do not require termination of the treatment.
5. Skin disruptions – rash, cheilites, urticaria, Lyell syndrome. At around 0.5–1% of the treated patients is observed manifestation of alopecia which is reversed with the termination of the intake of statins,
6. Diabetes mellitus – gathering the results of randomized clinical trials with different types of statins it is established the risk of having a newly onset of diabetes at around 13% of the patients,
7. Tendinitis and arthritis,
8. Peripheral neuropathy,
9. Dysfunction of central nervous system – headache, depression and sleep disruption
10. Erectile dysfunction,
11. Long term decreasing of the effect of the medications

In a view of this it is important to find new methods and strategies for alternative treatment of hypercholesterolemia with no or minimal side effects for the patient. One possible solution is the use of supplements containing naturally occurring monacolin K derived from the red yeast rice (Monascus purpureus). In the present study we analyze the results of the three months' treatment with an alternative supplement containing monacolin K, applied to patients with high levels of cholesterol, using a standard statistical analysis and the ICra approach. Materials and methods.

In order to avoid all of the stated ADR of statins, The Vascular Surgery Clinic of UMBAL Burgas has put through a randomized clinical trial of 50 patients with an alternative supplement called Linoprixol [12]. It is gluten-free natural food supplement containing Monacolin K from yeast cultured in red rice in combination with Co-enzyme Q10. The main ingredient supports normal levels of cholesterol in the blood, contributes to regular function of cardiovascular system and protects cells from oxidative stress. The mechanism of action is the same as statins' – inhibiting the enzyme HMG-CoA reductase. The main difference between them, however, is the absence of ADR on behalf of Linoprixol [12]. The object of the study was taking proof of the curative effect and advantages of Linoprixol versus statins in patients with hypercholesterolemia by selection of patients between 42 and 89 years' old who have or have not taken statins before and have elevated levels of cholesterol and triglycerides at around 2 units above the reference value. The treatment with Linoprixol consists of taking one tablet at night for the period of three months at least. The levels of cholesterol and triglycerides are measured monthly. The clinical trial was conducted in 2021 on 24 male and 26 female patients and the mean age of the patients was 65 years of age. The obtained data from the study was analyzed via the standard statistical methods and the relatively new method of ICra. The method of ICra uses two mathematical tools - Indexed Matrices (IMs) [2], and Intuitionistic Fuzzy Sets (IFSs) [3], thus rendering account of the effects of uncertainty. Originally, ICA was being proposed in [4].

2 Results and Discussion

After the data analysis we found that 92% of the patients treated with Linoprixol have lowered their levels of cholesterol and triglycerides in the trial period. The patients show statistically a decrease (Table 1) both in the levels of the total cholesterol, low-density lipoprotein (LDL) and the triglycerides (Fig. 1) after the 3th month of the trial period (**p < .00001*** mean levels of total cholesterol = **5.6** mmol/l, < **.00001*** mean levels of LDL = **2** mmol/l and **p = 0.00009*** mean levels triglycerides = **1.5** mmol/l) compared to the levels in the begging of the trial period (mean levels of total cholesterol = **6.29** mmol/l, mean levels of LDL = **2.69** mmol/l and mean levels of triglycerides = **2.3** mmol/l).

After data processing with ICrA software we obtain membership part (Table 2, Table 4 and Table 6) and non-membership part (Table 3, Table 5 and Table 7) of the intuitionistic fuzzy pairs [5] that represent an intuitionistic fuzzy evaluation of the relations between every pair of criteria (total cholesterol levels, LDL levels and triglycerides levels during the 3 months' trial period). Following [4, 6] in order to categorize all the values of the resultant n (n – 1)/2 pairs of criteria, we need to define two thresholds, α and β, for the positive and for the negative consonance, respectively. The threshold values α and

Table 1. P values from the statistical T-test analysis of the levels of total cholesterol, LDL and triglycerides in the beginning and at the end of the trial period of Linoprixol treatment. ("*" marks statistically significant result at $p < .05$)

Results	**Total Cholesterol**	**LDL**	**Triglycerides**
In the beginning and after the 3th month of the trial	$p < .00001*$	$p < .00001*$	$p = 0.00009*$

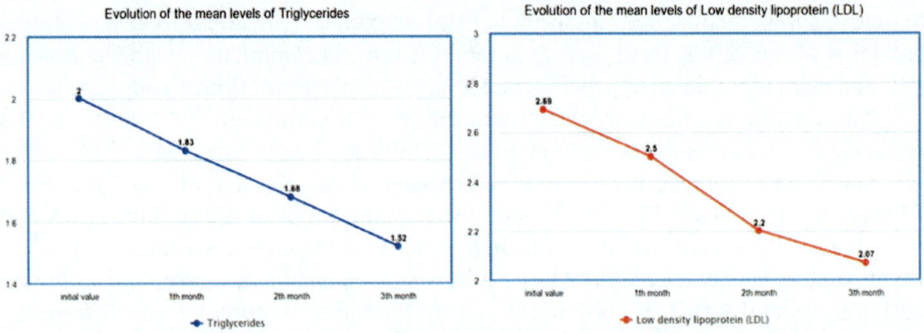

Fig. 1. The evolution of the triglycerides and the LDL mean levels

β are values on the [0; 1]- scale, changing with a precision step of 0.1. In our case the respective values, connected with the consonance/dissonance scale are as follows: strong positive consonance (0,95; 1), positive consonance (0,85; 0,95), weak positive consonance (0,75; 0,85). There are dependencies in the total cholesterol levels with a strong consonance $\langle 0.90, 0.098\rangle$ between the first month of the treatment and the second month of the treatment as well as the second month of the treatment and the third month of the treatment $\langle 0.88, 0.89\rangle$. The same tendency with strong positive consonance during the trial period are showing also the levels of the LDL ($\langle 0.86, 0.11\rangle$, $\langle 0.91, 0.07\rangle$) and the triglycerides ($\langle 0.92, 0.06\rangle$, $\langle 0.90, 0.07\rangle$) respectively.

Table 2. The membership parts of the Intuitionistic fuzzy pairs of the relations between the values of the total cholesterol levels during the trial period.

Degree of membership μ	initial levels	1th month	2th month	3th month
initial levels	1	0.7754	0.7355	0.7174
1th month	0.7754	1	0.9058	0.8442
2th month	0.7355	0.9058	1	0.8804
3th month	0.7174	0.8442	0.8804	1

Table 3. The non-membership parts of the Intuitionistic fuzzy pairs of the relations between the values of the total cholesterol levels during the trial period.

Degree of non-membership υ	initial levels	1th month	2th month	3th month
initial levels	0	0.1812	0.2269	0.2555
1th month	0.1812	0	0.098	0.1412
2th month	0.2269	0.098	0	0.0898
3th month	0.2555	0.1412	0.0898	0

Table 4. The membership parts of the Intuitionistic fuzzy pairs of the relations between the values of the LDL levels during the trial period.

Degree of membership μ	initial levels	1th month	2th month	3th month
initial levels	1	0.8424	0.8555	0.8294
1th month	0.8424	1	0.8686	0.8735
2th month	0.8555	0.8686	1	0.9151
3th month	0.8294	0.8735	0.9151	1

Table 5. The non-membership parts of the Intuitionistic fuzzy pairs of the relations between the values of the LDL levels during the trial period.

Degree of non-membership υ	initial levels	1th month	2th month	3th month
initial levels	0	0.1478	0.1322	0.16
1th month	0.1478	0	0.1192	0.1176
2th month	0.1322	0.1192	0	0.0735
3th month	0.16	0.1176	0.0735	0

That tendency means that Linoprixol supplement treatment gradually affects the levels of THL, LDL, triglycerides and the minimal treatment period should be at least 3 months.

Table 6. The membership parts of the Intuitionistic fuzzy pairs of the relations between the values of the triglycerides levels during the trial period.

Degree of membership μ	initial levels	1th month	2th month	3th month
initial levels	1	0.8122	0.8122	0.8033
1th month	0.8122	1	0.9208	0.8931
2th month	0.8122	0.9208	1	0.9094
3th month	0.8033	0.8931	0.9094	1

Table 7. The non-membership parts of the Intuitionistic fuzzy pairs of the relations between the values of the triglycerides levels during the trial period.

Degree of non-membership υ	initial state	1th month	2th month	3th month
initial state	0	0.1739	0.1763	0.178
1th month	0.1739	0	0.062	0.0808
2th month	0.1763	0.062	0	0.0702
3th month	0.178	0.0808	0.0702	0

3 Conclusions

The alternative treatment of hypercholesterolemia has its goal to reduce statin usage, respectively its ADRs. Based on the results of this study all of this can be achieved by using the alternative food supplement known as Linoprixol. The results of the analysis of the data show a definite decrease in the lipid values and a strong increase in dependences between the separate periods of taking the Linoprixol supplement during the three-month trial period. Naturally, given the relatively small size of the considered data it is not possible to claim with absolute certainty that our interpretations are doubtlessly valid but they provide a starting point for further investigations of the effect of the Linoprixol.

Acknowledgements. The authors are grateful for the support provided by the Bulgarian National Science Fund under Grant Ref. No. KP-06-N22/1/2018 "Theoretical research and applications of InterCriteria Analysis". The authors declare that there is no conflict of interest regarding the publication of this paper.

References

1. Altun, I., et al.: Effect of statins on endothelial function in patients with acute coronary syndrome: a prospective study using adhesion molecules and flow-mediated dilatation. J. Clin. Med. Res. **6**(5), 354–61 (2014). Epub 2014 Jul 28. PMID: 25110539; PMCID: PMC4125330. https://doi.org/10.14740/jocmr1863w

2. Atanassov, K.: On Intuitionistic Fuzzy Sets Theory. Springer, Berlin (2012). https://doi.org/10.1007/978-3-642-29127-2
3. Atanassov, K.: Index Matrices: Towards an Augmented Matrix Calculus. Springer, Cham (2014). https://doi.org/10.1007/978-3-319-10945-9
4. Atanassov, K., Mavrov, D., Atanassova, V.: Intercriteria decision making: a new approach for multicriteria decision making, based on index matrices and intuitionistic fuzzy sets. Issues Intuit. Fuzzy Sets Gen. Nets **11**, 1–8 (2014)
5. Atanassov, K., Atanassova, V., Gluhchev, G.: InterCriteria analysis: ideas and problems. Notes Intuit. Fuzzy Sets **21**(1), 81–88 (2015)
6. Atanassov, K., Szmidt, E., Kacprzyk, J.: On intuitionistic fuzzy pairs. Notes Intuit. Fuzzy Sets **19**(3), 1–13 (2013)
7. Atanassova, V., Mavrov, D., Doukovska, L., Atanassov, K.: Discussion on the threshold values in the intercriteria decision making approach. Int. J. Notes Intuit. Fuzzy Sets **20**(2), 94–99 (2014). ISSN 1310-4926
8. Bittencourt, M.S., Cerci, R.J.: Statin effects on atherosclerotic plaques: regression or healing? BMC Med. **13**, 260 (2015). PMID:26449405; PMCID:PMC4599025.https://doi.org/10.1186/s12916-015-0499-9
9. Espinheira, M., Vasconcelos, C., Medeiros, A., Alves, A., Bourbon, M., Guerra, A.: Hypercholesterolemia – a disease with expression since childhood. Revista Portuguesa de Cardiologia (English Edition) **32**, 379–386 (2013). https://doi.org/10.1016/j.repce.2012.09.013
10. Ibrahim, M.A., Asuka, E., Jialal, I.: Hypercholesterolemia. [Updated 2022 Jun 19]. In: StatPearls [Internet]. Treasure Island (FL): StatPearls Publishing, January 2022. https://www.ncbi.nlm.nih.gov/books/NBK459188/
11. https://www.eas-society.org/page/lp_a_consensus
12. https://sopharmacy.bg/bg/product/000000000030013802

ABO System Blood Groups Distribution in Bulgaria, Based on a Dataset of the Patients of the University Hospital "Saint Anna", Sofia, Bulgaria, from 2015 to 2021

Vassia Atanassova[1](✉), Nikolay Andreev[2], and Angel Dimitriev[1]

[1] Institute of Biophysics and Biomedical Engineering, Bulgarian Academy of Sciences, 105 Acad. Georgi Bonchev Str., 1113 Sofia, Bulgaria
vassia.atanassova@gmail.com

[2] Transfusion Hematology Department, University Hospital "Saint Anna", Sofia, 1 D. Mollov Str., Mladost 1, 1750 Sofia, Bulgaria

Abstract. The paper presents the analysis of a dataset containing the records of 47562 Bulgarian individuals, patients of the University Hospital "Saint Anna", Sofia, Bulgaria in the period from 1 January 2015 to 31 December 2021, and aims at the establishing the distribution of blood groups of the ABO system and the frequencies of the A_1 and A_2 subgroups and the Rh(D) antigen. In the frames of the conducted research, a series of data cleansing and data extraction procedures have been applied. The statistical analysis established the following frequencies of the ABO system blood groups in this sample of the Bulgarian population: A – 43.66%, B – 16.36%, O – 31.87% and AB – 8.11% . The prevalence of Rh(D) antigen determined was 86.38%. The distributions of A_1 and A_2 subgroups in the A group were determined as 91.92% and 8.09%, respectively, and the distributions of A_1B and A_2B subgroups in the AB group were established as 88.52% and 11.48%. These results are compared to the findings of other researchers regarding the distribution of ABO blood system blood antigens among the Bulgarian population, and slightly updates the results from the dataset of the patients of that Hospital from the period 2015–2020.

Keywords: ABO system · Blood group distribution · Rh(D) antigen distribution

1 Introduction

Ever since its discovery, the ABO system of blood groups and blood group antigens and their frequencies in population have concentrated researchers' interest. The concept of ABO blood types is crucially important for transfusion medicine, as transfusion of incompatible ABO type of blood is the most common cause of death from blood transfusion [8]. Antigens furthermore appear to play a significant role in human evolution because the frequencies of different ABO blood types exhibit significant variability among various human populations [7, 8, 11, 12], suggesting that a particular blood type conferred a selection advantage like resistance against an infectious disease, etc. [8].

S. Sotirov et al. (Eds.): BioInfoMed 2022, LNNS 658, pp. 72–83, 2023.
https://doi.org/10.1007/978-3-031-31069-0_9

Human migration further outlines the importance of the study of the ABO blood group antigens distribution, since the changing antigen proportions may complicate the provision of the most appropriate blood for blood transfusion, especially at places with lacking or weak blood donation culture.

It is important that every country maintains a regular, up-to-date record regarding the blood group distribution profile of its population. In this sense, the information about Bulgaria according to various publicly available online sources, needs concretization, update and additional references in order to be considered reliable. For instance, in a top-ranking web source dated to 2013 [7], the blood group distribution in Bulgarian population is given as 44% of A, 15% of B, 32% of O and 8% of AB, and one of the most popular websites Wikipedia has provided in [14] as of June 2017 (without citing a particular source) the following frequencies: A Rh(+) 37.4%, A Rh(−) 6.6%, B Rh(+) 12.8%, B Rh(−) 2.2%, O Rh(+) 28.0%, O Rh(−) 5.0%, AB Rh(+) 6.8%, AB Rh(−) 1.2%, summing respectively to the percentages in [7]. The same figures, as rounded to the nearest integer, are reported in another public database dated to 2019 [12]. Another available research from 2015 [13] reports these distributions for the Bulgarian population as differently as 39.96% of A, 16.84% of B, 35.80% of O, and 7.60% of AB. These observations show the definite need for conducting an up-to-date and detailed study aimed at generating recent, accurate and detailed data for Bulgaria, which has been the motivation behind the present research.

The present paper is structured as follows: in Sect. 2, we give the background of our study: source of the patients' data and methods of data cleansing and retrieval. Section 3 contains the results of the study, namely the established frequencies the ABO system blood groups and subgroups, Rh(D) antigen frequencies, and respective maintained ratios, all of these given across the patients' sex. The subsequent Sect. 4 makes a comparison with previous studies. Finally, Sect. 5 gives conclusion and directions for future research.

2 Materials and Methods

The research was conducted in the Centre of Transfusion Hematology in University Hospital "Saint Anna", Sofia, over a dataset of hospital patients registered in the period between January 1, 2015, and December, 31, 2021.

At the stage of primary data collection, we obtained the tabular data of: 6469 samples in 2015, 13716 samples in 2016, 8748 samples in 2017, 10214 samples in 2018, 10534 samples in 2019, 8769 samples in 2020, and 8968 samples in 2021.

We conducted the following procedure of cleansing the primary data, including:

1. Removal of records with incomplete or inaccurate entered personal data about the patients or data about their blood groups (e.g. recorded valued 'A' instead of 'A_1' or 'AB' instead of 'A_1B', missing value of Rh(D) antigen, etc.),
2. Removal of records for patients of foreign nationality (where the personal identity number is not in the "EGN" (unique citizenship number) format for Bulgarian nationals but in the "LNC" format for resident people of foreign nationality),

3. Removal of duplicate records with respect to the "EGN" (including numerous records of the blood samples of one and the same patient within a calendar year, as well as across the full six-year period),
4. Extraction from the "EGN" of the data about the patient's sex,
5. Anonymization of the data.

As a result of the data cleansing procedures, the total number of 67418 records in the primary input was reduced and the dataset over which our investigation was conducted and results are reported, includes data of 47562 unique individuals, including 22194 men and 25368 women. In 2019 and 2021, two similar studies, with a part of this data locked in the period 2015–2019, and 2015–2020, respectively were reported in [2] and [3].

In determining the blood group characteristics, all requirements have been met regarding the blood samples, test reagents and test erythrocytes, as postulated in Ordinance No. 18/2004 of the Bulgarian Ministry of Health. In the immune-hematological diagnostics, there has been included: determination of the ABO blood group, the A_1 and A_2 antigen subgroups, and the Rh(D) antigen. The determination of the blood groups from the ABO system has been performed by the crossmatching method. The blood test results have been recorded in the "Saint Anna" Hospital Information System. The present research does not contain data of the investigated erythrocyte antigens beyond the ABO and Rhesus factor systems.

The atomic approach to data adopted as early as the research design phase allowed the detailed breakdown of data simultaneously in blood group/subgroup, rhesus factor, and sex, and thus gave us the possibility to group and extract the data in various ways and formulate different conclusions.

3 Results

In this section, the results from the conducted research will be presented, from the most general to the most particular results. Thus, on the basis of the 47562 unique patients' records, we have established the following frequencies of the ABO system blood groups: A – 43.66%, B – 16.36%, O – 31.87%, and AB – 8.11% of all patients in the study (Table 1).

Table 1. Established frequencies of the ABO system blood groups

Blood group	Number of samples	Frequency,%
A	20767	43.66
B	7780	16.36
O	15156	31.87
AB	3859	8.11
Total	**47562**	**100.00**

The distribution of the blood groups after determining the Rh(D) antigen is the following: A Rh(+) – 37.70%; A Rh(−) – 5.96%; B Rh(+) – 14.19%; B Rh(−) – 2.17%; O Rh(+) – 27.45%; O Rh(−) – 4.41%; AB Rh(+) – 7.04%; AB Rh(−) – 1.07% of all investigated patients (Table 2).

Table 2. Established frequencies of the ABO system blood groups with the Rh(D) antigen determined.

Blood group	Number of samples	Frequency,%	Rh(D) Frequency within the group,%
A Rh(+)	17931	37.70	86.34
A Rh(−)	2836	5.96	13.66
Total A	**20767**	**43.66**	–
B Rh(+)	6748	14.19	86.74
B Rh(−)	1032	2.17	13.26
Total B	**7780**	**16.36**	–
O Rh(+)	13057	27.46	86.15
O Rh(−)	2099	4.41	13.85
Total O	**15156**	**31.87**	–
AB Rh(+)	3350	7.04	86.81
AB Rh(−)	509	1.07	13.19
Total AB	**3859**	**8.11**	–

Additionally, in the last column of Table 2, we have calculated the frequency with which both Rh(D) antigens occur in each blood group, respectively: A Rh(+) are 86.34% and A Rh(−) are 13.66% of all A patients; B Rh(+) are 86.74% and B Rh(−) are 13.26% of all B patients; O Rh(+) are 86.15% and O Rh(−) are 13.85% of all O patients; AB Rh(+) are 86.81% and AB Rh(−) are 13.19% of all AB patients.

Thus the following ratios have been determined: A Rh(+)/A Rh(−) = 6.32; B Rh(+)/B Rh(−) = 6.54; O Rh(+)/O Rh(−) = 6.22; AB Rh(+)/AB Rh(−) = 6.58. Hence, totally, the distribution of the Rh(D) antigen, according to our investigation, is 86.38% (41086/47562), while 13.62% (6476/47562) of the tested individuals have been determined as Rh(D) negative, and the ratio between all patients with the Rh(D) antigen and those without it is: 41086/6476 = 6.34. Compared to our previous research [3] based on the respective patients' dataset from the period 2015–2020, this marks in the year 2021 an increase in the ratio from 6.31 to 6.34.

The characterization of the subgroups of the A and AB blood groups is presented in the next Table 3: 40.13% of all 47562 patients are with the A_1 blood subgroup, 3.53% are with the A_2 blood subgroup, 7.18% are with the A_1B blood subgroup and 0.93% of all patients are with the A_2B blood subgroup. Going into deeper detail regarding the Rh(D) antigen, we have the following picture: A_1 Rh(+) – 34.72%; A_2 Rh(+) – 2.98%; A_1 Rh(−) – 5.41%; A_2 Rh(−) – 0.55%; A_1B Rh(+) – 6.21%; A_2B Rh(+) – 0.83%; A_1B Rh(−) – 0.97; A_2B Rh(−) – 0.10% of all the 47562 studied patients.

Table 3. Established frequencies of the A and AB blood group subgroups with the Rh(D) antigen determined

Blood subgroup	Number of samples	Frequency,%	A_1/A_2 Frequency within the blood group,%	Rh(D) Frequency within the blood subgroup,%
A_1 Rh(+)	16513	34.72	79.51	86.51
A_1 Rh(−)	2575	5.41	12.40	13.49
Total A_1	**19088**	**40.13**	**91.91**	–
A_2 Rh(+)	1418	2.98	6.83	84.45
A_2 Rh(−)	261	0.55	1.26	15.55
Total A_2	**1679**	**3.53**	**8.09**	–
A_1B Rh(+)	2956	6.21	76.60	86.53
A_1B Rh(−)	460	0.97	11.92	13.47
Total A_1B	**3416**	**7.18**	**88.52**	–
A_2B Rh(+)	394	0.83	10.21	88.94
A_2B Rh(−)	49	0.10	1.27	11.06
Total A_2B	**443**	**0.93**	**11.48**	–

The frequencies of the A_1 and A_2 subgroups within blood group A in this study has been determined, respectively, as 91.91% (19088/20767) and 8.09% (1679/20767) of all A patients. The frequencies of the A_1B and A_2B subgroups against blood group AB has been determined, respectively, as 88.52% (3416/3859) and 11.48% (443/3859) of all AB patients.

These findings are visualized on Fig. 1.

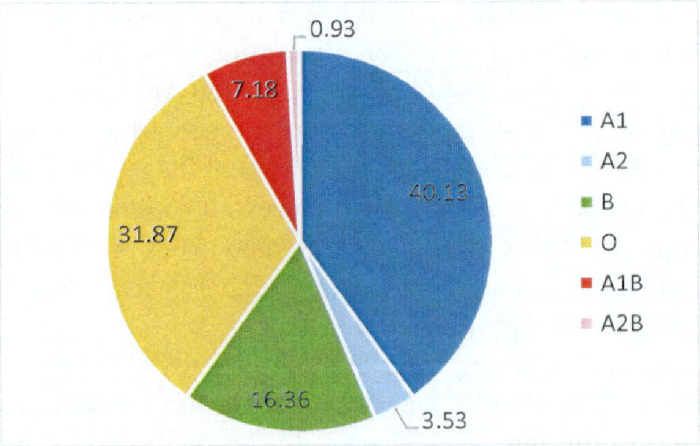

Fig. 1. Distribution of the ABO system blood groups and subgroups

The next Table 4 presents the determined frequencies of the ABO system blood groups in both sexes. The ratio of the blood samples of male to female individuals is 46.66% to 53.34%.

Table 4. Frequencies of the ABO system blood groups – males and females

Blood group	Number of samples		Frequency, %	
	M	F	M	F
A	9572	11195	43.13	44.13
B	3690	4090	16.62	16.12
O	7099	8057	31.99	31.76
AB	1833	2026	8.26	7.99
Total	**22194**	**25368**	**100.00**	**100.00**

The data show the following distribution within the group of male individuals: A – 43.13%, B – 16.62%, O – 31.99%, and AB – 8.26% of all studied male patients; and within the group of female individuals: A – 44.13%, B – 16.12%, O – 31.76%, and AB – 7.99% of all female patients in the study.

The distribution of the blood groups after determining the Rh(D) antigen is the following for males and females (Table 5). The distribution of the Rh(D) antigen in males, according to our investigation, is 86.99% (19306/22194), and in females is 85.86% (21780/25368). Respectively, determined as Rh(D) negative have been 13.01% of the tested males (2888/22194), and 14.14% of the tested females (3588/25368). In more details, the last two columns of Table 5 present the frequencies of the Rh(D) antigen within each of the blood groups for both sexes.

Table 5. Established frequencies of the ABO system blood groups with the Rh(D) antigen determined – males and females

Blood group	Number of samples		Rh(D) frequency within the group,%	
	M	F	M	F
A Rh(+)	8307	9624	86.78	85.97
A Rh(−)	1265	1571	13.22	14.03
Total A	**9572**	**11195**	–	–
B Rh(+)	3201	3547	86.75	86.72
B Rh(−)	489	543	13.25	13.28
Total B	**3690**	**4090**	–	–
O Rh(+)	6183	6874	87.10	85.32
O Rh(−)	916	1183	12.90	14.68
Total O	**7099**	**8057**	–	–

(*continued*)

Table 5. (*continued*)

Blood group	Number of samples		Rh(D) frequency within the group,%	
	M	F	M	F
AB Rh(+)	1615	1735	88.11	85.64
AB Rh(−)	218	291	11.89	14.36
Total AB	**1833**	**2026**	–	–

As presented in Table 3 (for all tested patients), the distributions of the A_1 and A_2 antigen in patients with the A blood group is as follows: $A_1/A = 91.87\%$ (19088/20767) and $A_2/A = 8.13\%$ (1679/20767). Respectively, these distributions in the group of males and the group of females are the following: Males: $A_1/A = 91.50\%$ (8758/9572); $A_2/A = 8.50\%$ (814/9572); Females: $A_1/A = 92.27\%$ (10330/11195); $A_2/A = 7.73\%$ (865/11195).

Again, as presented in Table 3 (for all tested patients), the distributions of the A_1 and A_2 antigen in patients with the AB blood group is as follows: $A_1B/AB = 88.52\%$ (3416/3859); $A_2B/AB = 11.48\%$ (443/3859). Respectively, these distributions in the group of males and the group of females are the following: Males: $A_1B/AB = 88.33\%$ (1619/1833); $A_2B/AB = 11.67\%$ (214/1833); Females: $A_1B/AB = 88.70\%$ (1797/2026); $A_2B/AB = 11.30\%$ (229/2026).

In details the information about the distributions of the A_1 and A_2 antigen in the patients with A and AB blood groups is presented in the Table 6. The following Fig. 2 graphically represents the hitherto presented data.

Table 6. Established frequencies of the A and AB blood group subgroups with the Rh(D) antigen determined – males and females

Blood subgroup	Number of samples		A_1/A_2 Frequency within the blood group,%		Rh(D) Frequency within the blood subgroup,%	
	M	F	M	F	M	F
A_1 Rh(+)	7603	8910	79.43	79.59	86.81	86.25
A_1 Rh(−)	1155	1420	12.07	12.68	13.19	13.75
Total A_1	**8758**	**10330**	**91.50**	**92.27**	–	–
A_2 Rh(+)	704	714	7.35	6.38	86.49	82.54
A_2 Rh(−)	110	151	1.15	1.35	13.51	17.46
Total A_2	**814**	**865**	**8.50**	**7.73**	–	–
A_1B Rh(+)	1423	1533	77.63	75.67	87.89	85.31
A_1B Rh(−)	196	264	10.69	13.03	12.11	14.69
Total A_1B	**1619**	**1797**	**88.32**	**88.70**	–	–

(*continued*)

Table 6. (*continued*)

Blood subgroup	Number of samples		A_1/A_2 Frequency within the blood group,%		Rh(D) Frequency within the blood subgroup,%	
	M	F	M	F	M	F
A_2B Rh(+)	192	202	10.47	9.97	89.72	88.21
A_2B Rh(−)	22	27	1.20	1.33	10.28	11.79
Total A_2B	**214**	**229**	**11.67**	**11.30**	–	–

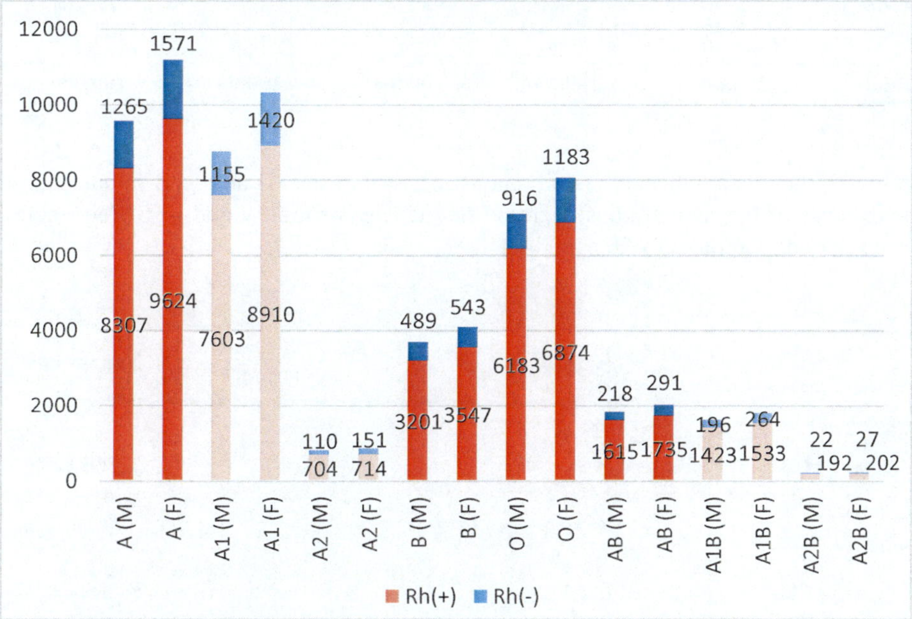

Fig. 2. ABO system blood and subgroups distribution among males and females

4 Comparison with Previous Studies

In the next Table 7, we present a comparison of the results obtained in the course of our present investigation with the results from similar investigations of the blood group distributions in Bulgarian population, conducted by other authors in the past, which are found in the available Bulgarian literature [1, 9, 11]. As our aim has been to present all the available data in the most comparable way, wherever necessary, missing table entries have been populated on the basis of appropriate computations [2].

As presented graphically on Fig. 3, we can outline the most notable trend of increase of the AB blood group frequency at the expense of lesser changes of the frequencies of the A, B and O blood groups. At the same time, according to the available data, visualized

Table 7. Comparison of the presented results (2022) with results from previous studies (1957, 1977, 2012, 2021)

	1957, [8]	1977, [1]	2012, [9]	2021 [3]	2022
A	**44.80**	**43.30**	**43.20**	**43.62**	**43.66**
A_1/A	*87.73*	*86.77*	*90.28*	*91.42*	*92.27*
A_2/A	*12.27*	*13.23*	*9.72*	*8.58*	*7.73*
B	**16.80**	**16.60**	**16.10**	**16.42**	**16.36**
O	**32.10**	**32.50**	**32.60**	**31.79**	**31.87**
AB	**6.30**	**7.60**	**8.10**	**8.18**	**8.11**
A_1B/AB	*76.40*	*n/a*	*83.95*	*87.56*	*88.70*
A_2B/AB	*23.60*	*n/a*	*16.05*	*12.44*	*11.30*
Total	**100.00**	**100.00**	**100.00**	**100.00**	**100.00**

on Fig. 4, there is a noticeable monotonous increase of the A_1 and A_1B frequencies at the expense of the monotonous decrease frequencies of the A_2 and A_2B frequencies, over a period of about 65 years.

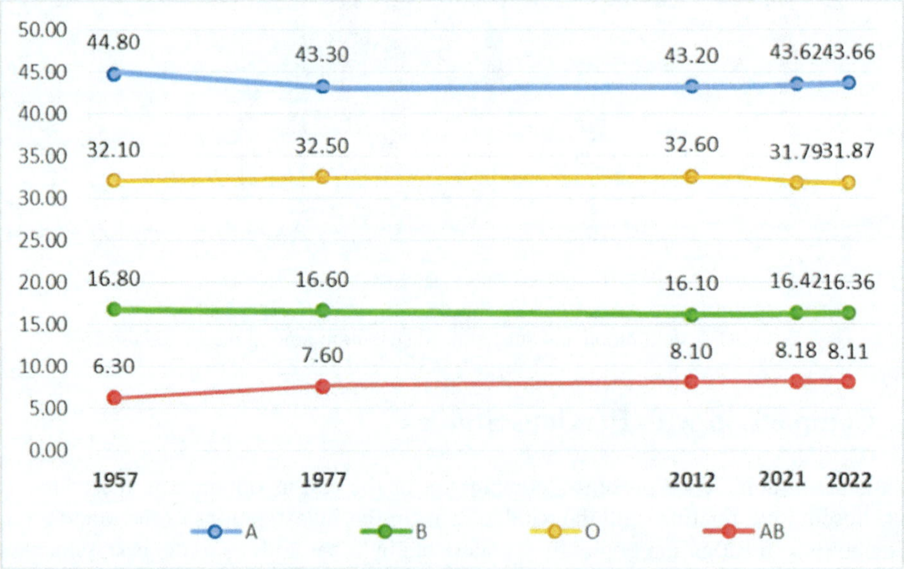

Fig. 3. Temporal change of the distributions of the A, B, O and AB blood groups (in years 1957, 1977, 2012, 2021, 2022)

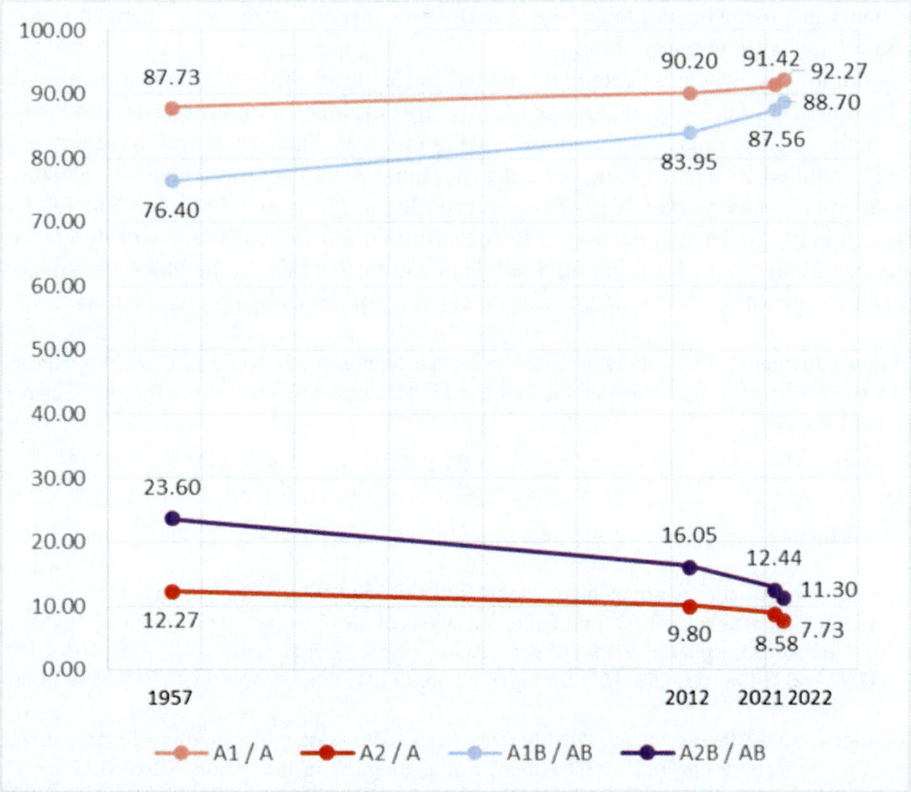

Fig. 4. Distributions of the A_1 and A_2 antigens in the A and AB blood groups (in years 1957, 1977, 2012, 2021, 2022)

5 Conclusions and Further Research

The results from the present research are informative for the specialists of transfusion hematology in Bulgaria regarding the actual state-of-the-art of the ABO system blood groups distributions, as well as the transfusion capacity of Bulgarian population.

On the basis of the available cleansed dataset of patients of the University Hospital "Saint Anna" collected in the period from 2015 to 2021 year, there have been envisaged at least two new steps of further research in this direction. From the patients' records, we have been able to extract automatically information about their year of birth (which varies in the range from 1913 to 2021 year), as well as information about their district place of birth. Thus we will be able to trace the temporal trends in blood group distribution in a sample of Bulgarian population that has lived for a period of a bit more than a century. Second, the patients' personal identity numbers contain information about the birthplace locations of the patients. Extracting this information would be potentially able to outline some additional patterns in the geographical distribution of the blood groups over the Bulgarian territory. Any significant finding in this latest direction of research would be informative for the easier detection and optimized routing of blood banks across different regional transfusion hematology centres in Bulgaria, especially in cases when rarer types of blood are being urgently sought.

Additionally, when collected and treated in this level of detail, the results provide ample opportunities for application of the recently developed multicriteria decision making method called InterCriteria Analysis (ICA) [6, 10]. Such a research has been previously applied to a part of the dataset concerning patient in the University Hospital "Saint Anna" in the period 2015–2019, as reported in [2, 3], and developed in a related research in [4, 5]. The application of ICA can further allow the discovery of trends and patterns of knowledge from the reported data, and further benefit the blood transfusion specialists regarding the blood transfusion capacity of the Bulgarian population.

Acknowledgement. The authors are grateful for the support provided under Grant No. KP-06-N-22/1 "Theoretical research and applications of InterCriteria Analysis" of the National Science Fund of Bulgaria.

References

1. Anastasov, A., et al.: Immunohematology, Sofia, p. 55 (1977). (in Bulgarian)
2. Andreev, N., Atanassova, V.: InterCriteria analysis of the blood group distribution of patients of Saint Anna hospital in 2015–2019. In: Atanassov, K.T., et al. (eds.) IWIFSGN 2019 2019. AISC, vol. 1308, pp. 158–165. Springer, Cham (2021). https://doi.org/10.1007/978-3-030-77716-6_14
3. Andreev, N., Atanassova, V.: Distribution of the ABO system blood groups in the patients of the university hospital "Saint Anna", Sofia, Bulgaria in the period 2015–2020. In: 1st Scientific Conference on Transfusion Hematology, Varna, Bulgaria, 24–26 September 2021 (2021). (in Bulgarian)
4. Andreev, N., Sotirova, E., Ribagin, S.: InterCriteria analysis of data from the centres for transfusion haematology in Bulgaria. Comptes rendus de l'Acade'mie bulgare des Sci. **72**(7), 982–990 (2019)
5. Andreev, N., Vassilev, P., Ribagin, S., Sotirov, S.: InterCriteria analysis of data for blood collection in the transfusion hematology department, university hospital "St. Anna", Sofia. Notes Intuitionistic Fuzzy Sets **25**(2), 88–95 (2019). https://doi.org/10.7546/nifs.2019.25.2.88-95
6. Atanassov, K., Mavrov, D., Atanassova, V.: InterCriteria decision making: a new approach for multicriteria decision making, based on index matrices and intuitionistic fuzzy sets. Issues Intuitionistic Fuzzy Sets Gen. Nets **11**, 1–8 (2014)
7. BloodBook.com: Racial & ethnic distribution of ABO blood types (2013). http://www.bloodbook.com/world-abo.html. Accessed 14 Oct 2022
8. Dean, L.: Blood groups and red cell antigens. National Center for Biotechnology Information (US), Bethesda (MD), pp. 31–33 (2005). https://www.ncbi.nlm.nih.gov/books/n/rbcantigen/pdf/. Accessed 14 Oct 2022
9. Dobreva, A., Doychinova, N., Vasilev, N. (eds.): Transfusion Hematology, p. 68, 114–115. State Publishing House "Medicina I Fizkultura", Sofia (1988). (in Bulgarian)
10. Doukovska, L., Atanassova, V., Sotirova, E., Vardeva, I., Radeva, I.: Defining consonance thresholds in InterCriteria analysis: an overview. In: Hadjiski, M., Atanassov, K.T. (eds.) Intuitionistic Fuzziness and Other Intelligent Theories and Their Applications. SCI, vol. 757, pp. 161–179. Springer, Cham (2019). https://doi.org/10.1007/978-3-319-78931-6_11

11. Popov, R., Petrov, N., Vaseva, V.: Distribution of blood groups of the ABO system in the Military Medical Academy immunohematological diagnostics. Bulg. Med. J. **VI**(2), 45–48 (2012). (in Bulgarian)
12. RhesusNegative.net: Blood type frequencies by country including the Rh factor (2019). http://www.rhesusnegative.net/themission/bloodtypefrequencies/. Accessed 14 Oct 2022
13. Salduz, Z., et al.: ABO and Rh blood group distribution in Istanbul province (Turkey). Istanb. Med. J. **16**, 98–100 (2015). https://doi.org/10.5152/imj.2015.14890
14. Wikipedia Contributors: Blood type distribution by country. In: Wikipedia, The Free Encyclopedia (2017). https://en.wikipedia.org/w/index.php?title=Blood_type_distribution_by_country&oldid=786847072

InterCriteria Analysis of the Geographic Distribution of the ABO System Blood Groups in the Patients of the University Hospital "Saint Anna", Sofia, Bulgaria, from 2015 to 2021

Vassia Atanassova[1](✉), Nikolay Andreev[2], and Angel Dimitriev[1]

[1] Institute of Biophysics and Biomedical Engineering, Bulgarian Academy of Sciences, 105 Acad. Georgi Bonchev Str., 1113 Sofia, Bulgaria
vassia.atanassova@gmail.com

[2] Transfusion Hematology Department, University Hospital "Saint Anna", Sofia 1 D. Mollov Str., Mladost 1, 1750 Sofia, Bulgaria

Abstract. The paper presents the results of the application of the method of InterCriteria Analysis on a dataset of the blood group of 47562 Bulgarian individuals, patients of the University Hospital "Saint Anna", Sofia, Bulgaria collected in the period from 2015 to 2021. Apart of the ABO system blood groups, the recorded data contain information about the patients' A_1 and A_2 subgroups and the Rh(D) antigen. In the current leg of research, in addition to the easily extracted information about the patients' year of birth and sex, here for the first time, the patients' birthplace regions are algorithmically extracted on the basis of the recorded personal identification numbers. This allows us to have a deeper and more insightful picture of the regional distribution of the frequencies of the ABO system, A_1 and A_2 subgroups and the Rh(D) antigen in the frames of the Bulgarian population, which is a level of detail that has not been yet reported in the existing literature and is considered an important contribution of the current research work to the state-of-the-art. The conducted InterCriteria Analysis on these data gives additional details that may help decision makers on national and regional level with respect to regional and cross-regional demand and supply of blood and blood products in Bulgaria.

Keywords: ABO system · Blood group distribution · Rh(D) antigen distribution · InterCriteria Analysis · Intuitionistic fuzzy sets

1 Introduction

Ever since the discovery of the ABO system of blood groups and blood group antigens, their frequency in the human population has been a subject of research study. The ABO system's significance for transfusion medicine is due to the fact that transfusion of an incompatible ABO type of blood is the most common cause of death from blood transfusion [15]. Antigens furthermore appear to play a significant role in human evolution because the frequencies of different ABO blood types exhibit significant variability

S. Sotirov et al. (Eds.): BioInfoMed 2022, LNNS 658, pp. 84–97, 2023.
https://doi.org/10.1007/978-3-031-31069-0_10

among various human populations [12, 15, 19, 20], suggesting that a particular blood type conferred a selection advantage like resistance against an infectious disease, etc. [15].

Every country needs to maintain a regular, up-to-date record with respect to the blood group distribution profile of its population. In this respect, various publicly available online sources contain rather outdated or approximate information about the blood group frequencies of Bulgarian population, in order to be considered reliable and usable. For example, in a top-ranking web source dated to 2013 [12], the data for the Bulgarian population's blood group distribution is roughly given as 44% of A, 15% of B, 32% of O and 8% of AB, and one of the most popular reference websites Wikipedia has provided in [22] as of June 2017 (without citing a particular source) the following frequencies: A Rh(+) 37.4%, A Rh(−) 6.6%, B Rh(+) 12.8%, B Rh(−) 2.2%, O Rh(+) 28.0%, O Rh(−) 5.0%, AB Rh(+) 6.8%, AB Rh(−) 1.2%, summing respectively to the percentages in [12]. Identical numbers, rounded to the nearest integer, are contained in another public database dated to 2019 [20]. A rather different picture is presented in a 2015 research [21] where the blood group distributions for the Bulgarian population are reported as 39.96% of A, 16.84% of B, 35.80% of O, and 7.60% of AB. Literature review of the blood group distributions in Bulgarian population also comprises sources from the years 1977 (Anastasov et al. [1]), 1988 (Dobreva et al. [16]), and 2012 (Popov et al. [19]), which we have discussed in comparison with our recent findings in [3, 4, 10].

These observations of available online information sources demonstrate the particular need for a detailed contemporary investigation of the problem and establishing an accurate database for Bulgaria, which motivates the present research. In addition, here a next level of detail regarding the geographic distribution of ABO blood groups on subnational/regional level is presented for the first time for the country's population.

The present paper is structured as follows: in Sect. 2, we give the background of our study: source of the patients' data and methods of data cleansing and retrieval. Section 3 contains the results of the study, namely the established frequencies the ABO system blood groups and subgroups, Rh(D) antigen frequencies, and respective maintained ratios, categorized by and analyzed in the light of the patients' birthplace location. The subsequent Sect. 4 analyzes the reported data with the apparatus of the intuitionistic fuzzy sets based method of InterCriteria Analysis. Finally, Sect. 5 gives conclusion and directions for future research.

2 Materials and Methods

The data collection for the present research is performed by the Centre of Transfusion Hematology in University Hospital "Saint Anna", Sofia, which is the source of the dataset of hospital patients registered in the period between January 1, 2015, and December, 31, 2021. In determining the blood group characteristics, all legal requirements related to the blood samples collection, test reagents and test erythrocytes, have been addressed, as stipulated in Ordinance No. 18/2004 of the Bulgarian Ministry of Health. The immune-hematological diagnostics comprises determination of the ABO blood group by the crossmatching method, determination of the A_1 and A_2 antigen subgroups, and determination of the Rh(D) antigen, followed by recording these data in the "Saint Anna" Hospital internal information system. The present research does not contain data of the investigated erythrocyte antigens beyond the ABO and Rhesus factor systems.

We are specifically noting the fact that, being one of the largest multispecialty hospitals in the capital city of Bulgaria, Sofia, University Hospital "Saint Anna" (locally known as '*The District Hospital*') treats both city residents and patients from all regional districts of Bulgaria. This fact served as an additional motivation for our research, as it suggested the idea to extract from the patients' data the information about their birthplace and collect as an additional parameter (column) of the dataset.

At the stage of primary data collection, we obtained the tabular data of: 6469 samples in 2015, 13716 samples in 2016, 8748 samples in 2017, 10214 samples in 2018, 10534 samples in 2019, 8769 samples in 2020, and 8968 samples in 2021.

We conducted the following procedure of cleansing the primary data, including:

1. Removal of records with incomplete or inaccurate entered personal data about the patients or data about their blood groups (e.g. recorded valued 'A' instead of 'A1' or 'AB', missing value of Rh(D) antigen, etc.),
2. Removal of records for patients of foreign nationality (where the personal identification number is not in the "EGN" (Unique citizenship number) format for Bulgarian nationals but in the "LNC" format for resident people of foreign nationality),
3. Removal of duplicate records with respect to the "EGN" (including numerous records of the blood samples of one and the same patient within a calendar year, as well as across the full six-year period),
4. Extraction from the "EGN" of the data about the patient's sex, year of birth and birthplace location (region) (notably, this is the novel aspect of the present research),
5. Anonymization of the data.

As a result of the data cleansing procedure (Steps 1–3), the total number of 67418 records in the primary input was reduced and the dataset over which our investigation was conducted and results are reported, includes data of 47562 unique individuals. In 2019 and 2021, two similar studies, with a part of this data locked in the period 2015–2019, and 2015–2020, respectively were reported in [3] and [4]. The full results from the authors' research of the dataset from 2015–2021 are provided in [10], as which continuation the present paper is standing.

We will comment here specifically on Step 4 of the followed procedure. Extraction from the unique citizenship number ("EGN") of the information about the person's sex, year of birth and place of birth (on district level, Bulgaria having 28 regional districts) is an algorithmically easy job given the public nature of the algorithm itself. For the readers' reference and convenience, we will note that a detailed explanation for the algorithm is provided in [18], and an online tool for checking the validity of the EGNs, supported with thorough information and source code is provided on the webpage [13]. For the needs of the present research, the software needed to extract the necessary relevant information has been originally developed by the third author.

On the basis of the extraction of the data regarding patients' place of birth, we can report the following geographic profile of the patients of the "Saint Anna" Hospital in Sofia, see Fig. 1.

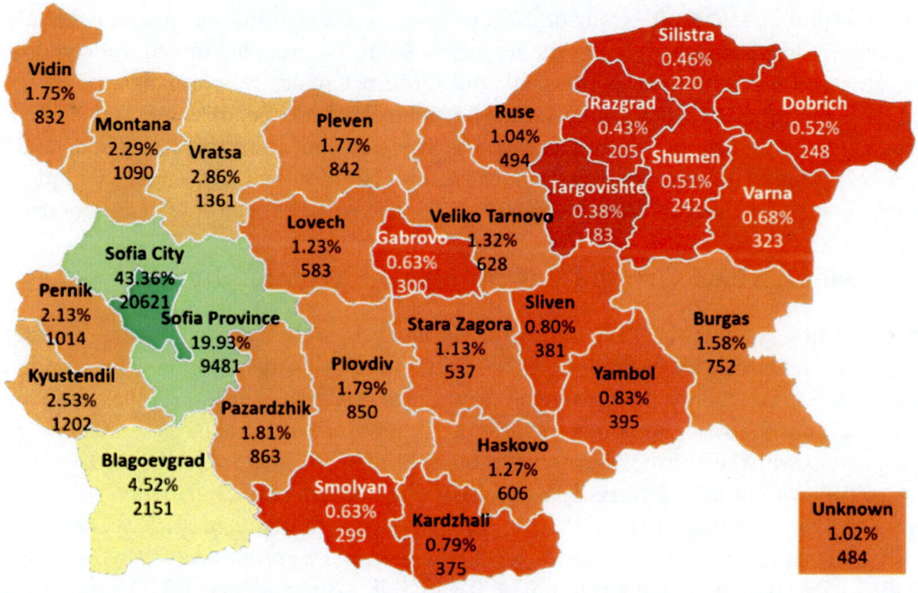

Fig. 1. Map of the distribution of "Saint Anna" Hospital patients from the period 2015–2021, covered in the study of ABO blood groups distribution

We shall immediately note that the map on Fig. 1 above should be regarded with a certain level of conditionality, for several reasons that will be listed below.

First, it is not based on all of the hospital patients but on that portion of them, who are Bulgarian nationals, and whose EGN and blood related data were completely and correctly input in the hospital information system, after removing duplication of patients with multiple entries across the years, which means it is based on the 47562 unique individuals out of the 67418 data records from the system. Moreover, the statistics above does not contain information about patients who did not have their blood checked as a part of the routine of emergency diagnostics.

Second, the map reflects the patients' birthplace, not the place of residence – and specifically for a large Sofia-based hospital it is of importance that many residents of Sofia are actually not born in the city or in the province of Sofia. So, while the map cannot directly suggest the actual transfusion capacity of the population by place of residence (which is the one that would make sense), it can be – to a certain extent – informative regarding the regional blood group profiles in the different parts of Bulgaria, regardless of the internal migration.

Our third notable consideration is that our capability of making certain conclusions on national level on the basis of the so derived data, suffers from the limitation related to the disproportionate representation of all the regions in the presented statistics. Expectedly, for various reasons (geographic, logistic, financial, etc.), the largest share of patients of "Saint Anna" are people from the city and the region of Sofia, with several neighbouring regions from Western Bulgaria being far better represented compared to regions from the Eastern Bulgaria. Patients from these regions get most of their treatment in the local town and provincial hospitals, so our data collected in "Saint Anna" hospital is evidently skewed towards the patients from the region of Sofia. We have attempted to overcome this shortcoming of comparing datasets with different scales by normalization of the data, i.e. having all of its variations falling between zero and one, or as represented here in percentages. Indisputably, local statistics derived using the same rigorous and detailed data collection, data cleansing and data extraction methodology will surely yield more representative results for those provinces than the one we have been able to present here.

3 Results: Blood Group Distributions on Regional Level

In this section, the findings from the conducted analysis of the cleansed patients' dataset will be presented, from the most general to the most particular results. Thus, on the basis of the 47562 unique patients' records, we have established in Table 1 the following frequencies of the ABO system blood groups per province of birthplace (see also [10]). The patients with their birthplace province, number and percentage of representativeness, are as follows (sorted by representativeness): Sofia–City – 20621 patients (43.36%), Sofia–Province – 9481 patients (19.93%), Blagoevgrad – 2151 patients (4.52%), Vratsa – 1361 patients (2.86%), Kyustendil – 1202 patients (2.53%), Montana – 1090 patients (2.29%), Pernik – 1014 patients (2.13%), Pazardzhik – 863 patients (1.81%), Plovdiv – 850 patients (1.79%), Pleven – 842 patients (1.77%), Vidin – 832 patients (1.75%), Burgas – 752 patients (1.58%), Veliko Tarnovo – 628 patients (1.32%), Haskovo – 606 patients (1.27%), Lovech – 583 patients (1.23%), Stara Zagora – 537 patients (1.13%), Ruse – 494 patients (1.04%), Yambol – 395 patients (0.83%), Sliven – 381 patients (0.80%), Kardzhali – 375 patients (0.79%), Varna – 323 patients (0.68%), Gabrovo – 300 patients (0.63%), Smolyan – 299 patients (0.63%), Dobrich (ex. Tolbuhin) – 248 patients (0.52%), Shumen – 242 patients (0.51%), Silistra – 220 patients (0.46%), Razgrad – 205 patients (0.43%), Targovishte – 183 patients (0.38%), and finally "Unknown birthplace province" – 484 patients (1.02%). Due to the high level of detail achieved with presenting the data on regional level, we will not additionally slice it here by male/female sex, although we have this information in hand.

We will note that for a better comprehension and evaluation of the data presented in Table 1, the provinces are given with the percentage of representativeness in the statistics. The rows with data for patients with unknown birthplace and the total national have been left uncoloured. The colour codes of the table cells (from blue to red) are defined as follows: provinces recording lowest frequencies of a given blood group compared to all the other provinces are in blue colour, and provinces recording the highest frequencies of a given blood group compared to all the other provinces are in red colour. For instance, the province with the highest frequency of the AB blood group is Silistra (10.00%) compared to Varna with the lowest frequency (6.19%), all this on the basis of the dataset of registered patients in Sofia's "Saint Anna" Hospital.

The distribution of the blood groups after determining the Rh(D) antigen is given in the following Table 2. Again, the frequencies per province will be presented disregarding the frequencies per sex, with the data about males and females given *en gross*. The colour codes for the cells in Table 2 are identical to those described for Table 1.

The characterization of the subgroups of the A and AB blood groups is presented in the next Table 3.

Table 1. Established frequencies of the ABO system blood groups (A, B, O, AB), on provincial and national level.

Province in Bulgaria (% of data) \ Blood group		A	B	O	AB
Sofia City	43.36	44.51	16.21	31.05	8.23
Sofia Province	19.93	43.85	16.66	31.48	8.01
Blagoevgrad	4.52	43.75	15.43	33.80	7.02
Vratsa	2.86	42.40	17.49	31.45	8.67
Kyustendil	2.53	44.01	16.64	31.11	8.24
Montana	2.29	43.39	17.80	30.73	8.07
Pernik	2.13	42.31	15.68	33.33	8.68
Pazardzhik	1.81	43.22	14.95	34.99	6.84
Plovdiv	1.79	41.53	17.65	34.47	6.35
Pleven	1.77	43.59	18.65	28.98	8.79
Vidin	1.75	41.11	14.78	33.17	10.94
Burgas	1.58	41.62	15.96	34.44	7.98
Veliko Tarnovo	1.32	42.83	16.40	33.12	7.64
Haskovo	1.27	41.91	14.52	36.30	7.26
Lovech	1.23	40.65	17.15	32.59	9.61
Stara Zagora	1.13	41.15	15.27	35.38	8.19
Ruse	1.04	42.51	14.57	36.64	6.28
Yambol	0.83	44.56	18.23	30.89	6.33
Sliven	0.80	41.47	18.37	32.02	8.14
Kardzhali	0.79	41.33	16.80	32.27	9.60
Varna	0.68	42.72	16.41	34.67	6.19
Gabrovo	0.63	42.00	15.33	34.67	8.00
Smolyan	0.63	38.46	16.05	36.79	8.70
Dobrich	0.52	42.74	13.31	35.89	8.06
Shumen	0.51	44.21	12.81	35.12	7.85
Silistra	0.46	48.18	13.64	26.82	11.36
Razgrad	0.43	36.10	20.49	36.10	7.32
Targovishte	0.38	48.09	14.21	30.05	7.65
-- *Unknown*	1.02	*40.29*	*19.83*	*30.99*	*8.88*
Bulgaria TOTAL	**100.00**	**43.85**	**16.36**	**31.87**	**8.11**

The frequencies of the A_1 and A_2 subgroups within blood group A in this study, as well as the frequencies of the A_1B and A_2B subgroups against blood group AB have been determined, for the patients in all provinces, as listed in Table 4. The color legend in the table is similar to the legends of Tables 1, 2 and 3: the higher the value of the frequencies of the A_1 subgroup in both the A and AB groups (2nd and 4th columns), the more intensive the red shade, the lower this frequency, the bluer the shade, yet since

Table 2. Established frequencies of the ABO system blood groups with the Rh(D) antigen determined, on provincial and national level.

Blood group and Rh(D) / Province in Bulgaria (% of data)		A		B		O		AB	
		Rh(+)	Rh(-)	Rh(+)	Rh(-)	Rh(+)	Rh(-)	Rh(+)	Rh(-)
Sofia City	43.36	38.27	6.24	13.93	2.28	26.75	4.30	7.09	1.13
Sofia Province	19.93	37.91	5.94	14.52	2.14	27.02	4.46	7.08	0.93
Blagoevgrad	4.52	38.08	5.67	13.62	1.81	29.43	4.37	6.09	0.93
Vratsa	2.86	37.25	5.14	15.06	2.42	26.67	4.78	8.01	0.66
Kyustendil	2.53	37.77	6.24	14.14	2.50	27.12	3.99	7.32	0.92
Montana	2.29	37.61	5.78	15.87	1.93	27.16	3.58	6.51	1.56
Pernik	2.13	37.38	4.93	14.10	1.58	28.50	4.83	7.40	1.28
Pazardzhik	1.81	37.43	5.79	12.75	2.20	31.17	3.82	6.03	0.81
Plovdiv	1.79	35.41	6.12	15.06	2.59	30.35	4.12	5.06	1.29
Pleven	1.77	38.24	5.34	17.10	1.54	24.11	4.87	7.96	0.83
Vidin	1.75	35.58	5.53	12.62	2.16	28.37	4.81	9.01	1.92
Burgas	1.58	36.70	4.92	14.36	1.60	28.46	5.98	7.45	0.53
Veliko Tarnovo	1.32	36.94	5.89	14.33	2.07	28.66	4.46	6.37	1.27
Haskovo	1.27	36.30	5.61	13.53	0.99	31.52	4.79	6.11	1.16
Lovech	1.23	34.65	6.00	14.24	2.92	29.16	3.43	8.06	1.54
Stara Zagora	1.13	36.50	4.66	13.97	1.30	29.80	5.59	6.89	1.30
Ruse	1.04	37.25	5.26	12.15	2.43	31.58	5.06	5.87	0.40
Yambol	0.83	38.23	6.33	16.71	1.52	25.82	5.06	5.82	0.51
Sliven	0.80	35.43	6.04	16.01	2.36	28.35	3.67	7.35	0.79
Kardzhali	0.79	35.47	5.87	14.67	2.13	29.07	3.20	8.80	0.80
Varna	0.68	37.15	5.57	13.62	2.79	29.72	4.95	5.88	0.31
Gabrovo	0.63	37.67	4.33	13.67	1.67	30.33	4.33	7.33	0.67
Smolyan	0.63	32.11	6.35	12.04	4.01	31.10	5.69	6.35	2.34
Dobrich	0.52	35.48	7.26	12.90	0.40	31.85	4.03	7.66	0.40
Shumen	0.51	40.08	4.13	12.81	0.00	29.75	5.37	6.20	1.65
Silistra	0.46	42.27	5.91	10.45	3.18	22.27	4.55	10.00	1.36
Razgrad	0.43	33.17	2.93	17.07	3.41	32.68	3.41	6.83	0.49
Targovishte	0.38	40.98	7.10	11.48	2.73	26.78	3.28	4.92	2.73
-- *Unknown*	1.02	*32.02*	*8.26*	*17.56*	*2.27*	*24.79*	*6.20*	*7.44*	*1.45*
Bulgaria TOTAL	**100.00**	**37.70**	**5.96**	**14.19**	**2.17**	**27.45**	**4.41**	**7.04**	**1.07**

the A_2 subgroup complements A_1, the color code of the 3rd and 5th columns, i.e., the frequencies A_2 subgroup in the A and AB groups, the logic in the colour legend is the opposite.

Table 3. Established frequencies of the A and AB blood group subgroups with the Rh(D) antigen determined, on provincial and national level.

Blood subgroup and Rh(D) / Province in Bulgaria (% of data)		A_1		A_2		A_1B		A_2B	
		Rh(+)	**Rh(-)**	**Rh(+)**	**Rh(-)**	**Rh(+)**	**Rh(-)**	**Rh(+)**	**Rh(-)**
Sofia City	43.36	35.38	5.66	2.89	0.57	6.28	1.01	0.82	0.13
Sofia Province	19.93	34.62	5.38	3.29	0.56	6.19	0.89	0.89	0.04
Blagoevgrad	4.52	35.43	5.07	2.65	0.60	5.72	0.79	0.37	0.14
Vratsa	2.86	33.73	4.63	3.53	0.51	7.05	0.59	0.96	0.07
Kyustendil	2.53	35.19	5.66	2.58	0.58	6.66	0.83	0.67	0.08
Montana	2.29	33.94	5.32	3.67	0.46	5.87	1.19	0.64	0.37
Pernik	2.13	35.60	4.44	1.78	0.49	6.21	1.18	1.18	0.10
Pazardzhik	1.81	34.18	5.56	3.24	0.23	4.98	0.70	1.04	0.12
Plovdiv	1.79	31.41	5.53	4.00	0.59	4.59	1.06	0.47	0.24
Pleven	1.77	34.80	4.99	3.44	0.36	7.01	0.71	0.95	0.12
Vidin	1.75	33.29	4.81	2.28	0.72	7.57	1.80	1.44	0.12
Burgas	1.58	34.18	4.52	2.53	0.40	5.85	0.53	1.60	0.00
Veliko Tarnovo	1.32	32.96	5.89	3.98	0.00	5.57	0.96	0.80	0.32
Haskovo	1.27	33.99	5.45	2.31	0.17	5.45	1.16	0.66	0.00
Lovech	1.23	31.90	4.97	2.74	1.03	7.55	1.54	0.51	0.00
Stara Zagora	1.13	31.66	4.28	4.84	0.37	6.15	1.30	0.74	0.00
Ruse	1.04	34.62	4.45	2.63	0.81	5.67	0.40	0.20	0.00
Yambol	0.83	35.44	5.82	2.78	0.51	5.06	0.51	0.76	0.00
Sliven	0.80	32.02	5.51	3.41	0.52	6.56	0.79	0.79	0.00
Kardzhali	0.79	33.87	5.87	1.60	0.00	8.00	0.80	0.80	0.00
Varna	0.68	35.29	5.26	1.86	0.31	4.64	0.31	1.24	0.00
Gabrovo	0.63	34.33	3.67	3.33	0.67	6.33	0.67	1.00	0.00
Smolyan	0.63	29.77	4.68	2.34	1.67	6.02	2.34	0.33	0.00
Dobrich	0.52	34.27	6.85	1.21	0.40	6.45	0.40	1.21	0.00
Shumen	0.51	38.84	2.89	1.24	1.24	5.79	1.24	0.41	0.41
Silistra	0.46	39.55	5.91	2.73	0.00	9.55	1.36	0.45	0.00
Razgrad	0.43	32.20	2.93	0.98	0.00	5.37	0.49	1.46	0.00
Targovishte	0.38	35.52	5.46	5.46	1.64	4.92	2.19	0.00	0.55
-- *Unknown*	1.02	*28.72*	*7.85*	*3.31*	*0.41*	*6.20*	*1.45*	*1.24*	*0.00*
Bulgaria TOTAL	**100.00**	**34.72**	**5.41**	**2.98**	**0.55**	**6.21**	**0.97**	**0.83**	**0.10**

Table 4. Established frequencies of the A and AB blood group subgroups with the Rh(D) antigen determined, on provincial and national level.

Subgroup / Group Frequency / Province in Bulgaria (% of data)		A_1/A	A_2/A	A_1B/AB	A_2B/AB
Sofia City	43.36	92.22	7.78	88.51	11.49
Sofia Province	19.93	91.22	8.78	88.41	11.59
Blagoevgrad	4.52	92.56	7.44	92.72	7.28
Vratsa	2.86	90.47	9.53	88.14	11.86
Kyustendil	2.53	92.82	7.18	90.91	9.09
Montana	2.29	90.49	9.51	87.50	12.50
Pernik	2.13	94.64	5.36	85.23	14.77
Pazardzhik	1.81	91.96	8.04	83.05	16.95
Plovdiv	1.79	88.95	11.05	88.89	11.11
Pleven	1.77	91.28	8.72	87.84	12.16
Vidin	1.75	92.69	7.31	85.71	14.29
Burgas	1.58	92.97	7.03	80.00	20.00
Veliko Tarnovo	1.32	90.71	9.29	85.42	14.58
Haskovo	1.27	94.09	5.91	90.91	9.09
Lovech	1.23	90.72	9.28	94.64	5.36
Stara Zagora	1.13	87.33	12.67	90.91	9.09
Ruse	1.04	91.90	8.10	96.77	3.23
Yambol	0.83	92.61	7.39	88.00	12.00
Sliven	0.80	90.51	9.49	90.32	9.68
Kardzhali	0.79	96.13	3.87	91.67	8.33
Varna	0.68	94.93	5.07	80.00	20.00
Gabrovo	0.63	90.48	9.52	87.50	12.50
Smolyan	0.63	89.57	10.43	96.15	3.85
Dobrich	0.52	96.23	3.77	85.00	15.00
Shumen	0.51	94.39	5.61	89.47	10.53
Silistra	0.46	94.34	5.66	96.00	4.00
Razgrad	0.43	97.30	2.70	80.00	20.00
Targovishte	0.38	85.23	14.77	92.86	7.14
-- Unknown	*1.02*	*90.77*	*9.23*	*86.05*	*13.95*
Bulgaria TOTAL	**100.00**	**91.91**	**8.09**	**88.52**	**11.48**

4 Results: InterCriteria Analysis

The presented findings from Sect. 3 are analyzed with the intuitionistic fuzzy sets based decision making approach of InterCriteria Analysis. Detailed theoretical presentation of the method is provided in a number of publications [7–9, 14, 17], and its applications to problems and datasets in the area of blood transfusion are contained in [2, 3, 5, 6].

The presented dataset contains 29 rows (28 Bulgarian provinces and 1 'Unknown', rare yet permissible by the 'EGN' algorithm) and 28 columns (standing respectively for the criteria: A, B, O, AB, A(+), A(−), B(+), B(−), O(+), O(−), AB(+), AB(−), A_1, A_2, A_1B, A_2B, A_1(+), A_1(−), A_2(+), A_2(−), A_1B(+), A_1B(−), A_2B(+), A_2B(−), A_1/A, A_2/A, A_1B/AB, A_2B/AB. After running the ICA algorithm on this dataset, the following results are obtained (Tables 5 and 6).

Table 5. Results of ICA analysis on the patients' dataset showing the links between the data sorted per district: μ values.

μ	Sofia - Cit	Sofia - Prc	Blagoevgr	Vratsa	Kyustendi	Montana	Pernik	Pazardzhi	Plovdiv	Pleven	Vidin	Burgas	Veliko Tar	Haskovo
Sofia - Cit	1.00	0.99	0.99	0.98	0.99	0.99	0.98	0.97	0.97	0.99	0.99	0.96	0.98	0.97
Sofia - Prc	0.99	1.00	0.98	0.99	0.98	0.98	0.98	0.97	0.97	0.99	0.97	0.96	0.97	0.96
Blagoevgr	0.99	0.98	1.00	0.97	0.98	0.98	0.96	0.96	0.97	0.97	0.97	0.94	0.97	0.96
Vratsa	0.98	0.99	0.97	1.00	0.98	0.98	0.98	0.98	0.96	0.99	0.97	0.97	0.97	0.96
Kyustendi	0.99	0.98	0.98	0.98	1.00	0.99	0.98	0.96	0.97	0.98	0.99	0.96	0.97	0.98
Montana	0.99	0.98	0.98	0.98	0.99	1.00	0.97	0.97	0.98	0.98	0.98	0.95	0.97	0.97
Pernik	0.98	0.98	0.96	0.98	0.98	0.97	1.00	0.96	0.95	0.98	0.99	0.96	0.96	0.96
Pazardzhi	0.97	0.97	0.96	0.98	0.96	0.97	0.96	1.00	0.97	0.98	0.96	0.97	0.98	0.96
Plovdiv	0.97	0.97	0.97	0.96	0.97	0.98	0.95	0.97	1.00	0.96	0.96	0.94	0.97	0.96
Pleven	0.99	0.99	0.97	0.99	0.98	0.98	0.98	0.98	0.96	1.00	0.97	0.97	0.97	0.96
Vidin	0.99	0.97	0.97	0.97	0.99	0.98	0.99	0.96	0.96	0.97	1.00	0.96	0.97	0.97
Burgas	0.96	0.96	0.94	0.97	0.96	0.95	0.96	0.97	0.94	0.97	0.96	1.00	0.95	0.95
Veliko Tar	0.98	0.97	0.97	0.97	0.97	0.97	0.96	0.98	0.97	0.97	0.97	0.95	1.00	0.96
Haskovo	0.97	0.96	0.96	0.96	0.98	0.97	0.96	0.96	0.96	0.96	0.97	0.95	0.96	1.00
Lovech	0.96	0.95	0.97	0.94	0.96	0.96	0.94	0.94	0.95	0.95	0.95	0.94	0.94	0.95
Stara Zagc	0.96	0.95	0.96	0.95	0.96	0.96	0.94	0.94	0.94	0.95	0.95	0.94	0.95	0.96
Ruse	0.95	0.94	0.96	0.94	0.94	0.95	0.93	0.93	0.94	0.94	0.94	0.93	0.94	0.94
Yambol	0.95	0.96	0.94	0.97	0.95	0.95	0.94	0.97	0.96	0.97	0.94	0.95	0.95	0.94
Sliven	0.97	0.98	0.97	0.97	0.97	0.98	0.97	0.96	0.96	0.97	0.96	0.96	0.96	0.96
Kardzhali	0.95	0.95	0.93	0.94	0.95	0.94	0.96	0.93	0.92	0.95	0.95	0.93	0.94	0.94
Varna	0.94	0.94	0.93	0.95	0.94	0.93	0.94	0.96	0.93	0.96	0.94	0.96	0.94	0.94
Gabrovo	0.96	0.97	0.96	0.98	0.96	0.96	0.96	0.97	0.95	0.98	0.95	0.97	0.96	0.95
Smolyan	0.93	0.92	0.94	0.92	0.92	0.93	0.91	0.91	0.93	0.92	0.92	0.91	0.92	0.93
Dobrich	0.92	0.92	0.91	0.93	0.93	0.91	0.93	0.95	0.92	0.93	0.93	0.94	0.92	0.94
Shumen	0.95	0.94	0.95	0.94	0.96	0.94	0.95	0.92	0.93	0.94	0.96	0.93	0.93	0.94
Silistra	0.94	0.93	0.94	0.92	0.94	0.93	0.93	0.91	0.91	0.93	0.94	0.91	0.93	0.93
Razgrad	0.91	0.91	0.90	0.92	0.92	0.90	0.93	0.92	0.89	0.92	0.92	0.93	0.92	0.91
Targovish	0.93	0.92	0.94	0.91	0.92	0.93	0.90	0.91	0.94	0.91	0.91	0.88	0.92	0.91
..Unknow	0.97	0.96	0.96	0.96	0.97	0.97	0.95	0.97	0.98	0.96	0.96	0.95	0.97	0.97

μ	Lovech	Stara Zagc	Ruse	Yambol	Sliven	Kardzhali	Varna	Gabrovo	Smolyan	Dobrich	Shumen	Silistra	Razgrad	Targovish	..Unknow
Sofia - Cit	0.96	0.96	0.95	0.95	0.97	0.95	0.94	0.96	0.93	0.92	0.95	0.94	0.91	0.93	0.97
Sofia - Prc	0.95	0.95	0.94	0.96	0.98	0.95	0.94	0.97	0.92	0.92	0.94	0.93	0.91	0.92	0.96
Blagoevgr	0.97	0.96	0.96	0.94	0.97	0.93	0.93	0.96	0.94	0.91	0.95	0.94	0.90	0.94	0.96
Vratsa	0.94	0.95	0.94	0.97	0.97	0.94	0.95	0.98	0.92	0.93	0.94	0.92	0.92	0.91	0.96
Kyustendi	0.96	0.96	0.94	0.95	0.97	0.95	0.94	0.96	0.92	0.93	0.96	0.94	0.92	0.92	0.97
Montana	0.96	0.96	0.95	0.95	0.98	0.94	0.93	0.96	0.93	0.91	0.94	0.93	0.90	0.93	0.97
Pernik	0.94	0.94	0.93	0.94	0.97	0.96	0.94	0.96	0.91	0.93	0.95	0.93	0.93	0.90	0.95
Pazardzhi	0.94	0.94	0.93	0.97	0.96	0.93	0.96	0.97	0.91	0.95	0.92	0.91	0.92	0.91	0.97
Plovdiv	0.95	0.94	0.94	0.96	0.96	0.92	0.93	0.95	0.93	0.92	0.93	0.91	0.89	0.94	0.98
Pleven	0.95	0.95	0.94	0.97	0.97	0.95	0.96	0.98	0.92	0.93	0.94	0.93	0.92	0.91	0.96
Vidin	0.95	0.95	0.94	0.94	0.96	0.95	0.94	0.95	0.92	0.93	0.96	0.94	0.92	0.91	0.96
Burgas	0.94	0.94	0.93	0.95	0.96	0.93	0.96	0.97	0.91	0.94	0.93	0.91	0.93	0.88	0.95
Veliko Tar	0.94	0.95	0.94	0.95	0.96	0.94	0.94	0.96	0.92	0.92	0.93	0.93	0.92	0.92	0.97
Haskovo	0.95	0.96	0.94	0.94	0.96	0.94	0.94	0.95	0.93	0.94	0.94	0.93	0.91	0.91	0.97
Lovech	1.00	0.95	0.98	0.93	0.96	0.94	0.92	0.94	0.96	0.90	0.92	0.96	0.89	0.92	0.95
Stara Zagc	0.95	1.00	0.95	0.93	0.95	0.92	0.91	0.96	0.93	0.91	0.93	0.92	0.90	0.92	0.95
Ruse	0.98	0.95	1.00	0.92	0.94	0.92	0.91	0.94	0.97	0.89	0.92	0.95	0.88	0.91	0.94
Yambol	0.93	0.93	0.92	1.00	0.96	0.92	0.97	0.97	0.90	0.94	0.91	0.90	0.90	0.90	0.97
Sliven	0.96	0.95	0.94	0.96	1.00	0.96	0.94	0.97	0.92	0.92	0.92	0.92	0.91	0.91	0.96
Kardzhali	0.94	0.92	0.92	0.92	0.96	1.00	0.94	0.93	0.91	0.92	0.91	0.97	0.93	0.88	0.93
Varna	0.92	0.91	0.91	0.97	0.94	0.94	1.00	0.95	0.89	0.96	0.90	0.92	0.92	0.88	0.95
Gabrovo	0.94	0.96	0.94	0.97	0.97	0.93	0.95	1.00	0.92	0.94	0.93	0.91	0.92	0.90	0.96
Smolyan	0.96	0.93	0.97	0.90	0.92	0.91	0.89	0.92	1.00	0.88	0.90	0.93	0.88	0.89	0.93
Dobrich	0.90	0.91	0.89	0.94	0.92	0.92	0.96	0.94	0.88	1.00	0.91	0.90	0.92	0.87	0.93
Shumen	0.92	0.93	0.92	0.91	0.92	0.91	0.90	0.93	0.90	0.91	1.00	0.90	0.89	0.89	0.92
Silistra	0.96	0.92	0.95	0.90	0.92	0.97	0.92	0.91	0.93	0.90	0.90	1.00	0.91	0.88	0.92
Razgrad	0.89	0.90	0.88	0.90	0.91	0.93	0.92	0.92	0.88	0.92	0.89	0.91	1.00	0.84	0.90
Targovish	0.92	0.92	0.91	0.90	0.91	0.88	0.88	0.90	0.89	0.87	0.89	0.88	0.84	1.00	0.93
..Unknow	0.95	0.95	0.94	0.97	0.96	0.93	0.95	0.96	0.93	0.93	0.92	0.92	0.90	0.93	1.00

Table 6. Results of ICA analysis on the patients' dataset showing the links between the data sorted per district: ν values.

ν	Sofia - Cit	Sofia - Prc	Blagoevgr	Vratsa	Kyustendi	Montana	Pernik	Pazardzhi	Plovdiv	Pleven	Vidin	Burgas	Veliko Tar	Haskovo
Sofia - Cit	0.00	0.01	0.01	0.02	0.01	0.01	0.02	0.03	0.02	0.01	0.01	0.03	0.02	0.02
Sofia - Prc	0.01	0.00	0.01	0.01	0.01	0.01	0.02	0.02	0.02	0.01	0.02	0.02	0.01	0.02
Blagoevgr	0.01	0.01	0.00	0.03	0.02	0.02	0.03	0.04	0.03	0.03	0.02	0.04	0.03	0.03
Vratsa	0.02	0.01	0.03	0.00	0.02	0.02	0.02	0.02	0.04	0.01	0.03	0.02	0.03	0.03
Kyustendi	0.01	0.01	0.02	0.02	0.00	0.01	0.01	0.04	0.03	0.02	0.01	0.03	0.02	0.01
Montana	0.01	0.01	0.02	0.02	0.01	0.00	0.02	0.03	0.02	0.02	0.02	0.04	0.02	0.02
Pernik	0.02	0.02	0.03	0.02	0.01	0.02	0.00	0.03	0.04	0.02	0.00	0.02	0.02	0.02
Pazardzhi	0.03	0.02	0.04	0.02	0.04	0.03	0.03	0.00	0.03	0.02	0.04	0.02	0.02	0.03
Plovdiv	0.02	0.02	0.03	0.04	0.03	0.02	0.04	0.03	0.00	0.04	0.03	0.05	0.02	0.03
Pleven	0.01	0.01	0.03	0.01	0.02	0.02	0.02	0.02	0.04	0.00	0.02	0.02	0.03	0.03
Vidin	0.01	0.02	0.02	0.03	0.01	0.02	0.00	0.04	0.03	0.02	0.00	0.03	0.02	0.02
Burgas	0.03	0.02	0.04	0.02	0.03	0.04	0.02	0.02	0.05	0.02	0.03	0.00	0.03	0.04
Veliko Tar	0.02	0.01	0.03	0.03	0.02	0.02	0.02	0.02	0.02	0.03	0.02	0.03	0.00	0.02
Haskovo	0.02	0.02	0.03	0.03	0.01	0.02	0.02	0.03	0.03	0.03	0.02	0.04	0.02	0.00
Lovech	0.03	0.04	0.02	0.05	0.04	0.03	0.05	0.06	0.04	0.04	0.04	0.06	0.04	0.04
Stara Zagc	0.03	0.03	0.03	0.04	0.03	0.03	0.04	0.05	0.04	0.04	0.04	0.04	0.04	0.03
Ruse	0.04	0.04	0.03	0.05	0.05	0.05	0.06	0.07	0.05	0.06	0.05	0.07	0.05	0.06
Yambol	0.03	0.02	0.04	0.02	0.03	0.03	0.03	0.01	0.02	0.02	0.04	0.03	0.03	0.04
Sliven	0.01	0.00	0.01	0.01	0.01	0.00	0.02	0.02	0.02	0.01	0.02	0.02	0.01	0.02
Kardzhali	0.03	0.03	0.04	0.03	0.02	0.03	0.02	0.04	0.05	0.03	0.02	0.04	0.04	0.03
Varna	0.04	0.04	0.06	0.04	0.04	0.05	0.04	0.02	0.05	0.03	0.04	0.03	0.04	0.05
Gabrovo	0.02	0.01	0.03	0.01	0.03	0.03	0.02	0.02	0.04	0.01	0.03	0.02	0.03	0.03
Smolyan	0.06	0.06	0.04	0.06	0.06	0.06	0.07	0.07	0.06	0.07	0.06	0.08	0.06	0.06
Dobrich	0.05	0.04	0.07	0.04	0.05	0.06	0.03	0.03	0.06	0.04	0.04	0.03	0.04	0.04
Shumen	0.04	0.04	0.04	0.05	0.03	0.04	0.03	0.07	0.06	0.05	0.03	0.05	0.06	0.03
Silistra	0.05	0.05	0.05	0.07	0.04	0.06	0.05	0.08	0.07	0.06	0.04	0.07	0.06	0.05
Razgrad	0.07	0.06	0.08	0.06	0.07	0.08	0.05	0.07	0.09	0.06	0.06	0.05	0.07	0.07
Targovish	0.05	0.05	0.04	0.07	0.06	0.04	0.07	0.07	0.03	0.07	0.06	0.08	0.06	0.06
..Unknow	0.02	0.02	0.03	0.03	0.03	0.02	0.04	0.03	0.01	0.03	0.03	0.04	0.02	0.02

ν	Lovech	Stara Zagc	Ruse	Yambol	Sliven	Kardzhali	Varna	Gabrovo	Smolyan	Dobrich	Shumen	Silistra	Razgrad	Targovish	..Unknow
Sofia - Cit	0.03	0.03	0.04	0.03	0.01	0.03	0.04	0.02	0.06	0.05	0.04	0.05	0.07	0.05	0.02
Sofia - Prc	0.04	0.03	0.04	0.02	0.00	0.03	0.04	0.01	0.06	0.04	0.04	0.05	0.06	0.05	0.02
Blagoevgr	0.02	0.03	0.03	0.04	0.01	0.04	0.06	0.03	0.04	0.07	0.04	0.05	0.08	0.04	0.03
Vratsa	0.05	0.04	0.05	0.02	0.01	0.03	0.04	0.01	0.06	0.04	0.05	0.07	0.06	0.07	0.03
Kyustendi	0.04	0.03	0.05	0.03	0.01	0.02	0.04	0.03	0.06	0.05	0.03	0.04	0.07	0.06	0.03
Montana	0.03	0.03	0.05	0.03	0.00	0.03	0.05	0.03	0.06	0.06	0.04	0.06	0.08	0.04	0.02
Pernik	0.05	0.04	0.06	0.03	0.02	0.02	0.04	0.02	0.07	0.03	0.03	0.05	0.05	0.07	0.04
Pazardzhi	0.06	0.05	0.07	0.01	0.02	0.04	0.02	0.02	0.07	0.03	0.07	0.08	0.07	0.07	0.03
Plovdiv	0.04	0.04	0.05	0.02	0.02	0.05	0.05	0.04	0.06	0.06	0.06	0.07	0.09	0.03	0.01
Pleven	0.04	0.04	0.06	0.02	0.01	0.03	0.03	0.01	0.07	0.04	0.05	0.06	0.06	0.07	0.03
Vidin	0.04	0.04	0.05	0.04	0.02	0.02	0.04	0.03	0.06	0.04	0.03	0.04	0.06	0.06	0.03
Burgas	0.06	0.04	0.07	0.03	0.02	0.04	0.03	0.02	0.08	0.03	0.05	0.07	0.05	0.08	0.04
Veliko Tar	0.04	0.04	0.05	0.03	0.01	0.04	0.04	0.03	0.06	0.04	0.06	0.06	0.07	0.06	0.02
Haskovo	0.04	0.03	0.06	0.04	0.02	0.03	0.05	0.03	0.06	0.04	0.03	0.05	0.07	0.06	0.02
Lovech	0.00	0.04	0.02	0.06	0.03	0.04	0.07	0.05	0.03	0.08	0.06	0.04	0.09	0.05	0.05
Stara Zagc	0.04	0.00	0.05	0.05	0.03	0.06	0.07	0.03	0.06	0.08	0.05	0.07	0.08	0.05	0.04
Ruse	0.02	0.05	0.00	0.06	0.04	0.06	0.08	0.05	0.02	0.09	0.06	0.04	0.11	0.07	0.06
Yambol	0.06	0.05	0.06	0.00	0.02	0.04	0.02	0.02	0.07	0.04	0.07	0.08	0.07	0.06	0.02
Sliven	0.03	0.03	0.04	0.02	0.00	0.03	0.04	0.01	0.05	0.04	0.04	0.05	0.06	0.05	0.02
Kardzhali	0.04	0.06	0.06	0.04	0.03	0.00	0.03	0.04	0.06	0.04	0.05	0.02	0.05	0.07	0.05
Varna	0.07	0.07	0.08	0.02	0.04	0.03	0.00	0.04	0.09	0.03	0.08	0.07	0.06	0.08	0.04
Gabrovo	0.05	0.03	0.05	0.02	0.01	0.04	0.04	0.00	0.06	0.04	0.05	0.07	0.06	0.07	0.03
Smolyan	0.03	0.06	0.02	0.07	0.05	0.06	0.09	0.06	0.00	0.09	0.07	0.05	0.09	0.07	0.06
Dobrich	0.08	0.08	0.09	0.04	0.04	0.04	0.03	0.04	0.09	0.00	0.06	0.07	0.04	0.09	0.05
Shumen	0.06	0.05	0.06	0.07	0.04	0.05	0.08	0.05	0.07	0.06	0.00	0.07	0.08	0.07	0.06
Silistra	0.04	0.07	0.04	0.08	0.05	0.02	0.07	0.07	0.05	0.07	0.07	0.00	0.08	0.08	0.07
Razgrad	0.09	0.08	0.11	0.07	0.06	0.05	0.06	0.06	0.09	0.04	0.08	0.08	0.00	0.12	0.09
Targovish	0.05	0.05	0.07	0.06	0.05	0.07	0.08	0.07	0.07	0.09	0.07	0.08	0.12	0.00	0.04
..Unknow	0.05	0.04	0.06	0.02	0.02	0.05	0.04	0.03	0.06	0.05	0.06	0.07	0.09	0.04	0.00

Notably, the closer relation of part of the districts and their better representation in the dataset results in the strong positive consonances detected (green), but also suggests that more precise conclusions will be possible in a dataset where all districts are more or less equally represented.

The same dataset has been transposed, thus allowing another insight on the data and determining the ICA relationships between the blood groups across the provinces (Tables 7 and 8).

Notably, yet not unexpected, here is the presence of higher positive consonance values in cells parallel to the main diagonal like the pairs between O and O(+), A_1B(+) and

Table 7. Results of ICA analysis on the transposed patients' dataset showing the links between the data sorted per blood indicator: μ values.

μ	0(-)	0(+)	A1(-)	A1(+)	A2(-)	A2(+)	B(-)	B(+)	A1B(-)	A1B(+)	A2B(-)	A2B(+)
0(-)	1.00	0.46	0.41	0.52	0.52	0.45	0.39	0.46	0.51	0.46	0.29	0.53
0(+)	0.46	1.00	0.36	0.39	0.50	0.35	0.47	0.35	0.42	0.35	0.30	0.48
A1(-)	0.41	0.36	1.00	0.50	0.37	0.55	0.52	0.53	0.51	0.52	0.38	0.49
A1(+)	0.52	0.39	0.50	1.00	0.52	0.42	0.41	0.36	0.43	0.47	0.46	0.45
A2(-)	0.52	0.50	0.37	0.52	1.00	0.51	0.57	0.37	0.60	0.50	0.46	0.32
A2(+)	0.45	0.35	0.55	0.42	0.51	1.00	0.52	0.58	0.58	0.47	0.48	0.42
B(-)	0.39	0.47	0.52	0.41	0.57	0.52	1.00	0.47	0.55	0.51	0.33	0.44
B(+)	0.46	0.35	0.53	0.36	0.37	0.58	0.47	1.00	0.40	0.53	0.34	0.61
A1B(-)	0.51	0.42	0.51	0.43	0.60	0.58	0.55	0.40	1.00	0.56	0.48	0.33
A1B(+)	0.46	0.35	0.52	0.47	0.50	0.47	0.51	0.53	0.56	1.00	0.30	0.55
A2B(-)	0.29	0.30	0.38	0.46	0.46	0.48	0.33	0.34	0.48	0.30	1.00	0.28
A2B(+)	0.53	0.48	0.49	0.45	0.32	0.42	0.44	0.61	0.33	0.55	0.28	1.00
0	0.57	0.89	0.34	0.39	0.51	0.34	0.45	0.34	0.42	0.39	0.27	0.52
A	0.47	0.34	0.61	0.83	0.51	0.55	0.44	0.41	0.47	0.47	0.53	0.38
A1	0.49	0.39	0.63	0.87	0.48	0.42	0.39	0.37	0.45	0.50	0.47	0.45
A2	0.47	0.41	0.48	0.38	0.62	0.90	0.55	0.53	0.62	0.46	0.47	0.37
B	0.45	0.36	0.54	0.34	0.41	0.59	0.59	0.88	0.43	0.52	0.33	0.59
AB	0.50	0.34	0.54	0.43	0.49	0.49	0.54	0.53	0.69	0.86	0.33	0.53
A1B	0.49	0.33	0.53	0.45	0.53	0.51	0.56	0.49	0.72	0.84	0.34	0.46
A2B	0.52	0.47	0.49	0.47	0.33	0.44	0.41	0.61	0.37	0.50	0.38	0.90
A(-)	0.40	0.38	0.84	0.46	0.53	0.58	0.62	0.50	0.59	0.51	0.37	0.40
A(+)	0.50	0.32	0.50	0.85	0.54	0.57	0.41	0.40	0.50	0.49	0.53	0.40
AB(-)	0.52	0.43	0.50	0.45	0.62	0.59	0.53	0.40	0.95	0.52	0.53	0.32
AB(+)	0.48	0.37	0.52	0.43	0.41	0.43	0.49	0.57	0.54	0.89	0.28	0.66
A1/A	0.51	0.57	0.53	0.64	0.34	0.13	0.42	0.48	0.37	0.53	0.29	0.63
A2/A	0.49	0.43	0.46	0.35	0.64	0.87	0.58	0.52	0.63	0.47	0.45	0.36
A1B/AB	0.46	0.48	0.50	0.52	0.66	0.54	0.59	0.37	0.66	0.56	0.37	0.14
A2B/AB	0.52	0.51	0.48	0.46	0.30	0.44	0.39	0.61	0.32	0.42	0.38	0.84

μ	0	A	A1	A2	B	AB	A1B	A2B	A(-)	A(+)	AB(-)	AB(+)	A1/A	A2/A	A1B/AB	A2B/AB
0(-)	0.57	0.47	0.49	0.47	0.45	0.50	0.49	0.52	0.40	0.50	0.52	0.48	0.51	0.49	0.46	0.52
0(+)	0.89	0.34	0.39	0.41	0.36	0.34	0.33	0.47	0.38	0.32	0.43	0.37	0.57	0.43	0.48	0.51
A1(-)	0.34	0.61	0.63	0.48	0.54	0.54	0.53	0.49	0.84	0.50	0.50	0.52	0.53	0.46	0.50	0.48
A1(+)	0.39	0.83	0.87	0.38	0.34	0.43	0.45	0.47	0.46	0.85	0.45	0.43	0.64	0.35	0.52	0.46
A2(-)	0.51	0.51	0.48	0.62	0.41	0.49	0.53	0.33	0.53	0.54	0.62	0.41	0.34	0.64	0.66	0.30
A2(+)	0.34	0.55	0.42	0.90	0.59	0.49	0.51	0.44	0.58	0.57	0.59	0.43	0.13	0.87	0.54	0.44
B(-)	0.45	0.44	0.39	0.55	0.59	0.54	0.56	0.41	0.62	0.41	0.53	0.49	0.42	0.58	0.59	0.39
B(+)	0.34	0.41	0.37	0.53	0.88	0.53	0.49	0.61	0.50	0.40	0.40	0.57	0.48	0.52	0.37	0.61
A1B(-)	0.42	0.47	0.45	0.62	0.43	0.69	0.72	0.37	0.59	0.50	0.95	0.54	0.37	0.63	0.66	0.32
A1B(+)	0.39	0.47	0.50	0.46	0.52	0.86	0.84	0.50	0.51	0.49	0.52	0.89	0.53	0.47	0.56	0.42
A2B(-)	0.27	0.53	0.47	0.47	0.33	0.33	0.34	0.38	0.37	0.53	0.53	0.28	0.29	0.45	0.37	0.38
A2B(+)	0.52	0.38	0.45	0.37	0.59	0.53	0.46	0.90	0.40	0.40	0.32	0.66	0.63	0.36	0.14	0.84
0	1.00	0.33	0.37	0.41	0.34	0.36	0.36	0.50	0.34	0.33	0.42	0.40	0.56	0.44	0.46	0.52
A	0.33	1.00	0.87	0.49	0.40	0.42	0.45	0.41	0.58	0.88	0.48	0.42	0.54	0.46	0.54	0.44
A1	0.37	0.87	1.00	0.36	0.35	0.44	0.45	0.48	0.57	0.81	0.46	0.46	0.68	0.32	0.50	0.48
A2	0.41	0.49	0.36	1.00	0.57	0.49	0.53	0.39	0.58	0.52	0.63	0.41	0.04	0.96	0.59	0.39
B	0.34	0.40	0.35	0.57	1.00	0.54	0.50	0.57	0.55	0.40	0.42	0.54	0.44	0.56	0.41	0.58
AB	0.36	0.42	0.44	0.49	0.54	1.00	0.92	0.52	0.56	0.45	0.66	0.84	0.49	0.51	0.57	0.42
A1B	0.36	0.45	0.45	0.53	0.50	0.92	1.00	0.44	0.55	0.48	0.68	0.80	0.45	0.55	0.64	0.34
A2B	0.50	0.41	0.48	0.39	0.57	0.52	0.44	1.00	0.39	0.43	0.39	0.61	0.62	0.38	0.08	0.89
A(-)	0.34	0.58	0.57	0.58	0.55	0.56	0.55	0.39	1.00	0.46	0.57	0.48	0.42	0.58	0.58	0.40
A(+)	0.33	0.88	0.81	0.52	0.40	0.45	0.48	0.43	0.46	1.00	0.52	0.43	0.50	0.49	0.55	0.43
AB(-)	0.42	0.48	0.46	0.63	0.42	0.66	0.68	0.39	0.57	0.52	1.00	0.50	0.36	0.64	0.64	0.34
AB(+)	0.40	0.42	0.46	0.41	0.54	0.84	0.80	0.61	0.48	0.43	0.50	1.00	0.58	0.42	0.46	0.52
A1/A	0.56	0.54	0.68	0.04	0.44	0.49	0.45	0.62	0.42	0.50	0.36	0.58	1.00	0.00	0.37	0.61
A2/A	0.44	0.46	0.32	0.96	0.56	0.51	0.55	0.38	0.58	0.49	0.64	0.42	0.00	1.00	0.61	0.37
A1B/AB	0.46	0.54	0.50	0.59	0.41	0.57	0.64	0.08	0.58	0.55	0.64	0.46	0.37	0.61	1.00	0.02
A2B/AB	0.52	0.44	0.48	0.39	0.58	0.42	0.34	0.89	0.40	0.43	0.34	0.52	0.61	0.37	0.02	1.00

Table 8. Results of ICA analysis on the transposed patients' dataset showing the links between the data sorted per blood indicator: ν values.

ν	0(-)	0(+)	A1(-)	A1(+)	A2(-)	A2(+)	B(-)	B(+)	A1B(-)	A1B(+)	A2B(-)	A2B(+)
0(-)	0.00	0.53	0.58	0.47	0.46	0.54	0.60	0.53	0.48	0.53	0.44	0.46
0(+)	0.53	0.00	0.64	0.60	0.48	0.65	0.53	0.64	0.58	0.65	0.43	0.52
A1(-)	0.58	0.64	0.00	0.49	0.61	0.45	0.48	0.46	0.48	0.47	0.35	0.50
A1(+)	0.47	0.60	0.49	0.00	0.46	0.57	0.58	0.63	0.56	0.52	0.27	0.54
A2(-)	0.46	0.48	0.61	0.46	0.00	0.47	0.41	0.61	0.37	0.48	0.27	0.66
A2(+)	0.54	0.65	0.45	0.57	0.47	0.00	0.48	0.42	0.42	0.53	0.25	0.57
B(-)	0.60	0.53	0.48	0.58	0.41	0.48	0.00	0.52	0.44	0.49	0.40	0.55
B(+)	0.53	0.64	0.46	0.63	0.61	0.42	0.52	0.00	0.59	0.47	0.39	0.38
A1B(-)	0.48	0.58	0.48	0.56	0.37	0.42	0.44	0.59	0.00	0.43	0.25	0.66
A1B(+)	0.53	0.65	0.47	0.52	0.48	0.53	0.49	0.47	0.43	0.00	0.43	0.45
A2B(-)	0.44	0.43	0.35	0.27	0.27	0.25	0.40	0.39	0.25	0.43	0.00	0.45
A2B(+)	0.46	0.52	0.50	0.54	0.66	0.57	0.55	0.38	0.66	0.45	0.45	0.00
0	0.42	0.11	0.66	0.60	0.46	0.66	0.55	0.66	0.57	0.61	0.47	0.48
A	0.53	0.66	0.38	0.17	0.47	0.45	0.56	0.58	0.53	0.53	0.20	0.62
A1	0.50	0.61	0.36	0.13	0.50	0.58	0.61	0.62	0.55	0.50	0.27	0.54
A2	0.52	0.58	0.51	0.61	0.36	0.10	0.44	0.46	0.37	0.54	0.26	0.63
B	0.54	0.64	0.45	0.66	0.57	0.41	0.41	0.11	0.57	0.48	0.40	0.41
AB	0.50	0.66	0.46	0.57	0.50	0.51	0.46	0.46	0.30	0.14	0.40	0.46
A1B	0.50	0.67	0.47	0.54	0.45	0.49	0.44	0.50	0.27	0.16	0.39	0.53
A2B	0.47	0.53	0.50	0.52	0.64	0.55	0.59	0.38	0.62	0.49	0.35	0.10
A(-)	0.59	0.62	0.16	0.54	0.45	0.42	0.38	0.49	0.40	0.49	0.36	0.60
A(+)	0.49	0.67	0.49	0.14	0.44	0.43	0.58	0.59	0.50	0.51	0.20	0.59
AB(-)	0.47	0.57	0.49	0.54	0.35	0.40	0.47	0.59	0.05	0.47	0.20	0.67
AB(+)	0.51	0.63	0.47	0.56	0.57	0.57	0.51	0.42	0.46	0.11	0.46	0.34
A1/A	0.49	0.43	0.46	0.35	0.64	0.87	0.58	0.52	0.63	0.47	0.45	0.36
A2/A	0.51	0.57	0.53	0.64	0.34	0.13	0.42	0.48	0.37	0.53	0.29	0.63
A1B/AB	0.52	0.51	0.48	0.46	0.30	0.44	0.39	0.61	0.32	0.42	0.37	0.84
A2B/AB	0.46	0.48	0.50	0.52	0.66	0.54	0.59	0.37	0.66	0.56	0.36	0.14

ν	0.00	A	A1	A2	B	AB	A1B	A2B	A(-)	A(+)	AB(-)	AB(+)	A1/A	A2/A	A1B/AB	A2B/AB
0(-)	0.42	0.53	0.50	0.52	0.54	0.50	0.50	0.47	0.59	0.49	0.47	0.51	0.49	0.51	0.52	0.46
0(+)	0.11	0.66	0.61	0.58	0.64	0.66	0.67	0.53	0.62	0.67	0.57	0.63	0.43	0.57	0.51	0.48
A1(-)	0.66	0.38	0.36	0.51	0.45	0.46	0.47	0.50	0.16	0.49	0.49	0.47	0.46	0.53	0.48	0.50
A1(+)	0.60	0.17	0.13	0.61	0.66	0.57	0.54	0.52	0.54	0.14	0.54	0.56	0.35	0.64	0.46	0.52
A2(-)	0.46	0.47	0.50	0.36	0.57	0.50	0.45	0.64	0.45	0.44	0.35	0.57	0.64	0.34	0.30	0.66
A2(+)	0.66	0.45	0.58	0.10	0.41	0.51	0.49	0.55	0.42	0.43	0.40	0.57	0.87	0.13	0.44	0.54
B(-)	0.55	0.56	0.61	0.44	0.41	0.46	0.44	0.59	0.38	0.58	0.47	0.51	0.58	0.42	0.39	0.59
B(+)	0.66	0.58	0.62	0.46	0.11	0.46	0.50	0.38	0.49	0.59	0.59	0.42	0.52	0.48	0.61	0.37
A1B(-)	0.57	0.53	0.55	0.37	0.57	0.30	0.27	0.62	0.40	0.50	0.05	0.46	0.63	0.37	0.32	0.66
A1B(+)	0.61	0.53	0.50	0.54	0.48	0.14	0.16	0.49	0.49	0.51	0.47	0.11	0.47	0.53	0.42	0.56
A2B(-)	0.47	0.20	0.27	0.26	0.40	0.40	0.39	0.35	0.36	0.20	0.20	0.46	0.45	0.29	0.37	0.36
A2B(+)	0.48	0.62	0.54	0.63	0.41	0.46	0.53	0.10	0.60	0.59	0.67	0.34	0.36	0.63	0.84	0.14
0	0.00	0.67	0.63	0.58	0.65	0.64	0.64	0.50	0.65	0.67	0.57	0.60	0.44	0.56	0.52	0.46
A	0.67	0.00	0.13	0.50	0.60	0.58	0.55	0.59	0.41	0.12	0.51	0.58	0.46	0.54	0.44	0.54
A1	0.63	0.13	0.00	0.64	0.65	0.56	0.54	0.52	0.42	0.18	0.53	0.54	0.32	0.68	0.48	0.50
A2	0.58	0.50	0.64	0.00	0.43	0.50	0.46	0.60	0.42	0.47	0.36	0.59	0.96	0.04	0.39	0.59
B	0.65	0.60	0.65	0.43	0.00	0.46	0.50	0.42	0.45	0.60	0.57	0.46	0.56	0.44	0.58	0.41
AB	0.64	0.58	0.56	0.50	0.46	0.00	0.07	0.48	0.44	0.55	0.34	0.16	0.51	0.49	0.42	0.57
A1B	0.64	0.55	0.54	0.46	0.50	0.07	0.00	0.55	0.44	0.51	0.32	0.20	0.55	0.45	0.34	0.64
A2B	0.50	0.59	0.52	0.60	0.42	0.48	0.55	0.00	0.60	0.57	0.60	0.39	0.38	0.62	0.89	0.08
A(-)	0.65	0.41	0.42	0.42	0.45	0.44	0.44	0.60	0.00	0.53	0.42	0.52	0.58	0.42	0.40	0.58
A(+)	0.67	0.12	0.18	0.47	0.60	0.55	0.51	0.57	0.53	0.00	0.47	0.56	0.49	0.50	0.43	0.55
AB(-)	0.57	0.51	0.53	0.36	0.57	0.34	0.32	0.60	0.42	0.47	0.00	0.50	0.64	0.36	0.34	0.64
AB(+)	0.60	0.58	0.54	0.59	0.46	0.16	0.20	0.39	0.52	0.56	0.50	0.00	0.42	0.58	0.52	0.46
A1/A	0.44	0.46	0.32	0.96	0.56	0.51	0.55	0.38	0.58	0.49	0.64	0.42	0.00	1.00	0.61	0.37
A2/A	0.56	0.54	0.68	0.04	0.44	0.49	0.45	0.62	0.42	0.50	0.36	0.58	1.00	0.00	0.37	0.61
A1B/AB	0.52	0.44	0.48	0.39	0.58	0.42	0.34	0.89	0.40	0.43	0.34	0.52	0.61	0.37	0.00	0.98
A2B/AB	0.46	0.54	0.50	0.59	0.41	0.57	0.64	0.08	0.58	0.55	0.64	0.46	0.37	0.61	0.98	0.00

AB and AB(+). Again, a more detailed a meaningful analysis would be possible when ICA is performed on a dataset where all the districts are more equally represented, thus producing reliable links between the pairs of blood indicators. We can though visualize the results on the IFS interpretation triangle, and Tables 5 and 6 will be collectively exhibited on Fig. 2, and Tables 7 and 8–on Fig. 3.

From Fig. 2, we specifically notice that all the ICA results plotted as IF points are condensed in a rather small zone, locked between the set's *Interior* $\langle \mu_{\min}, \nu_{\max} \rangle = \langle 0.84,\ 0.12 \rangle$ and the set's *Closure* $\langle \mu_{\min}, \nu_{\max} \rangle = \langle 1,\ 0 \rangle$. Depending on the user selected strategy for traversing the set [11] and on the user selected scale for estimation of the degrees of consonance or dissonance between criteria [7], the results can be interpreted not on the [0,1]-scale but scaled down to the interval defined by the *Interior* and the *Closure* operators, ranked and traversed in a set of alternative procedures, detailed in [11].

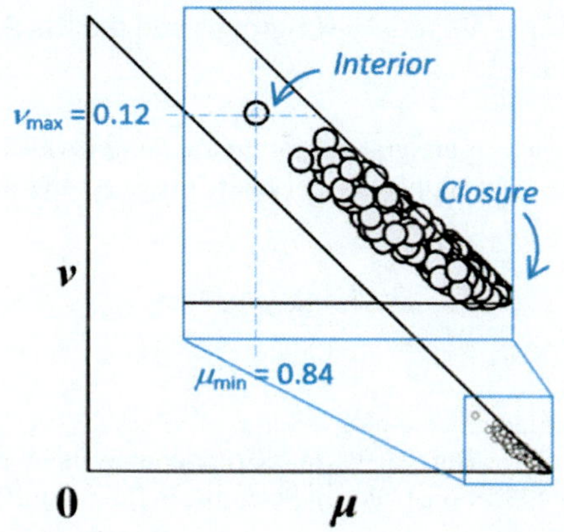

Fig. 2. The ICA results from Tables 5 and 6 visualized on the IF interpretation triangle

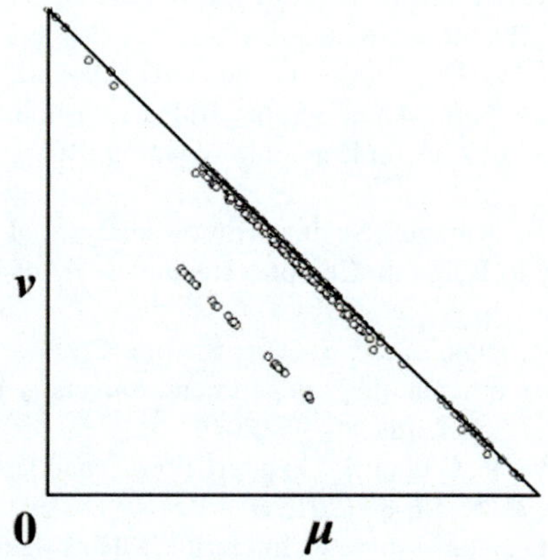

Fig. 3. The ICA results from Tables 7 and 8 visualized on the IF interpretation triangle

5 Discussions and Conclusions

While the results of the ICA analysis reported in Sect. 4, were not fully conclusive, they are promising on a conceptual level and the method will prove fully applicable on a larger dataset featuring better representability of patients. The crucial element of the herewith presented research is the adoption as early as the research's design phase of that atomic approach to data, which has allowed then the detailed breakdown of data simultaneously in blood group/subgroup, rhesus factor, sex, year of birth and (for the first time) region of birth. The presented data collection and data extraction methodology however can be adopted in the hospital units in other regions in the country in order to assemble a wider, more representative picture of the problem considered. Notwithstanding, this atomic approach to data has allowed us for the first time, in contrast to the available literature, to group and slice the data about the blood profile of Bulgarian population, and formulate in the next sections results and conclusions regarding the geographic distribution of the different ABO blood groups, A_1 and A_2 subgroups and the Rh(D) antigen in the frames on national and regional level.

Acknowledgements. The authors are grateful for the support provided under Grant No. KP-06-N-22/1 "Theoretical research and applications of InterCriteria Analysis" of the National Science Fund of Bulgaria.

References

1. Anastasov, A., et al.: Immunohematology. Sofia, p. 55 (1977). (in Bulgarian)
2. Andreev, N., Atanassov, K., Bureva, V.: InterCriteria analysis on data for blood collection. Annual of the Informatics Section, Union of Scientists in Bulgaria, 10, pp. 30–53 (2019–2020). (in Bulgarian)
3. Andreev, N., Atanassova, V.: InterCriteria analysis of the blood group distribution of patients of Saint Anna hospital in 2015–2019. In: Atanassov, K.T., et al. (eds.) Advances and New Developments in Fuzzy Logic and Technology (IWIFSGN 2019). AISC, vol. 1308, pp. 158–165. Springer, Cham (2021). https://doi.org/10.1007/978-3-030-77716-6_14
4. Andreev, N., Atanassova, V.: Distribution of the ABO system blood groups in the patients of the university hospital "Saint Anna", Sofia, Bulgaria in the period 2015–2020. In: 1st Scientific Conference on Transfusion Hematology, Varna, Bulgaria, 24–26 September 2021 (2021). (in Bulgarian)
5. Andreev, N., Sotirova, E., Ribagin, S.: Intercriteria analysis of data from the centers for transfusion haematology in Bulgaria. Comptes Rendus de l'Academie Bulgare des Science **72**(7), 982–990 (2018)
6. Andreev, N., Vassilev, P., Ribagin, S., Sotirov, S.: InterCriteria analysis of data for blood collection in the transfusion hematology department, university hospital "St. Anna", Sofia. Notes Intuitionistic Fuzzy Sets **25**(2), 88–95 (2019)
7. Atanassov, K., Atanassova, V., Gluhchev, G.: InterCriteria analysis: ideas and problems. Notes Intuitionistic Fuzzy Sets **21**(1), 81–88 (2015)
8. Atanassov, K., Mavrov, D., Atanassova, V.: Intercriteria decision making: a new approach for multicriteria decision making, based on index matrices and intuitionistic fuzzy sets. Issues Intuitionistic Fuzzy Sets Generalized Nets **11**, 1–8 (2014)
9. Atanassova, V.: Interpretation in the intuitionistic fuzzy triangle of the results, obtained by the InterCriteria analysis. In: Proceedings of 16th World Congress of IFSA, 9th Conference of EUSFLAT, June 2015, Gijon, Spain, pp. 1369–1374 (2015)
10. Atanassova, V., Andreev, N., Dimitriev, A.: ABO system blood groups distribution in Bulgaria, based on a dataset of the patients of the university hospital "Saint Anna", Sofia, Bulgaria, from 2015 to 2021. In: Sotirov, S., Pencheva, T., Kacprzyk, J., Atanassov, K., Sotirova, E., Ribagin, S. (eds.) Recent Contributions to Bioinformatics and Biomedical Sciences and Engineering, pp. xx–yy. Springer, Cham (2023)
11. Atanassova, V., Vardeva, I., Sotirova, E., Doukovska, L.: Traversing and ranking of elements of an intuitionistic fuzzy set in the intuitionistic fuzzy interpretation triangle. In: Atanassov, K.T., et al. (eds.) Novel Developments in Uncertainty Representation and Processing. AISC, vol. 401, pp 161–174. Springer, Cham (2016). https://doi.org/10.1007/978-3-319-26211-6_14
12. BloodBook.com: Racial & ethnic distribution of ABO blood types (2013). http://www.bloodbook.com/world-abo.html. Accessed 14 Oct 2022
13. Chorbadzhiyski, G.: Information, check and generator of unique citizen numbers (EGN). Source code, version 1.50 (30 Sep 2006) (2006). https://georgi.unixsol.org/programs/egn.php/view/

14. Chorukova, E., Marinov, P., Umlenski, I.: Survey on theory and applications of InterCriteria analysis approach. In: Atanassov, K.T. (ed.) Research in Computer Science in the Bulgarian Academy of Sciences. SCI, vol. 934, pp. 453–469. Springer, Cham (2021). https://doi.org/10.1007/978-3-030-72284-5_20
15. Dean, L.: Blood groups and red cell antigens. National Center for Biotechnology Information (US), Bethesda (MD), pp. 31–33 (2005). https://www.ncbi.nlm.nih.gov/books/n/rbcantigen/pdf/. Accessed 14 Oct 2022
16. Dobreva, A., Doychinova, N., Vasilev, N. (eds.): Transfusion Hematology, p. 68, 114–115. State Publishing House "Medicina I Fizkultura", Sofia (1988). (in Bulgarian)
17. Doukovska, L., Atanassova, V., Sotirova, E., Vardeva, I., Radeva, I.: Defining consonance thresholds in InterCriteria analysis: an overview. In: Hadjiski, M., Atanassov, K.T. (eds.) Intuitionistic Fuzziness and Other Intelligent Theories and Their Applications. SCI, vol. 757, pp. 161–179. Springer, Cham (2019). https://doi.org/10.1007/978-3-319-78931-6_11
18. Kohler, I., Kaltchev, J., Dimova, M.: Integrated information system for demographic statistics 'ESGRAON-TDS' in Bulgaria. Demogr. Res. **6**, 325–354 (2002)
19. Popov, R., Petrov, N., Vaseva, V.: Distribution of blood groups of the ABO system in the Military Medical Academy immunohematological diagnostics. Bulg. Med. J. **VI**(2), 45–48 (2012). (in Bulgarian)
20. RhesusNegative.net: Blood type frequencies by country including the Rh factor (2019). http://www.rhesusnegative.net/themission/bloodtypefrequencies/. Accessed 14 Oct 2022
21. Salduz, Z., et al.: ABO and Rh blood group distribution in Istanbul province (Turkey). Istanb. Med. J. **16**, 98–100 (2015)
22. Wikipedia Contributors: Blood type distribution by country. In: Wikipedia, The Free Encyclopedia (2017). https://en.wikipedia.org/w/index.php?title=Blood_type_distribution_by_country&oldid=786847072

Comparison of Docking Scoring Functions by InterCriteria Analysis on a Set of Protein Targets Related to Alzheimer and Parkinson Diseases

Petko Alov(✉), Ilza Pajeva, Ivanka Tsakovska, and Tania Pencheva(✉)

Department of QSAR and Molecular Modelling, Institute of Biophysics and Biomedical Engineering, Bulgarian Academy of Sciences, Acad. G. Bonchev Str., Bl.105, 1113 Sofia, Bulgaria

petko.alov@biophys.bas.bg, {pajeva,itsakovska, tania.pencheva}@biomed.bas.bg

Abstract. Nowadays, the pharmaceutical industry extensively uses *in silico* drug design methods to deal with the enormously wide chemical space of druggable compounds and to reduce the R&D expenses. Having a significant number of tools for structure-based *in silico* screening, the aim of this study was to assess the putative consonance between the docking scores obtained by different scoring functions, which could provide a basis for reduction of the computational cost and/or for optimal choice of scoring functions.

The proteins used in the study were acetylcholine esterase (AChE), histone deacetylase 2 (HDAC2), and monoamine oxidase B (MAO-B), enzymes related to the treatment of symptoms of Alzheimer and Parkinson diseases and potential subjects for inclusion in "single drug – multiple targets" research. The 11085 small molecules (ligands) used in the study were selected from a database of more than 600000 commercially available drug-like compounds which docking scores obtained by a rigid docking were better than the re-docking scores of the co-crystalized reference ligands in the selected enzymes. To assess the differences in performance of the scoring functions implemented in different molecular docking software packages, docking of these ligands in the target proteins was performed using several widely used molecular modelling platforms: (i) rigid- and flexible-protein docking in MOE (v. 2019.01, https://www.chemcomp.com); (ii) rigid-protein docking in FlexX (v. 4.3) and ligand optimization and rescoring in HYDEscorer (v. 1.0), respectively (www.biosolveit.de); and (iii) rigid-protein docking in AutoDock Vina (http://vina.scripps.edu). Besides docking scores, the time for docking was also recorded and the computational costs were calculated for each of the studied docking protocol/scoring function pairs.

The binding energies estimated by the selected scoring functions were subjected to intercriteria analysis (ICrA). The ICrA approach relies on the formalisms of the intuitionistic fuzzy sets and index matrices, and attempts to uncover similarities in the behavior of criteria applied for evaluation of multiple objects. ICrA was employed as a potential tool to support the selection of an appropriate scoring function for ranking of more than 11 000 ligands identified to interact with all three proteins simultaneously. Further, an analysis of the intersections of the top

S. Sotirov et al. (Eds.): BioInfoMed 2022, LNNS 658, pp. 98–110, 2023.
https://doi.org/10.1007/978-3-031-31069-0_11

1000 ligands for each target ranked by different scoring functions, was performed. ICrA analysis revealed consonances between FlexX and AutoDock Vina scoring functions in all studied proteins, and additionally between MOE flexible-protein docking and AutoDock Vina in AChE. Analysis of intersection results was to a great extent in line with the intercriteria relations.

The results indicate that a precise selection of scoring functions and docking protocols, confirmed by the available knowledge of the studied objects, is needed. This analysis suggests also the possibility for optimization of *in silico* screening campaigns by avoidance of computationally expensive docking protocols that are highly consonant with the less expensive ones.

Keywords: Molecular docking · Scoring functions · Intercriteria analysis · Computational costs

1 Introduction

Computer-aided drug design (CADD, or *in silico* drug design) is a collection of computational methods used to optimize time-consuming and costly processes of development of new drugs. The most commonly used among the structure-based *in silico* methods is molecular docking. A computational simulation of placement of small molecules (ligands) in different orientations and conformations within the active site of a biomacromolecule (usually a protein) is performed, determining the optimal binding mode through the calculation of the ligand-receptor binding energy. This simulation follows the principle of the complementarity in the protein-ligand interaction and results in the estimation of free energy of their binding using a scoring function [1]. The selection of the best poses (the pose refers to both ligand conformation and orientation) to be subjected to further investigations is usually based on the calculated docking scores.

Recently, the main paradigm in pharmaceutical chemistry was "single drug – single target", which required multidrug therapy for patients with multi-symptom diseases and increases the risk of undesirable side and toxic effects. Nowadays, there is a paradigm shift towards "single drug – multiple targets", which could both increase the therapeutic efficiency and decrease the risk of side effects in these patients [2, 3]. An appropriate case for such studies are the neurodegenerative disorders (e.g. Alzheimer and Parkinson diseases) due to the fact that current therapies aim at several different brain enzyme systems [4, 5]. For example, a drug acting simultaneously on acetylcholine esterase (AChE), histone deacetylase 2 (HDAC2), and monoamine oxidase B (MAO-B), could produce a combined additive or synergistic effect. Molecular docking could be successfully used in studies searching for ligands able to interact specifically with multiple protein targets. Having this in mind, we selected AChE, HDAC2, and MAO-B as protein targets in our study.

There is a variety of docking algorithms developed, which performance is strongly dependent on the scoring functions and a lot of comparative studies on scoring functions have been reported [6–10]. However, it is not a trivial task to select a suitable scoring function to simulate the interactions of a particular set of ligands with a particular receptor(s). There is also some comparative studies of the docking software packages/platforms available either with commercial license or with free access [11–13]. However, the question which molecular docking software and protocol to select is still open. Intending to assist in this selection, the potential of InterCriteria analysis (ICrA) [14] to assess the performance of various scoring functions on a set of 11085 compounds selected in previous experiment [15, 16] as capable to interact with AChE, HDAC2, and MAO-B, is here explored.

ICrA is developed to discern possible relations in the behavior of criteria based on multiple objects observations [14]. ICrA brings together two mathematical formalisms – of the index matrices (IM) [17, 18] and the intuitionistic fuzzy sets (IFS) [19], which are an extension of Zadeh's fuzzy sets [20] Thus, ICrA pursues to obtain new information based on the criteria involved in the evaluation processes. The eventual revealing of any existing relations between the criteria themselves may support the decision-making process significantly. So far, ICrA has been applied in a variety of research topics aiming to compare predefined criteria and the objects estimated by them or to discover some dependencies between the criteria themselves. The ICrA successful applications, e.g. in medicine [21], ecology [22], artificial intelligence and metaheuristics algorithms performance [23], and e-learning [24], illustrate the approach's ability to reveal new relations and dependencies between considered criteria, thus allowing a new view on the analyzed data. ICrA has been successfully tested in the field of *in silico* drug design and computational toxicology [25] as well.

In this investigation ICrA is applied to assess the performance of a selection of scoring functions implemented in molecular docking software packages with commercial license or with free access. Five scoring functions from three software packages are used in the molecular docking on a compiled dataset of commercially-available substances in the active sites of AChE, HDAC2, and MAO-B.

2 Materials and Methods

2.1 InterCriteria Analysis Approach

The ICrA approach proposed by Atanassov et al. in 2014 [14] combines the fundamental mathematical concepts of index matrices (IM) [17, 18] and intuitionistic fuzzy sets (IFS) [19]. A brief overview of the ICrA's theoretical background is outlined below.

IFS are introduced by Atanassov [19] as an extension of Zadeh's fuzzy sets (FS) [20]. FS work with a membership function, taking values in the interval [0, 1] for each element in a set with boundary conditions 0 when an element does not belong to the set and 1 when it does. The IFS extend this idea further by introducing a degree of non-membership, taking values also in the closed unit interval, adding the requirement the sum of the two degrees to fall in the closed unit interval as well. The IFS A^* has the following formal definition:

$$A^* = \{\langle x, \mu_A(x), \nu_A(x)\rangle | x \in E\},$$

where E is a universe set, $A \subseteq E$; $\mu_A(x), \nu_A(x): E \to [0, 1]$ are, respectively, the degree of membership and the degree of non-membership for each element $x \in E$ to the fixed subset A of E, satisfying $0 \leq \mu_A(x) + \nu_A(x) \leq 1$. An ordered pair $\langle \mu, \nu \rangle$, such that

$\mu, \nu \in [0, 1]$ and $\mu + \nu \leq 1$ [26], is called an intuitionistic fuzzy pair (IFP). The IFPs are among the most important elements of ICrA, as an evaluation of the similarity of behavior of two criteria over a set of objects is obtained as IFP.

ICrA uses as input a two-dimensional IM, represented by sets of row and column indexes and elements corresponding to the combination of the aforementioned indexes. Thus, each 2D IM may be denoted as $[K, L, a_{k,l}]$, where K is the set of row indexes (labels), L – the set of column indexes (labels), and $a_{k,l}$ ($k \in K, l \in L$) – the element corresponding to row index k and column index l. In terms of ICrA, the input IM represents the objects evaluations by different criteria and, thus, may be denoted as $[O, C, e_{O,C}]$, where O is the index set of rows and corresponds to the distinct evaluated objects, C – the index set of columns and corresponds to different evaluating criteria, while $e_{O,C}$ – the evaluation assigned to the object "O" towards the criterion "C". In case of n criteria and m objects, the input 2D IM for ICrA is represented as follows:

	C_1	...	C_k	...	C_n
O_1	e_{O_1,C_1}	...	e_{O_1,C_k}	...	e_{O_1,C_n}
...	...	...	...	...	...
O_i	e_{O_i,C_1}	...	e_{O_i,C_k}	...	e_{O_i,C_n}
...	...	...	...	...	...
O_m	e_{O_m,C_1}	...	e_{O_m,C_k}	...	e_{O_m,C_n}

In such a manner, ICrA starts with an $m \times n$ table and performs a comparison between every two criteria over multiple objects' evaluations considering the respective elements relations, as shown in Fig. 1. After processing, ICrA results in $n \times n$ table of IFPs corresponding to InterCriteria relationships.

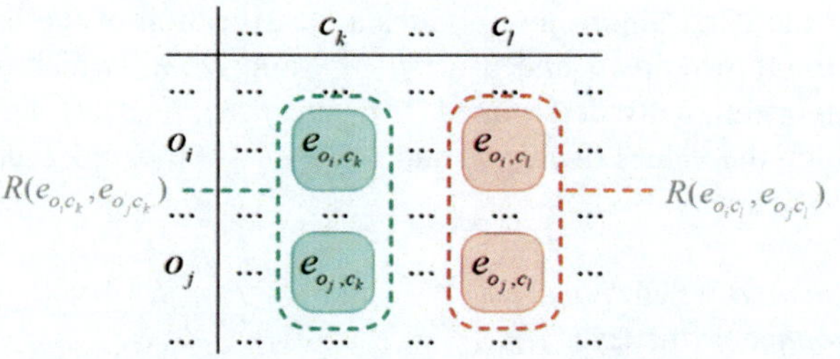

Fig. 1. Pairwise comparisons of the objects' evaluations against the criteria

There are several possible cases depending on the input elements' nature. In the case of coincidence of two relations $R(e_{O_i,C_k}, e_{O_j,C_k})$, $R(e_{O_i,C_l}, e_{O_j,C_l})$, where $R \in \{<, =, >\}$, the counter $S_{k,l}^{\mu}$ for agreement (or similarity) is increased. If they differ (one of the relations is the dual of the other, e.g., $R(e_{O_i,C_k}, e_{O_j,C_k})$ is the relation "<" and $R(e_{O_i,C_l}, e_{O_j,C_l})$ is the relation ">" or vice versa), the counter $S_{k,l}^{\nu}$ for disagreement (dissimilarity) is increased. Otherwise, the contribution is assigned to the score of uncertainty (indeterminacy).

The pairwise comparisons between the objects total to $m(m-1)/2$, therefore $0 \leq S^{\mu}_{k,l} + S^{\nu}_{k,l} \leq \frac{m(m-1)}{2}$ holds true. The following normalized values are obtained from counters $S^{\mu}_{k,l}$ and $S^{\nu}_{k,l}$ (for every k, l, such as $1 \leq k \leq l \leq m$ and $m \geq 2$):

- $\mu_{C_k,C_l} = 2\frac{S^{\mu}_{k,l}}{m(m-1)}$, called *degree of agreement* in terms of ICrA;
- $\nu_{C_k,C_l} = 2\frac{S^{\nu}_{k,l}}{m(m-1)}$, called *degree of disagreement* in terms of ICrA.

Due to the fact that $\mu_{C_k,C_l} = \mu_{C_l,C_k}$ and $\nu_{C_k,C_l} = \nu_{C_l,C_k}$, the degrees of *agreement/disagreement* between two criteria need to be calculated only once.

Obviously, $\langle \mu_{C_k,C_l}, \nu_{C_k,C_l} \rangle$ is an IFP. The value $\pi_{C_k,C_l} = 1 - \mu_{C_k,C_l} - \nu_{C_k,C_l}$ corresponds to the *degree of uncertainty.*

The resulting after ICrA application IM has all the collected IFPs $\langle \mu_{C_k,C_l}, \nu_{C_k,C_l} \rangle$, which may be viewed as an intuitionistic fuzzy evaluation of the relations between any two criteria C_k and C_l:

	C_1	...	C_k	...	C_n
C_1	$\langle 1, 0 \rangle$	...	$\langle \mu_{C_1,C_k}, \nu_{C_1,C_k} \rangle$	...	$\langle \mu_{C_1,C_n}, \nu_{C_1,C_n} \rangle$
...	...	...	...	...	...
C_k	$\langle \mu_{C_k,C_l}, \nu_{C_k,C_l} \rangle$	...	$\langle 1, 0 \rangle$	...	$\langle \mu_{C_k,C_n}, \nu_{C_k,C_n} \rangle$
...	...	...	...	...	...
C_n	$\langle \mu_{C_n,C_1}, \nu_{C_n,C_1} \rangle$	...	$\langle \mu_{C_n,C_k}, \nu_{C_n,C_k} \rangle$	...	$\langle 1, 0 \rangle$

In practice, it is more convenient to consider two distinct index matrices IM_{μ} and IM_{ν}, rather than the IM of IFPs as presented above. The detailed mathematical considerations may be found in [14].

The result of the ICrA algorithm provides a classification of the InterCriteria relations, based on the thresholds α and β for μ_{C_k,C_l} and ν_{C_k,C_l}, which might be chosen by the user or algorithmically determined. If α and β ($\alpha, \beta \in [0, 1], \alpha > \beta$) are the thresholds to which the values of μ_{C_k,C_l} and ν_{C_k,C_l} are compared, then the criteria C_k and C_l are said to be in:

- *positive consonance*, whenever $\mu_{C_k,C_l} > \alpha$ and $\nu_{C_k,C_l} < \beta$;
- *negative consonance*, whenever $\mu_{C_k,C_l} < \beta$ and $\nu_{C_k,C_l} > \alpha$;
- *dissonance*, otherwise.

Figure 2 presents the intuitionistic fuzzy triangle with the zones of *positive consonance* and *negative consonance* in the case of $\alpha = 0.75$ and $\beta = 0.25$, which are the thresholds default values.

The authors in [27] present a finer scale for types of *consonance* or *dissonance* between criteria pairs, as well as the results' interpretation with respect to the *degrees of agreement*, *disagreement*, and *uncertainty*. Later on, the authors in [28] have formulated and presented the rules from [27] as different algorithms for the calculation of ICrA

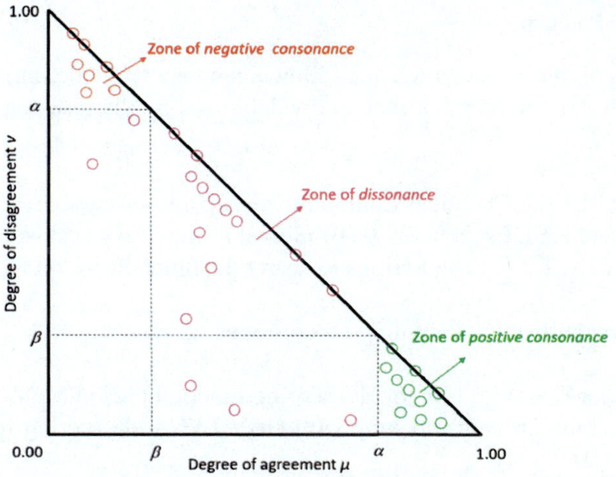

Fig. 2. Intuitionistic fuzzy triangle with zones of *positive consonance*, *dissonance*, and *negative consonance* for $\alpha = 0.75$ and $\beta = 0.25$

relations (with their pseudo-codes), called *μ-biased*, *balanced*, *ν-biased*, *unbiased*, and *weighted,* respectively.

ICrA was performed using the ICrAData v.2.5 software, which is freely available at http://intercriteria.net/software/.

2.2 Protein Targets and Small Ligand Dataset

For the purposes of this study the all available in Protein Data Bank (PDB, https://www.rcsb.org, [29]) crystal structures of the investigated here AChE, HDAC2 and MAO-B have been analyzed. The following protein-ligand complexes have been selected as reference structures for the purposes of molecular docking studies: AChE with donepezil (PDB ID 4EY7), HDAC2 with a benzamide derivative (PDB ID 4LY1), and MAO-B with safinamide (PDB ID 2V5Z). For the purposes of the subsequent molecular docking the B chains have been selected from 4EY7 for AChE and from 2V5Z for MAO-B, which are homodimers, and the C chain from 4LY1 for HDAC2, which is homotrimer. Flavin, in case of 2V5Z, and zinc, in case of 4LY1, have been kept during the molecular docking studies [30, 31].

The dataset consists of 11085 compounds, a subset of more than 600 000 commercially-available substances, compiled by Chemical Computing Group ULC (CCG) and OpenEye Scientific, and provided with CCG's Molecular Operating Environment (MOE) subscription (https://www.chemcomp.com/). The selection criterion was the compounds to possess rigid-protein docking scores better than the re-docking scores of the co-crystalized reference ligands in the studied enzymes.

2.3 Molecular Docking

For the purposes of this investigation, the following molecular modelling platforms have been selected, due to the fact that they are widely used by the research community in the field of *in silico* drug design:

- MOE v. 2019.01 (The Chemical Computing Group, http://www.chemcomp.com);
- FlexX v. 4.3 and HYDEscorer v. 1.0 (BioSolveIT, https://www.biosolveit.de);
- AutoDock Vina v. 1.2.0. (The Scripps Research Institute, https://vina.scripps.edu).

The following molecular docking protocols have been used:

- rigid-protein docking with London dG scoring function [32] in MOE, v. 2019.01;
- induced fit (flexible-protein) docking with GBVI/WSA dG scoring function [33] in MOE, v. 2019.01;
- rigid-protein docking in FlexX v. 4.3 with Böhm's scoring function [34, 35];
- ligand optimization and rescoring with HYDEscorer v. 1.0 and HYDE scoring function [36, 37];
- rigid-protein docking with AutoDock Vina and AutoDock scoring function [38, 39].

Water molecules have been removed from the binding sites of the studied proteins. The default placement algorithm in each of the used molecular docking packages has been applied. Up to 30 poses have been kept after docking. For each of the target proteins, the compounds not docked successfully in any of the docking tools have not been subjected to further analysis. As such, the compound libraries have been reduced to 10622 compounds in case of AChE, 8413 in case of HDAC2, and 8376 in case of MAO-B.

The results from the performed molecular docking studies have been processed and prepared as spreadsheets, in a format suitable for the subsequent ICrA analysis. All spreadsheets have been constructed in an identical way: the compounds (drug-like molecules) are considered as objects in terms of ICrA, while the docking scores calculated by the scoring functions, are considered as criteria.

3 Results and Discussion

Molecular docking studies have been performed in three protein targets, respectively AChE, HDAC2 and MAO-B. As an indicator of protein-ligand binding affinity, binding energies calculated by different scoring functions (for all scoring functions, lower scores indicate more favorable poses; the unit is kcal/mol) have been collected.

For the purposes of the current investigation, the values of the binding energy for the best out of 30 saved docking poses, for each of the tested compounds in each of the studied proteins, have been subjected to ICrA. The ICrA results did not outline any significant relations between the binding energies calculated by different scoring functions for none of the studied protein targets. Figure 3 presents the results obtained by ICrA application in case of docking in MAO-B. As seen in the figure, the user interface of ICrAData consists of left panel for the input data, the central panel for the result of the ICrA implementation in the form of a table with colored values, and the right panel with the graphical visualization of the results in the intuitionistic fuzzy interpretation triangle with points (circles) colored correspondingly to the results shown in the central panel. For a better visualization, cell colors have been introduced in accordance with the rules below and the user-defined thresholds α and β:

- the results are colored in green in the case of *positive consonance*;
- the results are colored in red in the case of *negative consonance*;
- otherwise, the results are colored in magenta when there is *dissonance*.

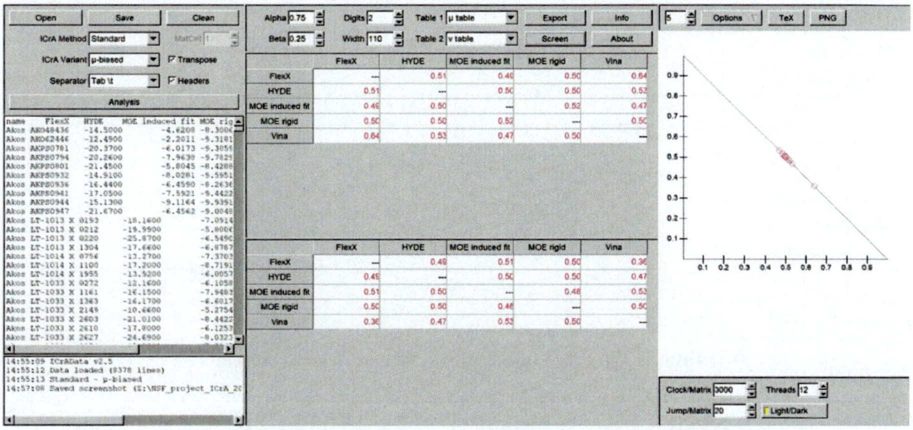

Fig. 3. ICrAData view on MAO-B docking results

The results presented in Fig. 3 show that each scoring function brings unique and inequivalent information in terms of ICrA.

Table 1 presents μ-tables from ICrA implementation in case of the three protein targets. As mentioned above, the thresholds α and β are set according to the user's choice. If one considers a lower value of the threshold α (0.7 or 0.6), a *positive consonance* will be identified between the following pairs of scoring functions:

- FlexX and AutoDock Vina in case of docking in MAO-B (Table 1a);
- Again between FlexX and AutoDock Vina, with the highest *degree of agreement* observed ($\mu = 0.68$) in case of docking in HDAC2 (Table 1b);
- Again between FlexX and AutoDock Vina, but also between MOE induced fit and AutoDock Vina in case of docking in AChE (Table 1c).

ICrA analysis revealed marginal consonances between FlexX and AutoDock Vina scoring functions in all studied proteins, and additionally between MOE flexible-protein docking and AutoDock Vina in AChE (Table 1c). The latter observation might be considered valuable, since the simulations employing GBVI/WSA dG scoring are significantly

Table 1. μ-tables from ICrA implementation for: (a) MAO-B; (b) HDAC2; and (c) AChE.

Docking tool / variant	FlexX	HYDEscorer	MOE induced fit	MOE rigid	AutoDock Vina
			(a)		
FlexX	–	0.51	0.49	0.50	0.64
HYDEscorer		–	0.50	0.50	0.53
MOE induced fit			–	0.52	0.47
MOE rigid				–	0.50
AutoDock Vina					–
			(b)		
FlexX	–	0.59	0.53	0.48	0.68
HYDEscorer		–	0.52	0.49	0.57
MOE induced fit			–	0.46	0.55
MOE rigid				–	0.47
AutoDock Vina					–
			(c)		
FlexX	–	0.46	0.54	0.49	0.65
HYDEscorer		–	0.52	0.51	0.46
MOE induced fit			–	0.50	0.61
MOE rigid				–	0.50
AutoDock Vina					–

Table 2. Computational costs of the docking studies.

Docking tool/variant	Single-CPU performance (s/molecule)	License or hardware CPU limit	Multi-CPU performance (s/molecule)	Total time (h)
FlexX	44	8 (license)	6	55
HYDEscorer*	23	6 (license)	4	37
MOE induced fit	1118	24 (license)	47	434
MOE rigid	17	6 (license)	3	28
AutoDock Vina	389	72 (hardware)	5	46

The total times shown are obtained on 33 255 instances (3 proteins, 11 085 compound/each).
*HYDEscorer performs a single pose optimization and rescoring.

more time-consuming than those with AutoDock Vina or any other scoring function (Table 2).

Further on, an analysis of the intersections between the top 1000 ligands for each of the three targets investigated, ranked by each of the five scoring functions, has been performed. Analysis of intersection results comes to a great extent in line with the intercriteria relations. Figure 4 presents the number of the common compounds in these 1000 hits for each of the scoring function pairs.

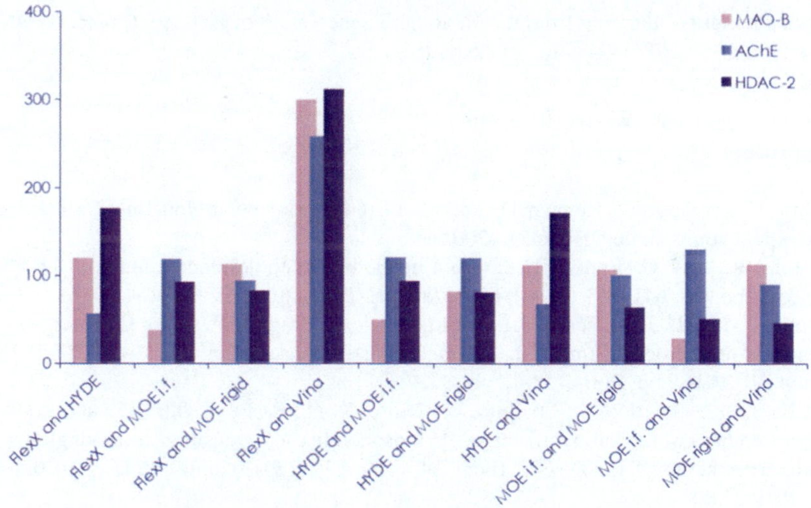

Fig. 4. Number of the common compounds in the 1000 top scored ligands for each of the scoring function pairs. *i.f. stands for induced fit.

The results about ICrA consonances (obtained for the full dataset) between FlexX and AutoDock Vina scoring for all protein targets and between MOE induced fit and AutoDock Vina scoring in case of AChE only, are in line with the observed broader intersections for the top scored 1000 compounds in these cases.

4 Conclusions

In this investigation, the capability of ICrA to assess the performance of the studied scoring functions available in MOE, FlexX, HYDEscorer, and AutoDock Vina was explored on the results obtained from docking simulations on a set of 11 085 drug-like compounds in the binding sites of the enzymes MAO-B, AChE and HDAC2, targeted to alleviate the symptoms of Alzheimer and Parkinson diseases. The performance of the scoring functions was evaluated based on the docking scores as approximations of the binding affinities. In general, ICrA reveals a consonance trend only between some pairs of the investigated scoring functions, namely FlexX/Böhm and AutoDock Vina, and GBVI/WSA dG and AutoDock Vina. These ICrA relations have been confirmed by the analysis of intersections between the compounds, top scored in different docking simulations.

The obtained results indicate that a precise selection of scoring functions and docking protocols is needed, supported by the available knowledge of the studied objects and/or a preliminary test on a small well-defined subset of compounds. The results suggest also the possibility for optimization of *in silico* screening campaigns by avoiding the computationally expensive docking protocols that are highly consonant with the less expensive ones.

Acknowledgements. Funding from the National Science Fund of Bulgaria (grants № DN 17/6 and № KP-06-OPR 03/8) is gratefully acknowledged.

References

1. Höltje, H.-D., Sippl, W., Rognan, D., Folkers, G. (eds.): Molecular Modeling: Basic Principles and Applications. Wiley, Hoboken (2008)
2. Morphy, R., Kay, C., Rankovic, Z.: From magic bullets to designed multiple ligands. Drug Discov. Today **9**, 641–651 (2004). https://doi.org/10.1016/S1359-6446(04)03163-0
3. Youdim, M.B.H., Buccafusco, J.J.: Multi-functional drugs for various CNS targets in the treatment of neurodegenerative disorders. Trends Pharmacol. Sci. **26**, 27–35 (2005). https://doi.org/10.1016/j.tips.2004.11.007
4. Cheong, S.L., Federico, S., Spalluto, G., Klotz, K.-N., Pastorin, G.: The current status of pharmacotherapy for the treatment of Parkinson's disease: transition from single-target to multitarget therapy. Drug Discov. Today **24**, 1769–1783 (2019). https://doi.org/10.1016/j.drudis.2019.05.003
5. Matos, M.J.: Multitarget therapeutic approaches for Alzheimer's and Parkinson's diseases: an opportunity or an illusion? Future Med. Chem. **13**, 1301–1309 (2021). https://doi.org/10.4155/fmc-2021-0119
6. Cheng, T., Li, X., Li, Y., Liu, Z., Wang, R.: Comparative assessment of scoring functions on a diverse test set. J. Chem. Inf. Model. **49**, 1079–1093 (2009). https://doi.org/10.1021/ci9000053
7. Li, Y., Han, L., Liu, Z., Wang, R.: Comparative assessment of scoring functions on an updated benchmark: 2. evaluation methods and general results. J. Chem. Inf. Model. **54**, 1717–1736 (2014). https://doi.org/10.1021/ci500081m
8. Su, M., et al.: Comparative assessment of scoring functions: the CASF-2016 update. J. Chem. Inf. Model. **59**, 895–913 (2019). https://doi.org/10.1021/acs.jcim.8b00545
9. Khamis, M.A., Gomaa, W.: Comparative assessment of machine-learning scoring functions on PDBbind 2013. Eng. Appl. Artif. Intell. **45**, 136–151 (2015). https://doi.org/10.1016/j.engappai.2015.06.021
10. Xu, W., Lucke, A.J., Fairlie, D.P.: Comparing sixteen scoring functions for predicting biological activities of ligands for protein targets. J. Mol. Graph. Model. **57**, 76–88 (2015). https://doi.org/10.1016/j.jmgm.2015.01.009
11. Wang, Z., et al.: Comprehensive evaluation of ten docking programs on a diverse set of protein–ligand complexes: the prediction accuracy of sampling power and scoring power. Phys. Chem. Chem. Phys. **18**, 12964–12975 (2016). https://doi.org/10.1039/C6CP01555G
12. Stanzione, F., Giangreco, I., Cole, J.C.: Chapter four - use of molecular docking computational tools in drug discovery. In: Witty, D.R., Cox, B. (eds.) Progress in Medicinal Chemistry, pp. 273–343. Elsevier (2021). https://doi.org/10.1016/bs.pmch.2021.01.004
13. Kalinowsky, L., Weber, J., Balasupramaniam, S., Baumann, K., Proschak, E.: A diverse benchmark based on 3D matched molecular pairs for validating scoring functions. ACS Omega **3**, 5704–5714 (2018). https://doi.org/10.1021/acsomega.7b01194
14. Atanassov, K., Mavrov, D., Atanassova, V.: A new approach for multicriteria decision making, based on index matrices and intuitionistic fuzzy sets. In: Issues in IFSs and GNs, pp. 1–8 (2014)
15. Alov, P., Tsakovska, I., Pencheva, T., Ruseva, M., Tzvetkov, N., Pajeva, I.: In search for multitarget leads for treating neurodegenerative disorders: Structure-based pharmacophore modelling and virtual screening. Presented at the Asian Federation for Medicinal Chemistry (AFMC) 12th International Symposium "New Avenues for Design and Development of Translational Medicine" (AIMECS 2019), Istanbul, Turky, 8 September 2019

16. Tzvetkov, N., Alov, P., Tsakovska, I., Pencheva, T., Pajeva, I.: From single to multi-target active ligands against neurodegenerative diseases: Multi-potency screening and virtual activity evaluation. Presented at the Asian Federation for Medicinal Chemistry (AFMC) 12th International Symposium "New Avenues for Design and Development of Translational Medicine" (AIMECS 2019), Istanbul, Turky, 8 September 2019
17. Atanassov, K.: Generalized index matrices. C. R. Acad. Bulg. Sci. **40**, 15–18 (1987)
18. Atanassov, K.: Index Matrices: Towards an Augmented Matrix Calculus. Springer International Publishing, Cham (2014). https://doi.org/10.1007/978-3-319-10945-9
19. Atanassov, K.: Intuitionistic Fuzzy Logics. Springer International Publishing, Cham (2017). https://doi.org/10.1007/978-3-319-48953-7
20. Zadeh, L.A.: Fuzzy sets. Inf. Control **8**, 338–353 (1965). https://doi.org/10.1016/S0019-9958(65)90241-X
21. Jekova, I., Vassilev, P., Stoyanov, T., Pencheva, T.: InterCriteria analysis: application for ECG data analysis. Mathematics **9**, 854 (2021). https://doi.org/10.3390/math9080854
22. Georgieva, V., Angelova, N., Roeva, O., Pencheva, T.: InterCriteria analysis of wastewater treatment quality. Ecol. Saf. **10**, 365–376 (2016)
23. Roeva, O., Fidanova, S.: Comparison of different metaheuristic algorithms based on InterCriteria analysis. J. Comput. Appl. Math. **340**, 615–628 (2018). https://doi.org/10.1016/j.cam.2017.07.028
24. Krawczak, M., Bureva, V., Sotirova, E., Szmidt, E.: Application of the InterCriteria decision making method to universities ranking. In: Atanassov, K.T., et al. (eds.) Novel Developments in Uncertainty Representation and Processing. AISC, vol. 401, pp. 365–372. Springer, Cham (2016). https://doi.org/10.1007/978-3-319-26211-6_31
25. Tsakovska, I., et al.: InterCriteria analysis implementation for exploration of the performance of various docking scoring functions. In: Dimov, I., Fidanova, S. (eds.) HPC 2019. SCI, vol. 902, pp. 88–98. Springer, Cham (2021). https://doi.org/10.1007/978-3-030-55347-0_8
26. Atanassov, K., Szmidt, E., Kacprzyk, J.: On intuitionistic fuzzy pairs. Notes Intuit. Fuzzy Sets **19**, 1–13 (2013)
27. Atanassov, K., Atanassova, V., Gluhchev, G.: InterCriteria analysis: ideas and problems. Notes Intuit. Fuzzy Sets **21**, 81–88 (2015)
28. Roeva, O., Vassilev, P., Ikonomov, N., Angelova, M., Su, J., Pencheva, T.: On different algorithms for intercriteria relations calculation. In: Hadjiski, M., Atanassov, K.T. (eds.) Intuitionistic Fuzziness and Other Intelligent Theories and Their Applications. SCI, vol. 757, pp. 143–160. Springer, Cham (2019). https://doi.org/10.1007/978-3-319-78931-6_10
29. Berman, H.M., et al.: The protein data bank. Nucl. Acids Res. **28**, 235–242 (2000). https://doi.org/10.1093/nar/28.1.235
30. Binda, C., et al.: Structures of human monoamine oxidase B complexes with selective noncovalent inhibitors: safinamide and coumarin analogs. J. Med. Chem. **50**, 5848–5852 (2007). https://doi.org/10.1021/jm070677y
31. Lauffer, B.E.L., et al.: Histone deacetylase (HDAC) inhibitor kinetic rate constants correlate with cellular histone acetylation but not transcription and cell viability. J. Biol. Chem. **288**, 26926–26943 (2013). https://doi.org/10.1074/jbc.M113.490706
32. Heitler, W., London, F.: Wechselwirkung neutraler Atome und homöopolare Bindung nach der Quantenmechanik. Z. Physik. **44**, 455–472 (1927). https://doi.org/10.1007/BF01397394
33. Labute, P.: The generalized Born/volume integral implicit solvent model: estimation of the free energy of hydration using London dispersion instead of atomic surface area. J. Comput. Chem. **29**, 1693–1698 (2008). https://doi.org/10.1002/jcc.20933
34. Rarey, M., Kramer, B., Lengauer, T., Klebe, G.: Fast flexible docking method using an incremental construction algorithm. J. Mol. Biol. **261**, 470–489 (1996). https://doi.org/10.1006/jmbi.1996.0477

35. Böhm, H.-J.: The development of a simple empirical scoring function to estimate the binding constant for a protein-ligand complex of known three-dimensional structure. J. Comput.-Aided Mol. Des. **8**, 243–256 (1994). https://doi.org/10.1007/BF00126743
36. Reulecke, I., Lange, G., Albrecht, J., Klein, R., Rarey, M.: Towards an integrated description of hydrogen bonding and dehydration: decreasing false positives in virtual screening with the HYDE scoring function. Chem. Med. Chem. **3**, 885–897 (2008). https://doi.org/10.1002/cmdc.200700319
37. Schneider, N., Lange, G., Hindle, S., Klein, R., Rarey, M.: A consistent description of HYdrogen bond and DEhydration energies in protein–ligand complexes: methods behind the HYDE scoring function. J. Comput. Aided Mol. Des. **27**, 15–29 (2013). https://doi.org/10.1007/s10822-012-9626-2
38. Trott, O., Olson, A.J.: AutoDock Vina: improving the speed and accuracy of docking with a new scoring function, efficient optimization, and multithreading. J. Comput. Chem. **31**, 455–461 (2010). https://doi.org/10.1002/jcc.21334
39. Koes, D.R., Baumgartner, M.P., Camacho, C.J.: Lessons learned in empirical scoring with smina from the CSAR 2011 benchmarking exercise. J. Chem. Inf. Model. **53**, 1893–1904 (2013). https://doi.org/10.1021/ci300604z

Multicriteria Analysis of Oncology Data During the Covid Pandemic

E. Sotirova[1(✉)], H. Bozov[1,2], S. Sotirov[1], G. Bozova[3], S. Ribagin[1], and V. Gonchev[1]

[1] Prof. Assen Zlatarov University, Prof. Yakimov Str., Burgas, Bulgaria
{esotirova,ssotirov}@btu.bg
[2] Oncology Complex Center - Burgas, 86 Demokratsiya Blvd, 8000 Burgas, Bulgaria
[3] Department of Nephrology, Military Medical Academy, St. George Sofiiski, Sofia, Bulgaria
greti@abv.bg

Abstract. In the article a data related to the treatment of Covid-19 patients for the period 28.10.2020–23.3.2022 were investigated. Some of the patients are with oncological diseases. The studied patients were admitted to a hospital in the city of Burgas, specialized in the treatment of cancer. The data contains information about the name of the disease according to the International Statistical Classification of Diseases and Health Problems (ICD), period of the patient's hospitalization, number of patients, medical staff by profile, etc. The method of InterCriteria Analysis was applied to study the dependencies between the parameters describing the patients and the medical staff. A self-organized neural network was developed for clustering the information related to the type of cancer.

Keywords: InterCriteria Analysis · Covid-19 · Oncological disease · Self-organized neural network · Intuitionistic Fuzzy Logic

1 Introduction

The data presented by the National Center for Public Health and Analysis show that the number of registered newly discovered diseases from malignant neoplasms in Bulgaria is about 30,000 people after 2000 [9, 10]. In the last decade, more and more emphasis has been placed on improving prevention, early diagnosis of oncological diseases and patient care in Bulgaria. Due to insufficiently good prevention, the diagnosis of cancer is very often accidental. There are over 350,000 patients with oncological diseases in Bulgaria.

Announced on March 11, 2020 by the World Health Organization (WHO), the pandemic of the SARS CoV-2 coronavirus, causing Covid-19, was a serious challenge for many risk groups of patients, including cancer patients. Specialized medical institutions for oncology care have also become involved in the care of patients with coronavirus infection [5, 11, 12].

In this study, data for patients with Covid-19 admitted for treatment in Oncology Complex Center in Burgas town for the period 28.10.2020–23.3.2022 was investigated.

S. Sotirov et al. (Eds.): BioInfoMed 2022, LNNS 658, pp. 111–118, 2023.
https://doi.org/10.1007/978-3-031-31069-0_12

The data contains information about different parameters: age of patients, name of the disease, according to International statistical classification of diseases and health problems, gender, marital status, data of the registration of the patient, etc.

For data analysis, the ICA method was applied. It is a new approach introduced in [3]. The ICA method supports multicriteria decision making by evaluating objects according to many criteria. It uses two mathematical tools - Indexed Matrices (IMs) [1] and Intuitionistic Fuzzy Sets (IFSs) [2]. The discovery of dependencies between criteria supports the decision-making process, especially in situations related to constraints such as time and resources. The use of intelligent tools to analyze data related to oncological diseases has become increasingly important in recent years. The authors applied the ICA approach for analysis of a data for patients with oncological diseases in [13–15].

In our investigation we use the ICA approach to study the dependencies between the parameters describing the patients and the medical staff. A self-organized neural network (SOM) was developed for clustering the information related to the type of patient's cancer.

2 Description of the Data

The ICA method is applied to real data The ICA method is applied to real data provided by the Complex Oncology Center Burgas. During the period 28.10.2020–23.3.2022, a total of 823 patients with Covid infection were hospitalized in the hospital. 789 patients were treated in the Covid Sector, of which 76 died. 34 patients were hospitalized in the Department of Anesthesiology and Intensive Care, of which 21 died.

The collected data are for 3 periods in which the Complex Oncology Center received patients with Covid-19 infection: I period: 28.10.2020–1.2.2021; II period 24.2.2021–7.5.2021 and III period 7.5.2021–23.3.2022.

The patients with Covid infection and malignant disease at the same time are 141. According to the ICD [8] their oncological diseases are in 25 groups: C00 "Malignant neoplasm of lip" (1 patient); C16 "Malignant neoplasm of stomach" (3 patients); C18 "Malignant neoplasm of colon" (8 patients); C20 "Malignant neoplasm of rectum" (13 patients); C21 "Malignant neoplasm of anus and anal canal" (1 patient); C22 "Malignant neoplasm of liver and intrahepatic bile ducts" (2 patients); C25 "Malignant neoplasm of pancreas" (6 patients); C32 "Malignant neoplasm of larynx" (2 patients); C34 "Malignant neoplasm of bronchus and lung" (15 patients); C43 "Malignant melanoma of skin" (4 patients); C44 "Other malignant neoplasms of skin" (4 patients); C50 "Malignant neoplasm of breast" (31 patients); C53 "Malignant neoplasm of cervix uteri" (10 patients); C54 "Malignant neoplasm of cervix uteri" (2 patients); C56 "Malignant neoplasm of ovary" (3 patients); C61 "Malignant neoplasm of prostate" (12 patients); C64 "Malignant neoplasm of kidney, except renal pelvis" (5 patients); C67 "Malignant neoplasm of bladder" (4 patients); C71 "Malignant neoplasm of brain" (1 patient); C75 "Malignant neoplasm of other endocrine glands and related structures" (1 patient); C78 "Secondary malignant neoplasm of respiratory and digestive organs" (7 patients); C79 "Secondary malignant neoplasm of other and unspecified sites" (2 patients); C85 "Other and unspecified types of non-Hodgkin lymphoma" (2 patients); C91 "Lymphoid leukaemia" (1 patient); C92 "Myeloid leukaemia" (1 patient).

For the same ICD 25 groups we observed the number of patients with malignant disease patients in the registry of the Complex Oncology Center Burgas in period mart 2029 - mart 2022: C00 - 25 patients; C16 - 272 patients; C18 - 446 patients; C20 - 304 patients; C21 - 8 patients; C22 - 137 patients; C25 - 208 patients; C32 - 79 patients; C34 - 600 patients; C43 - 64 patients; C44 - 1033 patients; C50 - 643 patients; C53 - 204 patients; C54 - 203 patients; C56 - 161 patients; C61 - 483 patients; C64 - 192 patients; C67 - 361 patients; C71 - 111 patients; C75 - 2 patients; C78 - patients; C79 - 0 patients; C85 - 20 patients; C91 - 15 patients; C92 - 10 patients.

Patients were cared for by medical staff that changed depending on the time period and the available number of beds for patients with Covid-19. The medical staff included 6 groups - Doctors, Nurses, X-ray laboratory assistant, Healthcare Assistants, Clinical laboratory assistant, Sanitarians. The maximum number of beds in the Covid sector for periods I, II and III were 42, 49 and 42, respectively. The medical staff was respectively 53 for the I period (15 Doctors, 21 Nurses, 1 X-ray laboratory assistant, 3 Healthcare Assistants, 13 Sanitarians), 55 for the II period (14 Doctors, 27 Nurses, 1 X-ray laboratory assistant, 2 Healthcare Assistants, 11 Sanitarians), and 39 for the III period (12 Doctors, 10 Nurses, 3 X-ray laboratory assistant, 2 Healthcare Assistants, 2 Clinical laboratory assistant, 10 Sanitarians).

3 The ICA Approach

3.1 Applying the ICA Approach to a Data for Patients with Oncological Disease and Covid-19 Infection

We use an index matrix with 6 rows (for 3 time periods: 28.10.2020–1.2.2021, 24.2.2021–7.5.2021 and 7.5.2021–23.3.2022, number of oncology patients treated for Covid-19 by location and number of patient in the register by location) and 26 columns (for 25 ICD groups for oncological disease of the Covid-19 patients).

After data processing with ICA software we obtain membership part (Table 1) and non-membership part (Table 2) of the intuitionistic fuzzy pairs [4] that represent an intuitionistic fuzzy evaluation of the relations between every pair of criteria.

The pair of criteria "III period of Covid-19 Infection-Oncology patients with Covid-19" is in weak positive consonance: $\langle 0,833; 0,017\rangle$. This means that for trend of the development of covid treatment, morbidity and mortality for the patients with malignant disease, the most decisive is period III (7.5.2021–23.3.2022). The other 9 pairs of criteria are in dissonance (2 pairs in weak dissonance, 4 pairs in dissonance, 3 pairs in strong dissonance), that means, that there is no consistent trend of development the mobility. The degree of uncertainty ($\pi = 1 - \mu - \nu$) is high, which is explained by the insufficiently large size of the analyzed data.

3.2 Applying the ICA Approach to a Data for Medical Staff

For this observation an index matrix with 8 rows (for every group of medical staff - Doctors, Nurses, X-ray laboratory assistant, Healthcare Assistants, Clinical laboratory assistant, Sanitarians, and number of beds) and 34 columns (for dates of opening of the

Table 1. Membership parts of the Intuitionistic fuzzy pairs of the relations

Degree of membership μ	I period	II period	III period	Oncology patients treated for Covid-19	Patients in the register
I period	1,000	0,420	0,573	0,563	0,397
II period	0,420	1,000	0,500	0,607	0,557
III period	0,573	0,500	1,000	0,833	0,693
Oncology patients treated for Covid	0,563	0,607	0,833	1,000	0,730
Patients in the register	0,397	0,557	0,693	0,730	1,000

Table 2. Non-membership parts of the Intuitionistic fuzzy pairs of the relations

Degree of non-membership ν	I period	II period	III period	Oncology patients with Covid-19	Patients in the register
I period	0,000	0,147	0,097	0,077	0,193
II period	0,147	0,000	0,163	0,107	0,153
III period	0,097	0,163	0,000	0,017	0,167
Oncology patients with Covid-19	0,077	0,107	0,017	0,000	0,173
Patients in the register	0,193	0,153	0,167	0,173	0,000

Covid sector and Department of Anesthesiology and Intensive Care, dates of closing of the Covid sector and Department of Anesthesiology and Intensive Care).

The membership part and non-membership part of the intuitionistic fuzzy pairs respectively are shown in Table 3 and Table 4.

One pair of criteria "Doctors - Nurses" is in strong positive consonance: $\langle 0, 911; 0, 063\rangle$. This shows the same tendency in distribution of doctors and nurses in the Covid sector and Department of Anesthesiology and Intensive Care. Two pairs are in positive consonance: "Doctors - Sanitarians" with evaluation $\langle 0, 888; 0, 068\rangle$ and "Nurses - Sanitarians" with evaluation $\langle 0, 867; 0, 108\rangle$, that means very similar tendency in the allocation of doctors, nurses and sanitarians. The other eighteen couples do not show a consistent trend.

Table 3. Membership parts of the Intuitionistic fuzzy pairs of the relations

Degree of membership μ	Beds	Doctors	Nurses	X-ray lab. assistant	Healthcare assistants	Clin.lab. assistant	Sanitarians
Number of beds	1,000	0,736	0,732	0,601	0,557	0,291	0,734
Doctors	0,790	1,000	0,911	0,663	0,617	0,307	0,888
Nurses	0,786	0,911	1,000	0,616	0,627	0,263	0,867
X-ray lab. assistant	0,631	0,659	0,611	1,000	0,696	0,645	0,609
Healthcare Assistants	0,589	0,634	0,637	0,742	1,000	0,368	0,641
Clin.lab. assistant	0,310	0,318	0,277	0,652	0,366	1,000	0,286
Sanitarians	0,749	0,864	0,839	0,575	0,611	0,251	1,000

Table 4. Non-membership parts of the Intuitionistic fuzzy pairs of the relations

Degree of non-membership ν	Beds	Doctors	Nurses	X-ray lab. assistant	Healthcare assistants	Clin.lab. assistant	Sanitarians
Number of beds	0,000	0,148	0,158	0,091	0,076	0,081	0,148
Doctors	0,114	0,000	0,063	0,095	0,072	0,095	0,068
Nurses	0,121	0,063	0,000	0,134	0,059	0,131	0,108
X-ray lab. assistant	0,075	0,101	0,143	0,000	0,097	0,000	0,137
Healthcare Assistants	0,062	0,075	0,067	0,077	0,000	0,077	0,052
Clin.lab. assistant	0,071	0,103	0,138	0,000	0,090	0,000	0,120
Sanitarians	0,138	0,083	0,131	0,159	0,060	0,133	0,000

4 Clustering of the Data

A self-organizing map (SOM) or self-organizing feature map (SOFM) is an unsupervised machine learning technique used to create an n-dimensional (usually two-dimensional) representation of a data set while preserving the topological structure of the data. The COM adaptively perceives the input data and, based on the numbers, spreads them into clusters of observations with similar values for the variables. These clusters can be visualized as a two-dimensional "map" that aggregates similar values from observations in the distal clusters. This can make high-dimensional data easier to visualize and analyze (Figs. 1, 2, 3 and 4).

For testing the SOM, all input vectors corresponding to individual MKBs are run for identification. In this way, the different clusters are identified to which of the cases they correspond and with which they are similar. The table shows the distribution of the individual vectors in the six clusters.

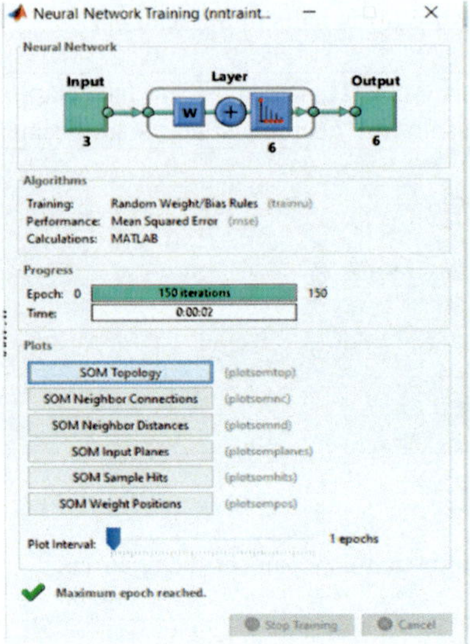

Fig. 1. Structure of the SOM

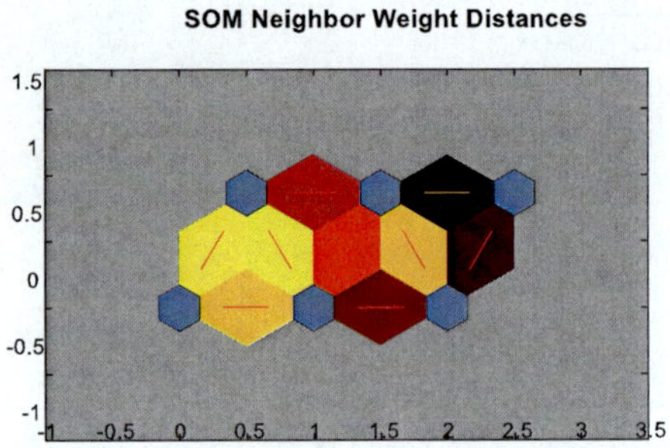

Fig. 2. Distances between the clusters in the SOM

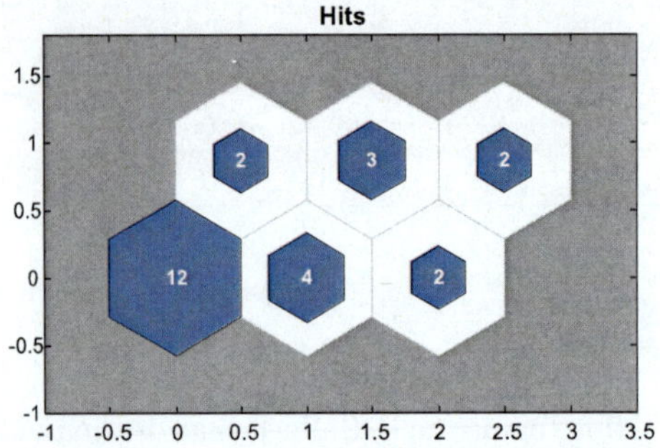

Fig. 3. Distribution of the individual vectors on the clusters

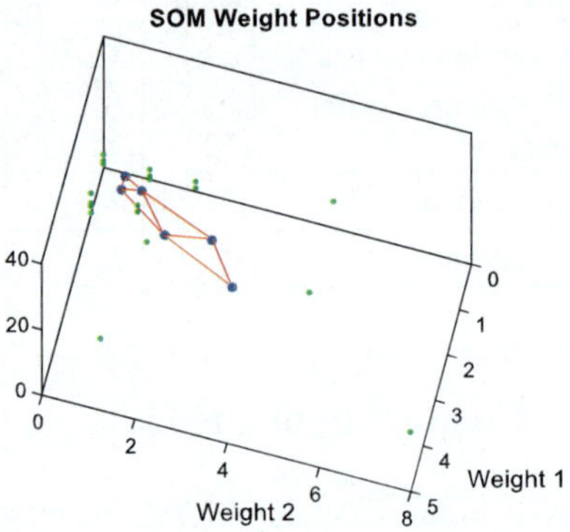

Fig. 4. Visualization of weights and cluster centers

	P1	P2	P3	P4	P5	P6	P7	P8	P9	P10	P11	P12	P13	P14	P15	P16	P17	P18	P19	P20	P21	P22	P23	P24	P25
C1	1	1	1	1	1	1	1	1	1	1	1	1	0	0	0	0	0	0	0	0	0	0	0	0	0
C2	0	0	0	0	0	0	0	0	0	0	0	0	0	0	1	1	1	1	0	0	0	0	0	0	0
C3	0	0	0	0	0	0	0	0	0	0	0	0	0	0	0	0	0	0	0	0	0	0	1	1	0
C4	0	0	0	0	0	0	0	0	0	0	0	0	1	1	0	0	0	0	0	0	0	0	0	0	0
C5	0	0	0	0	0	0	0	0	0	0	0	0	0	0	0	0	0	0	1	1	0	1	0	0	0
C6	0	0	0	0	0	0	0	0	0	0	0	0	0	0	0	0	0	0	0	0	1	0	0	0	1

Cluster 1 includes: Malignant neoplasm of the mouth, Malignant neoplasm of the anus and anal canal, Malignant neoplasm of the brain, Malignant neoplasm of other endocrine glands, Lymphoid leukemia, Myeloid leukemia, Malignant neoplasm of the liver and intrahepatic bile ducts, Malignant neoplasm of larynx, Malignant neoplasm of the body of the uterus, Secondary malignant neoplasm with other localizations, Other and unspecified types of Non-Hodgkin's lymphoma, Malignant neoplasm of the stomach.

Cluster 4 includes: Malignant neoplasm of the ovary, Malignant melanoma of the skin, Cluster 2 includes: Other malignant neoplasms of the skin, Malignant neoplasm of the bladder, Malignant neoplasm of the kidney, Malignant neoplasm of the pancreas.

Cluster 5 includes: Secondary malignant neoplasm of the respiratory and digestive systems, Malignant neoplasm of the colon, Malignant neoplasm of the prostate.

Cluster 6 includes: Malignant neoplasm of the cervix, Malignant neoplasm of the mammary gland.

Cluster 3 includes: Malignant neoplasm of the rectum, Malignant neoplasm of the bronchi and lung.

The location of the individual vectors responsible for the individual ICDs indicates the most common diseases and their classification only on the data available for processing. The COM itself is an adaptive structure that perceives all input vectors and adjusts itself. A very positive property of a neural network is that there is no predetermined goal to pursue. From this point of view, the system is adjusted independently and without additional external intervention.

5 Conclusion

Data summarization, determination, and clustering capabilities are very important tools when dealing with large amounts of data. In this case, various types of information processing tools were used. The data contain information about the name of the disease according to the ICD, the patient's hospitalization period, number of patients, medical staff by profile, etc. In our investigation, we used the method of InterCriteria Analysis, which was applied to study the dependencies between parameters describing patients and medical staff. In addition, a self-organized neural network was developed for clustering the information related to the type of cancer. The article examines data related to the treatment of patients with Covid-19 for the period 28.10.2020–23.03.2022. Some of the patients have oncological diseases. In the next observations the ICA approach will be apply to a patients with malignant melanoma in ovarian and primary cervical carcinoma registered in the Complex Oncology Center Burgas town [6, 7].

Acknowledgements. The authors are grateful for the support provided by the Bulgarian National Science Fund under Grant Ref. No. KP-06-N22/1/2018 "Theoretical research and applications of InterCriteria Analysis". The authors declare that there is no conflict of interest regarding the publication of this paper.

References

1. Atanassov, K.T.: Index matrices: towards an augmented matrix calculus. In: Studies in Computational Intelligence Series, vol. 573. Springer, Cham (2014). https://doi.org/10.1007/978-3-319-10945-9

2. Atanassov, K.T.: On Intuitionistic Fuzzy Sets Theory. Springer, Heidelberg (2012). https://doi.org/10.1007/978-3-642-29127-2
3. Atanassov, K., Mavrov, D., Atanassova, V.: Intercriteria decision making: a new approach for multicriteria decision making, based on index matrices and intuitionistic fuzzy sets. Issues Intuitionistic Fuzzy Sets Generalized Nets **11**(2014), 1–8 (2014)
4. Atanassov, K., Szmidt, E., Kacprzyk, J.: On intuitionistic fuzzy pairs. Notes Intuitionistic Fuzzy Sets **19**(3), 1–13 (2013)
5. Gosain, R., Abdou, Y., Singh, A., Rana, N., Puzanov, I., Ernstoff, M.S.: COVID-19 and Cancer: a Comprehensive Review. Curr. Oncol. Rep. **22**(5), 1–15 (2020). https://doi.org/10.1007/s11912-020-00934-7
6. Dobrev, P., Strashilov, S., Yordanov, A.: Metastasis of malignant melanoma in ovarian simulating primary ovarian cancer: a case report. Gazetta Medica Italiana – Archivio per le Scienze Medicine **180** (2021)
7. Dobrev, P., Yordanov, A., Strashilov, S.: Synchronous primary cervical carcinoma and ovarian fibroma: challenge in surgery. Gazzetta Medica Italiana-Archivio per le Scienze Mediche **179**(5), 375–377 (2020)
8. International Statistical Classification of Diseases and Related Health Problems 10th Revision (ICD-10)-WHO Version for 2019-covid-expanded. https://www.who.int/standards/classifications/classification-of-diseases
9. National Center of Public Health and Analyses, Annual information. http://ncpha.government.bg/index.php?lang=en
10. Ministry of Health, National Center of Public Health and Analyses. https://ncpha.government.bg/uploads/statistics/annual/health_B_3.pdf
11. Nekhlyudov, L., et al.: Addressing the needs of cancer survivors during the COVID-19 pandemic. J. Cancer Surviv. **14**(5), 601–606 (2020). https://doi.org/10.1007/s11764-020-00884-w
12. Raymond, E., Thieblemont, C., Alran, S., Faivre, S.: Impact of the COVID-19 outbreak on the management of patients with cancer. Target. Oncol. **15**(3), 249–259 (2020)
13. Sotirov, S., Petrova, Y., Bozov, H., Sotirova, E.: A hybrid algorithm for multilayer perceptron design with intuitionistic fuzzy logic using malignant melanoma disease data. In: Kahraman, C., Tolga, A.C., Cevik Onar, S., Cebi, S., Oztaysi, B., Sari, I.U. (eds.) Intelligent and Fuzzy Systems (INFUS 2022). LNNS, vol. 504, pp. 665–672. Springer, Cham (2022). https://doi.org/10.1007/978-3-031-09173-5_77
14. Sotirova, E., Vasilev, V., Bozova, G., Bozov, H., Sotirov, S.: Application of the InterCriteria analysis method to a dataset of malignant neoplasms of the digestive organs for the Burgas region for 2014–2018. In: 2019 Big Data, Knowledge and Control Systems Engineering (BdKCSE), pp. 1–6. IEEE (2019). https://doi.org/10.1109/BdKCSE48644.2019.9010609
15. Sotirova, E., Bozova, G., Bozov, H., Sotirov, S., Vasilev, V.: Application of the InterCriteria analysis method to a data of malignant melanoma disease for the Burgas region for 2014–2018. In: Atanassov, K.T., et al. (eds.) IWIFSGN 2019 2019. AISC, vol. 1308, pp. 166–174. Springer, Cham (2021). https://doi.org/10.1007/978-3-030-77716-6_15

Biomedical Approaches and Applications

Convolutional Autoencoder for Filtering of Power-Line Interference with Variable Amplitude and Frequency: Study of 12-Lead PTB-XL ECG Database

Kamen Ivanov, Irena Jekova, and Vessela Krasteva(✉)

Institute of Biophysics and Biomedical Engineering, Bulgarian Academy of Sciences, Acad. G. Bonchev str. bl. 105, 1113 Sofia, Bulgaria
vessika@biomed.bas.bg

Abstract. This study aims to explore a new deep learning strategy for electrocardiogram (ECG) denoising under adverse conditions of non-stationary power-line interference (PLI) with amplitude changes or nominal frequency deviations. The study presents an exhaustive training strategy of deep convolutional autoencoder (CAE), while input with one of the largest PhysioNet 12-lead ECG databases contaminated by simulated sinusoidal PLI noise with augmented settings. Twelve ECG leads (I, II, III, aVR, aVL, aVF, V1–V6) from 14890 PTB-XL records, divided patient-wise to training (50%, 7441 records), validation (20%, 2979 records) and test (30%, 4470 records) are superimposed by PLI with five signal-to-noise ratios (SNR) (–2.5, 0, 2.5, 5, 7.5 dB), nine frequencies (48, 48.5, 49, 49.5, 50, 50.5, 51, 51.5, 52 Hz), 12 amplitude slew rates ±(50, 100, 250, 500, 750, 1000 μV/s), and 14 frequency slew rates ±(0.01, 0.025, 0.05, 0.075, 0.1, 0.15, 0.2 Hz/s). CAE receptive field inputs one ECG lead with 1024 samples (2.048 s for 500 Hz sampling rate). CAE architecture is designed with seven 1D-convolutional layers, including three encoder layers (filters x kernel size = 16 × 8, 8 × 8, 8 × 8) and four decoder layers (8 × 8, 8 × 8, 16 × 8, 1 × 8) with linear activation function and same padding. CAE non-linear operations for max-pooling and up-sampling (pool size of 2) follow each encoder and decoder convolutional layers, respectively. Adam optimizer and mean squared error loss function are applied for CAE training over 250 epochs. The quality of clean ECG reconstruction in CAE output is evaluated by root-mean-square error (RMSE), percentage-root-mean-square difference (PRD) and improvement in signal-to-noise ratio (SNR_{imp}). Statistical test results for denoising of all 12 ECG leads present median RMSE = 5.3 μV, PRD = 3.5%, SNR_{imp} = 22–32 dB for SNR = –2.5 to 7.5 dB. The results do not substantially change for PLI frequencies 48–52 Hz, amplitude slew rates up to ±1000 μV/s and frequency slew rates up to ±0.2 Hz/s with median value divergence of $\Delta RMSE < 2$ μV, $\Delta PRD < 1.5\%$, $\Delta SNR_{imp} \leq 3$ dB. The observed performance stability justifies the deep learning strategy for training a CAE with generalizable application for denoising of ECG signals with non-stationary PLI.

Keywords: Electrocardiogram · Deep learning · Artificial neural networks · Digital filters · Denoising autoencoder · Signal-to-noise ratio · 50 Hz filtering

S. Sotirov et al. (Eds.): BioInfoMed 2022, LNNS 658, pp. 121–133, 2023.
https://doi.org/10.1007/978-3-031-31069-0_13

1 Introduction

The electrocardiogram (ECG) is a generic bio-signal used in non-invasive clinical examinations of cardiovascular pathologies [1]. Misinterpretation of ECG diagnostic components is highly susceptible to 50/60 Hz power-line interference (PLI) induced via parasitic capacitive couplings that typically occur between the body and power lines [2]. Although modern instrumentation amplifiers have high common mode rejection ratio, and that shielding and grounding are applied during ECG acquisition, it has been shown that an imbalance of input electrode impedances can induce differential PLI, which is amplified together with the diagnostic ECG [3]. PLI filtering is always mandatory in diagnostic ECG devices, which need to imply with international standards [4].

Different classical techniques for 50/60 Hz PLI filtering have been published. The simplest filter design uses low-pass comb filters for rejection of the fundamental PLI frequency and its harmonics [5, 6] at the cost of certain high-frequency QRS distortions. Comb filters perform well only when the PLI frequency is stable and fixed within the filter stop-band. A filter design for ECG signals is shown to preserve non-linear ECG components by subtracting a PLI estimate into linear segments, named the subtraction procedure [7, 8]. Adaptive filters follow continuously PLI variations by modifying the filter coefficients through least mean square algorithm [9]. They are shown to introduce distortions during QRS due to limited adaptation time and deficient estimation of the reference sinusoid, commonly extracted from the noisy input [9]. Various modifications by block based time-frequency domain adaptation algorithms [10] and cascaded multistage adaptive structures [11] present improved efficiency.

Existing ECG filtering techniques have certain limitations linked to the overlap of 50/60 Hz PLI and the ECG diagnostic bandwidth. Common filter distortions are QRS amplitude suppression and ringing after steep RS slopes. Therefore, new approaches are being sought in the field of deep neural networks as denoising autoencoders (DAE) assuming that they extract specific signal components in hidden encoder layers. A few studies have recently applied DAE for reduction of the baseline wander, muscle artifacts and electrode motion in ECG [12–15]. The designed DAE are based on dense, convolutional and hybrid architectures with convolutional and recurrent long-short-term-memory layers and residual blocks. To our knowledge none of these studies have provided extensive evaluation of PLI filtering, especially in non-stationary conditions.

This study aims to explore a new deep learning strategy for denoising of ECG signals under adverse conditions of non-stationary power-line interference (PLI) with amplitude changes or nominal frequency deviations. The study presents an exhaustive training strategy of deep convolutional autoencoder, while input with one of the largest PhysioNet 12-lead ECG databases contaminated by PLI sinusoidal noise with expanded settings, including constant amplitudes (signal-to-noise ratios from -2.5 to 7.5 dB), constant frequencies (range of ± 2 Hz around 50 Hz notch), linearly changing amplitudes (slew rates from 50 to ± 1000 μV/s) and frequencies (slew rates from 0.01 to ± 0.2 Hz/s). Using independent test set, the 12 ECG leads are compared in respect to the quality of the denoised ECG. The expected benefits would prove the ability of an extended deep learning strategy to train deep denoising autoencoder, which is able to sufficiently suppress broad PLI variations without additional adjustments or reference inputs.

2 Materials and Methods

2.1 ECG Database

This study uses the PhysioNet PTB-XL ECG database [16] (https://physionet.org/content/ptb-xl/1.0.1/), which is one of the largest freely available clinical 10s 12-lead ECG datasets, comprising 21837 records from 18885 patients. The dataset is balanced with respect to sex (52% male and 48% female) and covers the whole range of ages from 0 to 95 years (median 62, inter-quantile range of 22). The twelve standard ECG leads (I, II, III, aVL, aVR, aVF, V1–V6) are released at high resolution (1 μV/LSB, sampling frequency of 500 Hz) with the purpose for evaluation of machine learning algorithms, making certain that the whole database is a rich representative of ECG rhythms in norm and pathology. Although this study does not consider specific diagnostic labels, it is worthy to note that the reliability of the training strategy and evaluation results is justified with the diversity of the available clinical rhythms, including 9528 Normal ECGs, 5772 Conduction disturbances, 6886 Myocardial infarctions, 2819 Hypertrophies, 5788 ST/T changes (one record could have one or several diagnoses).

We excluded a total of 6947 records, applying criteria for duplicated records per patient (2952) and low signal quality (1607 with baseline drift, 2362 with static noise, 613 with burst noise, 30 with electrode problems, 293 with active pacemakers). Note that some records might present several of the exclusion conditions.

The included records are 14890, divided patient-wise to training (50%, 7441 records), validation (20%, 2979 records) and test (30%, 4470 records). Each 10 s record is divided into four sequential non-overlapping segments with duration of 1024 samples (2.048 s) as defined by the receptive field of the denoising autoencoder input (Sect. 2.2). Each lead is independently processed, therefore, the size of the database is computed by the multiplication: 4 segments * number of leads * number of records.

2.2 Convolutional Autoencoder Architecture

The nonlinear mapping qualities of autoencoders make them appropriate for signal denoising. In principle, the autoencoder provides a copy of the input data to the output. It can be trained to ignore certain noisy components of the input data when performing this operation. Such components can be the PLI harmonics in ECG. Autoencoders contain an encoder and a decoder modules (Fig. 1). The encoder maps the input sequence x into hidden representation ζ through a nonlinear transformation f. After that, through the nonlinear transformation g, the decoder maps the hidden representation ζ into the reconstructed output sequence $\hat{x}$. This process is formally described as follows [12, 13]

$$\zeta = f(xW + b) \tag{1}$$

$$\hat{x} = g\left(\zeta \hat{W} + \hat{b}\right) \tag{2}$$

W and b are weights and bias in the encoder, and $\hat{W}$ and $\hat{b}$ are the weights and bias in the decoder, respectively. During the training, the mean squared error is used to update the network parameter set $\theta = \left\{W, b, \hat{W}, \hat{b}\right\}$ by minimizing the loss function:

$$L(\theta) = \sum\nolimits_{n} \left\| x - \hat{x} \right\|^2 \tag{3}$$

A major decision on the autoencoder design is regarding the use of either fully-connected or convolutional layers. Convolutional layers are favored as they preserve the locally-spatial information from previous layers while fully-connected layers fail. Convolutional layers are associated with a smaller number of parameters, which also reduces the requirements on the hardware for training [12, 13]. Hence, in the present work, a convolutional autoencoder (CAE) architecture is proposed for the ECG denoising.

The architecture of designed CAE is illustrated in Fig. 1. CAE is applied on noisy input with a dimension of 1024 × 1 (2.048 s for sampling rate of 500 Hz) to reconstruct the clean ECG data at the output (size 1024 × 1). The architecture involves seven 1D-convolutional layers, including three encoder layers (filters x kernel size = 16 × 8, 8 × 8, 8 × 8) and four decoder layers (8 × 8, 8 × 8, 16 × 8, 1 × 8). Linear activation function and same padding are set up for convolutional layers as suitable for regression tasks. Each encoder convolutional layer is followed by a max-pooling operation with a pool size of 2 (shrinking the encoder feature map to 128 × 1), and their inversely symmetric decoder convolutional layers are followed by an upsampling operation with a size of 2.

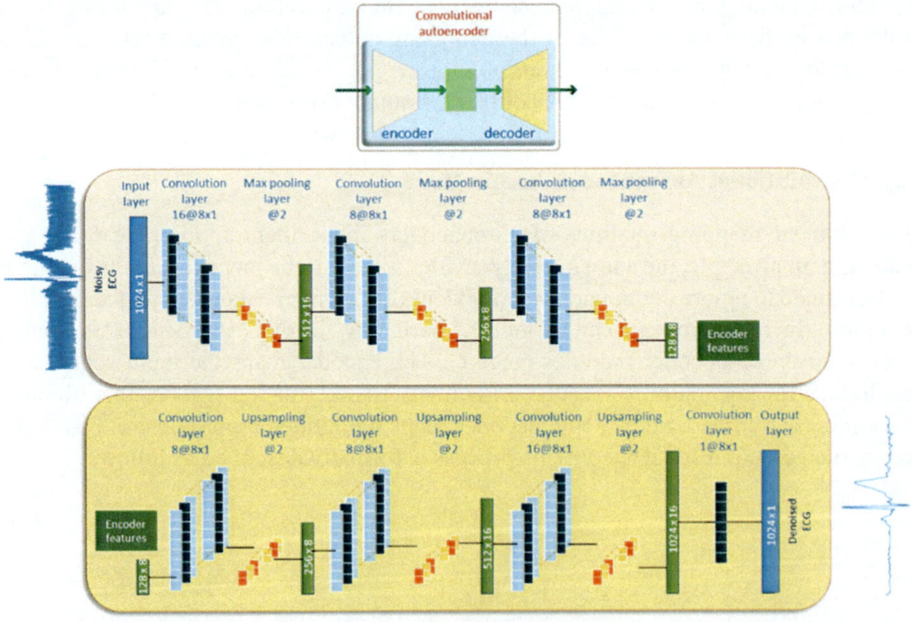

Fig. 1. The structure of the proposed convolutional autoencoder architecture.

2.3 Deep Learning Strategy

A deep learning strategy of CAE for ECG denoising with exhaustive number of artificially generated noisy ECG records (*nECG*) is designed. At the noisy input to CAE, an arithmetic sum of original ECG signals from databases (*oECG*, Sect. 2.1) and simulated PLI sinusoids with predefined amplitude (A_{PLI}) and frequency (F_{PLI}) is applied:

$$nECG = oECG(n) + A_{PLI}(n)sin(2\pi F_{PLI}(n)n/F_s), \tag{4}$$

where n denotes the sample number within the duration of the signal with N samples (n = 1,2,...N). A_{PLI} and F_{PLI} can be linearly changing values, starting from a baseline A_{PLI0} and F_{PLI0} with predefined slew rates, namely ΔA_{PLI} (μV/s) and ΔF_{PLI} (Hz/s), formally written as:

$$A_{PLI}(n) = A_{PLI0} + n\Delta A_{PLI}/F_s \tag{5}$$

$$F_{PLI}(n) = F_{PLI0} + n\Delta F_{PLI}/F_s \tag{6}$$

When $\Delta A_{PLI} = 0/s$ or $\Delta F_{PLI} = 0Hz/s$, the generated PLI sinusoid has a constant baseline amplitude or frequency, respectively.

The amplitude A_{PLI} should be rescaled independently for each *oECG* lead according to a predefined signal-to-noise ratio, computed for N = 5000 samples in 10 s:

$$SNR\,[dB] = 10log_{10}\left(\sqrt{\frac{1}{N}\sum\nolimits_{n=0}^{N-1}(oECG(n))^2}\Big/\sqrt{\frac{1}{N}\sum\nolimits_{n=0}^{N-1}(A_{PLI}(n))^2}\right) \tag{7}$$

Each original ECG lead is superimposed with four different PLI settings (Table 1) during the training and test process of CAE.

Table 1. Training and test strategy of CAE for filtering of non-stationary PLI. Note*: The positive or negative signs of ΔA_{PLI} and ΔF_{PLI} indicate increasing or decreasing trend for change of the PLI amplitude and frequency over time, respectively.

PLI setting	SNR (dB)	F_{PLI} (Hz)	ΔA_{PLI} (μV/s)*	ΔF_{PLI} (Hz/s)*
1: Constant PLI with different SNR	[−2.5, 0, 2.5, 5, 7.5]	50	0	0
2: Constant PLI with different frequencies	0	[48, 48.5, 49, 49.5, 50, 50.5, 51, 51.5, 52]	0	0
3: Non-stationary PLI with linearly changing ampl.	NA	50	±[50,100,250, 500,750,1000]	0
4: Non-stationary PLI with linearly changing frequency	0	50	0	±[0.01,0.025, 0.05,0.075, 0.1,0.15,0.2]

2.4 Evaluation of Denoised Signal Quality

CAE denoising performance is estimated by means of three benchmark metrics, namely root-mean-square error (*RMSE*), percentage root-mean-square difference (*PRD*) and improvement in the signal-to-noise ratio (SNR_{imp}) [12–14, 17].

RMSE evaluates the variance between the output, predicted by the denoising algorithm (*dECG*), and original ECG signal (*oECG*). A lower *RMSE* value corresponds to better performance of the denoising method. An advantage of *RMSE* is that it provides direct estimation of the output ECG deviations in original units, i.e., millivolts:

$$RMSE = \sqrt{\frac{1}{N}\sum\nolimits_{n=0}^{N-1}(oECG(n) - dECG(n))^2}, \quad [\text{mV}] \tag{8}$$

PRD evaluates the percent distortion of the denoised signal compared to the original one (Eq. 9). The lower the *PRD* value, the better the noise reduction. *PRD* is preferred because it is easy to calculate and takes into account the true amplitude of the input signal, since larger input amplitudes are usually associated with larger distortions. The *PRD* value depends on the zero-line offset in both *oECG* and *dECG*. For example, if an offset is present in the denominator of Eq. (9), then falsely lower *PRD* is obtained. Hence, *PRD* shall be applied to signals with subtracted mean:

$$PRD = 100\sqrt{\frac{\sum_{n=0}^{N-1}(oECG(n) - dECG(n))^2}{\sum_{n=0}^{N-1} oECG^2(n)}}, \quad [\%] \tag{9}$$

SNR_{imp} reflects the improvement in the SNR of the output denoised signal compared to the noisy input one:

$$SNR_{imp} = SNR_{out} - SNR_{in}, \ [dB] \tag{10}$$

where SNR_{in} and SNR_{out} are defined as follows:

$$SNR_{in} = 10\,log_{10}\left(\frac{\sum_{n=0}^{N-1} oECG^2(n)}{\sum_{n=0}^{N-1}(nECG - oECG(n))^2}\right), \ [dB] \tag{11}$$

$$SNR_{out} = 10\,log_{10}\left(\frac{\sum_{n=0}^{N-1} oECG^2(n)}{\sum_{n=0}^{N-1}(dECG(n) - oECG(n))^2}\right), \ [dB] \tag{12}$$

For preset SNR_{in}, better denoising performance is represented by the combination of a larger SNR_{imp} and both lower *RMSE* and *PRD*.

3 Results and Discussion

3.1 Training

CAE models are implemented in Python using Keras with Tensorflow backend. All experiments are conducted in a workstation PERSY Stinger with Intel CPU Xeon Silver 4214R @ 2.4 GHz (2 processors), 96 GB RAM, NVIDIA RTX A5000–24 GB GPU.

The model training is run for 250 epochs and the model with minimal validation loss computed according to Eq. 3 is stored for further evaluation in the test set. The learning curve in Fig. 2 indicates an optimal fit while both training and validation loss successively decrease down to similar levels. Our final choice is the minimal loss model at 229 epoch, highlighted over the curve plateau (Fig. 2), thereby no substantial improvement would be expected if more training epochs are applied.

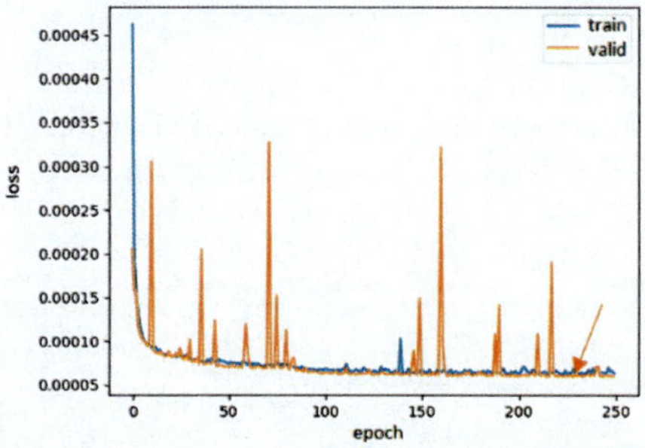

Fig. 2. CAE training process pursuing loss minimization in the training dataset. The observation of the minimal loss in the validation dataset is used to select an optimally trained model.

3.2 Test 1: Constant PLI with Different ECG Leads

The trained CAE model is applied for filtering of 12-lead ECG records in the test set of PTB-XL ECG database, with each lead filtered independently. The example in Fig. 3 illustrates that the PLI sinusoid (50 Hz, $SNR_{in} = 0$ dB) is completely rejected in the CAE output for all ECG leads, regardless of the beat waveform (normal or ventricular ectopic, presened with different amplitude polarities and slopes in 12 leads). Lower amplitude P, T-waves are also denoised without noticeable distortion.

Furhremore, this observation is confirmed in the statistical comparative study of all leads (Fig. 4), presenting narrow inter-lead span of the median values of *RMSE* by about 2.5 μV (4–6.5 μV), *PRD* by 4% points (2.5–6.5%), SNR_{imp} by 8 dB (25–33 dB). No significant differences are observed for CAE performance with different ECG leads, so statistics is reported for all ECG leads together in further tests (Sects. 3.3–3.6).

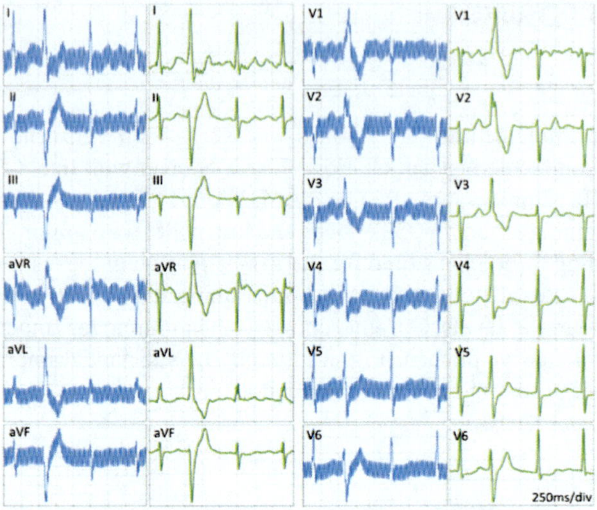

Fig. 3. Example for CAE filtering of standard 12-lead ECG (record '19391_hr'), showing: blue trace – noisy PLI input (2.048 s, 50 Hz, $SNR_{in} = 0$ dB), green trace – denoised ECG output.

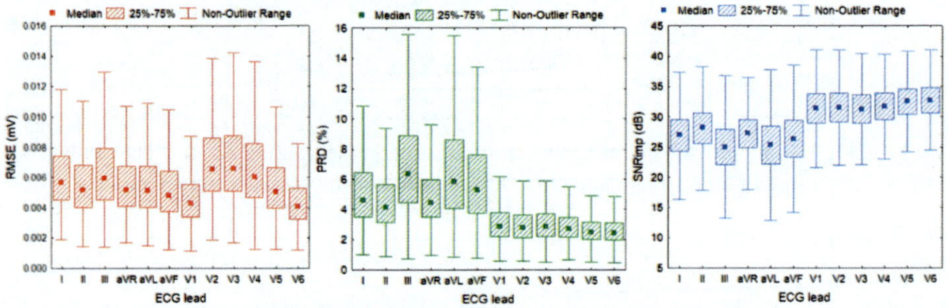

Fig. 4. Test performance for CAE filtering of constant PLI (50 Hz, 0 dB) in 12-lead ECG.

3.3 Test 1: Constant PLI with Different Input SNR

The example in Fig. 5 illustrates CAE denoising of constant PLI sinusoid (50 Hz) with low and high amplitudes preset by $SNR_{in} = 7.5$ and –2.5 dB, respectively. In both cases, PLI is sufficiently filtered, presenting notably the same output error (right trace) with equal estimations of $RMSE = 4.9$ μV and $PRD = 5.1\%$. Conversely, SNR_{imp} is equal to 17.8 dB (top) and 27.8 dB (bottom), noting 10 dB difference due to SNR_{in} span.

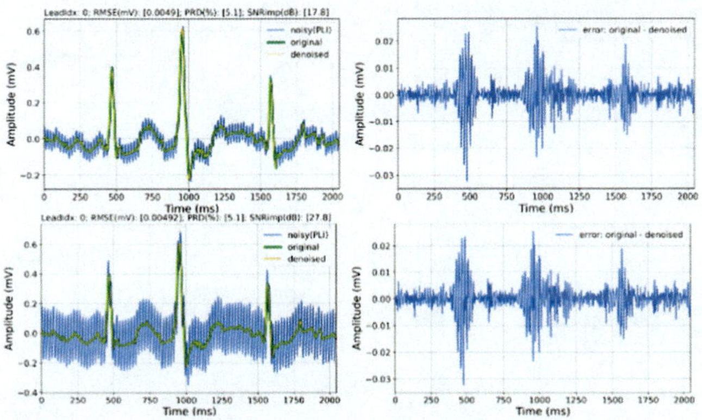

Fig. 5. ECG example (record '19391_hr', lead I, normal and ventricular ectopic beats) superimposed by PLI (50 Hz) with different $SNR_{in} = 7.5$ dB (top) and $SNR_{in} = -2.5$ dB (bottom). Focus is on CAE denoised output (left, orange trace) and error (right).

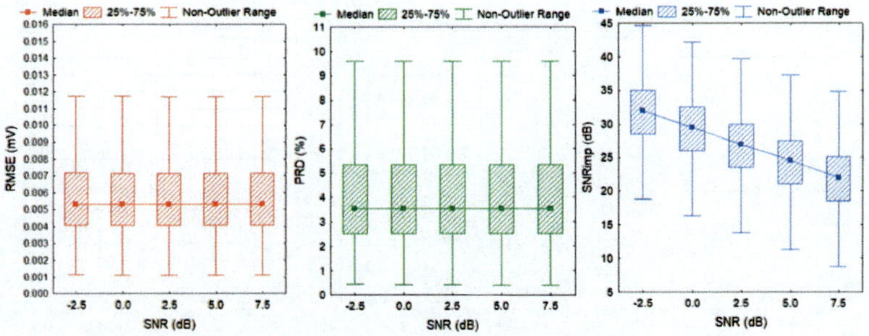

Fig. 6. Test performance for CAE filtering of constant PLI (50 Hz) added with five SNR_{in}.

Statistical test results for denoising of all 12 ECG leads in the case of PLI with constant frequency and amplitude (Fig. 6) show a substantially low and stable error for the overall test range of SNR_{in} (from −2.5 to 7.5 dB), presented as median value (quartile range, min-max range): $RMSE = 5.3$ µV (0.04–0.07, 0.001–0.012 µV), $PRD = 3.5\%$ (2.5–5.4, 0.4–9.7%). The median SNR_{imp} is from 22 to 32 dB (range from 8 to 45 dB).

3.4 Test 2: Constant PLI with Different Frequencies

The example in Fig. 7 illustrates CAE denoising output in the case of constant PLI with two borderline frequencies (48 and 52 Hz). In both cases, PLI sinusoid is completely filtered (left plot) with notably very similar output error (middle plot), estimated in the range $RMSE = 2.9$–3.4 µV, $PRD = 5.5$–6.3%, $SNR_{imp} = 25.4$–26.6 dB, considering the input setting of $SNR_{in} = 0$ dB. The FFT spectrum (right plot) justifies the rejection of the PLI spectral peak at 48 and 52 Hz, respectively, and shows an important CAE benefit to fully reproduce low frequency ECG components (<44 Hz) but also to keep high-frequency ECG components and especially those in the PLI bandwidth (48–52 Hz), which are usually fully suppressed in the stop-band of the conventional notch filters.

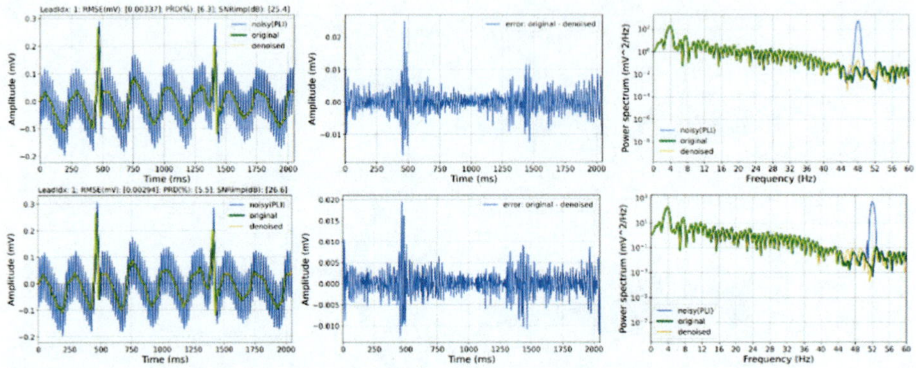

Fig. 7. ECG example (record '00706_hr', lead II, atrial flutter) superimposed by PLI noise (SNR = 0 dB) with two extreme frequencies: 48 Hz (top) and 52 Hz (bottom). PLI frequencies are seen by the peak in the spectrum (right, blue trace). Focus is on CAE denoised output (left, orange trace) and error (middle).

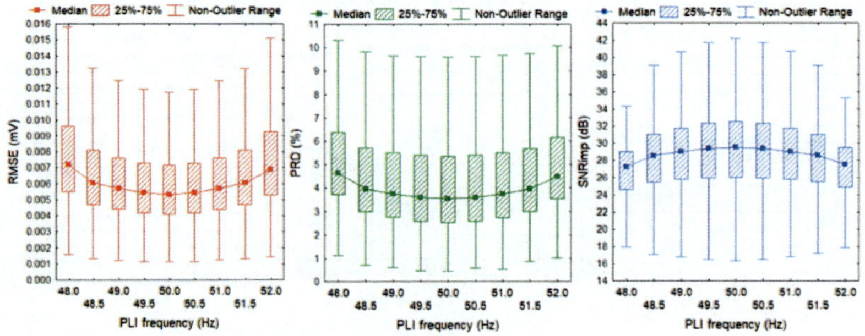

Fig. 8. Test performance for CAE filtering of constant PLI (SNR = 0 dB) with nine frequencies.

Statistical test results for denoising of all 12 ECG leads in the case of constant PLI with frequency settings in the range 48–52 Hz (Fig. 8) show a negligible influence of the PLI frequency on the CAE denoising performance, i.e. the median value variation is $< 2\ \mu V$ ($\Delta RMSE$), $< 1.5\%$ (ΔPRD), ≤ 3 dB (ΔSNR_{imp}).

3.5 Test 3: Non-stationary PLI with Linearly Changing Amplitude

The example in Fig. 9 illustrates CAE denoising output in the case of non-stationary PLI with linearly changing amplitude. PLI is sufficiently filtered (left plot) in both low (50 μV/s) and high slew rate ($-250\ \mu$V/s), presenting almost the same output error (right traces). Statistical test results (Fig. 10) show a consistently small output error (median $RMSE = 5.5\ \mu V$, $PRD = 4\%$, $SNR_{imp} = 29$–57 dB) up to extreme slew rates of $\pm 1000\ \mu$V/s.

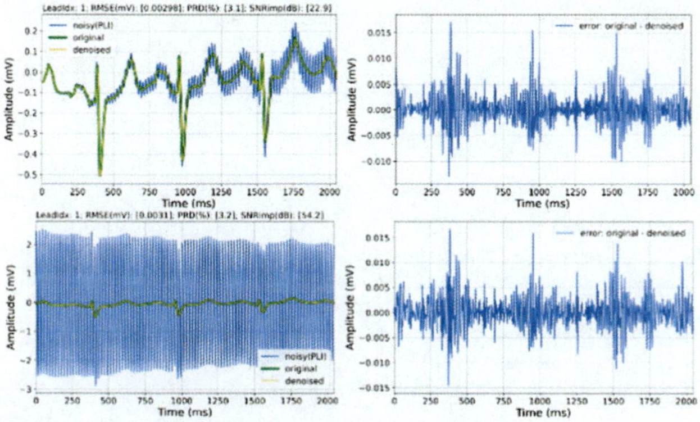

Fig. 9. ECG example (record '00103_hr', lead II, sinus tachycardia) superimposed by PLI noise (50 Hz) with two extreme amplitude slew rates: $\Delta A = 50$ uV/s (top) and $\Delta A = -250$ uV/s (bottom). Focus is on CAE denoised output (left, orange trace) and error (right).

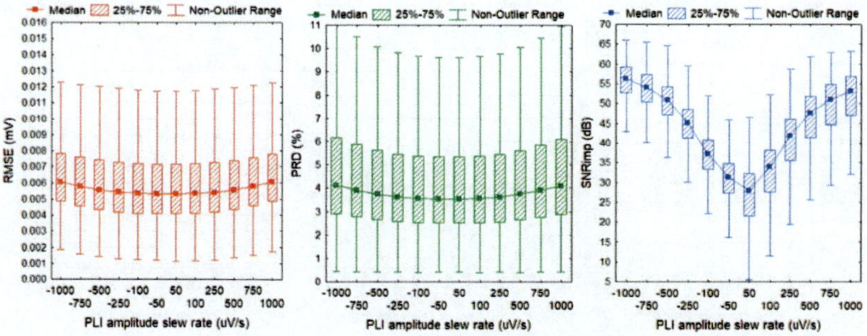

Fig. 10. Test performance for CAE filtering of PLI (50 Hz) with linearly changing amplitude defined by 12 amplitude slew rates in the range from –1000 to 1000 μV/s.

3.6 Test 4: Non-stationary PLI with Linearly Changing Frequency

The example in Fig. 11 illustrates that CAE sufficiently filters non-stationary PLI with high slew rates of frequency variation (±0.2 Hz/s). Both tests produce PLI spectral peaks above and below 50 Hz, which are rejected in the CAE output. Remarkably, the ECG spectral components remain undistorted. Statistical test results (Fig. 12) show a consistently small output error (median $RMSE = 5.5$ μV, $PRD = 4\%$, $SNR_{imp} = 29$ dB), which remains independent of the frequency variation rate.

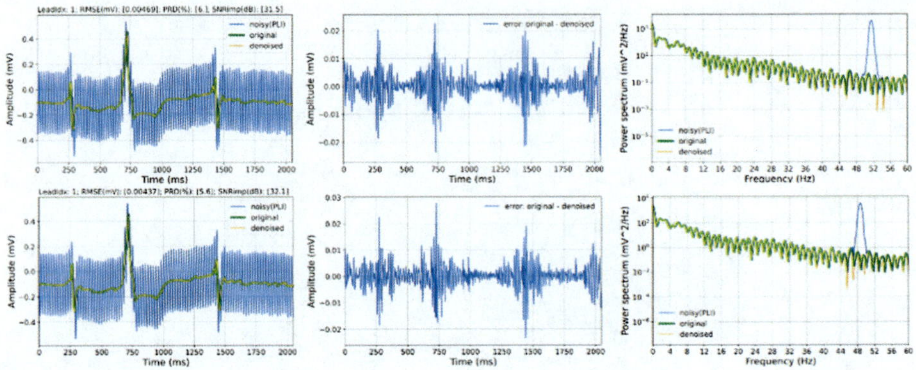

Fig. 11. ECG example (record '00674_hr', lead II, normal and ventricular ectopic beats) superimposed by PLI noise ($SNR = 0$ dB) with two extreme frequency slew rates: $\Delta F = 0.2$ Hz/s (top) and $\Delta F = -0.2$ Hz/s (bottom), seen as spectral peak above and below 50 Hz, respectively.

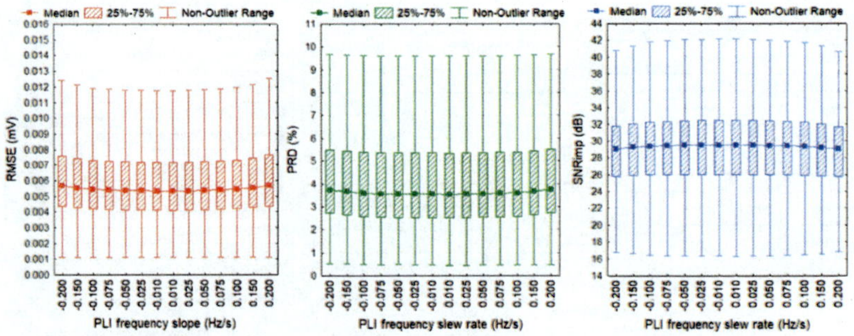

Fig. 12. Test performance for CAE filtering of PLI (48–52 Hz) with linearly changing frequency defined by 14 frequency slew rates in the range from –0.2 to 0.2 Hz/s.

4 Conclusions

Statistical test results for CAE denoising present median $RMSE = 5.3\ \mu$V, $PRD = 3.5\%$, $SNR_{imp} = 22$–32 dB for $SNR_{in} = -2.5$–7.5 dB (Fig. 6), without significant differences between 12 ECG leads (Fig. 4). The results do not substantially change for PLI frequencies 48–52 Hz (Fig. 8), amplitude slew rates up to $\pm 1000\ \mu$V/s (Fig. 10) and frequency slew rates up to ± 0.2 Hz/s (Fig. 12) with median value divergence of $\Delta RMSE < 2\ \mu$V, $\Delta PRD < 1.5\%$, $\Delta SNR_{imp} \leq 3$ dB. Remarkably, CAE preserves ECG spectral components without rejection in the PLI frequency band (Fig. 7, 11) as standard band-stop filters.

The observed performance stability justifies the deep learning strategy for training a CAE with generalizable application for ECG denoising with non-stationary PLI.

Acknowledgements. This work was supported by the Bulgarian National Science Fund, grant КП-06-Н42/3.

References

1. Bradley, W., et al.: Electrocardiogram in Clinical Medicine, 1st edn. John Wiley & Sons Ltd., Hoboken, NJ, US (2020). https://doi.org/10.1002/9781118754511
2. Sörnmo, L., Laguna P.: Chapter 6 – the electrocardiogram—a brief background. Bioelectr. Signal Process. Card. Neurol. Appl. **2005**, 411–452 (2005)
3. Dobrev, D., Neycheva, T.: High-quality biopotential acquisition without a reference electrode: power-line interference reduction by adaptive impedance balancing in a mixed analog–digital design. Med. Biol. Eng. Comput. **60**, 1801–1814 (2022)
4. IEC 60601-2-25:2011: Medical Electrical Equipment - Part 2-25: particular requirements for the basic safety and essential performance of electrocardiographs, 2nd edn., International Electrotechnical Commission, pp. 1–196 (2011)
5. Dobrev, D., Neycheva, T., Mudrov, N.: High-Q comb filter for mains interference suppression. Annu. J. Electron. **3**(1), 47–49 (2009)
6. Tabakov, S., Iliev, I., Krasteva, V.: Online digital filter and QRS detector applicable in low resource ECG monitoring systems. Ann. Biomed. Eng. **36**(11), 1805–1815 (2008)
7. Levkov, C., Mihov, G., Ivanov, R., et al.: Removal of power-line interference from the ECG: a review of the subtraction procedure. Biomed. Eng. OnLine **4**, 50 (2005). https://doi.org/10.1186/1475-925X-4-50
8. Dotsinsky, I., Stoyanov, T., Mihov, G.: Power-line interference removal from high sampled ECG signals using modified version of the subtraction procedure. Int. J. Bioautom. **24**(4), 381–392 (2020)
9. Stoyanov, T., Christov, I., Jekova, I., Krasteva, V.: Online adaptive filter for mains interference suppression in diagnostic electrocardiographs: cases of amplitude and frequency deviation. Annu. J. Electron. **4**, 150–153 (2010)
10. Rahman, M.Z.U., Karthik, G.V.S., Fathima, S.Y., Lay-Ekuakille, A.: An efficient cardiac signal enhancement using time–frequency realization of leaky adaptive noise cancelers for remote health monitoring systems. Measurement **46**(10), 3815–3835 (2013)
11. Faiz, M., Kale, I.: Removal of multiple artifacts from ECG signal using cascaded multistage adaptive noise cancellers. Array **14**, 100133 (2022)
12. Chiang, H., Hsieh, Y., Fu, S., et al.: Noise reduction in ECG signals using fully convolutional denoising autoencoders. IEEE Access **7**, 60806–60813 (2019)
13. Dasan, E., Panneerselvam, I.: A novel dimensionality reduction approach for ECG signal via convolutional denoising autoencoder with LSTM. Biomed. Signal Process. Control **63**, 102225 (2021)
14. Xiong, P., Wang, H., Liu, M., et al.: A stacked contractive denoising auto-encoder for ECG signal denoising. Physiol. Meas. **37**, 2214–2230 (2016)
15. Casas, L., Klimmek, A., Navab, N., Belagiannis, V.: Adversarial signal denoising with encoder-decoder networks. In: 28th European Signal Processing Conference, pp. 1467–1471. IEEE, Amsterdam, Netherlands (2021)
16. Wagner, P., Strodthoff, N., Bousseljot, R.D., et al.: A large publicly available ECG dataset. Sci. Data **7**, 154 (2020)
17. Němcová, A., Smíšek, R., Maršánová, L., Smital, L., Vítek, M.: A comparative analysis of methods for evaluation of ECG signal quality after compression. Biomed. Res. Int. **2018**, 1868519 (2018)

Lipid Order of Membranes Isolated from Erythrocytes of Patients with Coronary Artery Disease: Correlation with Biochemical Parameters

Vesela Yordanova[1](✉), Galya Staneva[1](✉), Plamen Krastev[2], Tania Markovska[1], Ana-Mariya Marinovska[1], Aneliya Kostadinova[1], Rusina Hazarosova[1], and Albena Momchilova[1]

[1] Institute of Biophysics and Biomedical Engineering, Bulgarian Academy of Sciences, Acad. G. Bonchev Str., bl. 21, 1113 Sofia, Bulgaria
vyordanova@biophys.bas.bg, gstaneva@obzor.bio21.bas.bg
[2] Cardiology Clinic, University Hospital "St. Ekaterina", 52A, P. Slaveykov Blvd., 1431 Sofia, Bulgaria

Abstract. Coronary artery disease (CAD) is a medical condition that is characterized by an inability of the arteries to supply the myocardium of the heart with enough blood, therefore affecting its function and leading to serious pathological changes. CAD development is caused by plaque buildup on the inner walls of the coronary arteries. The plasma and red blood cell lipids are involved in plaque formation. In this study, the fatty acid (FA) composition was determined in blood plasma and membranes isolated from erythrocytes of treated patients against CAD and healthy controls. Higher saturated/unsaturated FA ratio was found in the erythrocyte ghost membranes of patients compared to controls. Moreover, the triglyceride (TG) levels of CAD patients were also higher compared to the control group. No difference in total cholesterol and LDL-cholesterol levels was observed between patients and control group. Unexpectedly, despite the observed higher level of saturated FAs in the erythrocyte membranes, lower membrane lipid order was found for CAD group compared to the control one. Our results suggest that the serum TGs govern more significantly the lipid order of CAD ghost membranes than the saturated/unsaturated FA ratio. This finding is in accordance with the literature data stating an inverse relationship between TG levels and the lipid anisotropy in model and erythrocyte membranes. The present study implies that CAD treatment is able to decrease erythrocyte membrane lipid order, whereas higher levels of TGs in blood serum and saturated FAs in erythrocyte membranes still represent a critical potential risk for CAD patients compared to healthy individuals.

Keywords: Blood plasma · Erythrocyte ghosts · Fatty acids · Lipid profile · Membrane lipid order

V. Yordanova and G. Staneva—Both authors have equal contribution.

S. Sotirov et al. (Eds.): BioInfoMed 2022, LNNS 658, pp. 134–146, 2023.
https://doi.org/10.1007/978-3-031-31069-0_14

1 Introduction

Coronary artery disease (CAD), also known as ischemic heart disease, is the most common of the cardiovascular diseases (CVD) and one of the major causes of death worldwide [1]. Although it mainly affects the elderly, unfortunately its prevalence in young people increases. It is assumed to have complex etiology involving a complicated interaction between genetic predisposition and environmental factors [2]. CAD pathophysiology involves the reduction of blood flow to the heart muscle as a result of plaque buildup (atherosclerosis) in the coronary arteries leading to ischemia, unstable angina and ultimately to myocardial infarction (MI). Evidence has shown that red blood cells (RBCs) and plasma lipids play a key role in atherosclerotic plaque formation and indicated cholesterol as a component of the plaques revealing its involvement in the predominant pathogenic process that causes CAD [2–4].

Several studies have demonstrated an association of high blood cholesterol levels [5–7], elevated triglycerides (TGs) [8, 9] and fatty acid (FA) content [10–13] with cardiovascular events. High cholesterol as a major risk factor for CAD is related mainly to low-density lipoprotein (LDL)-cholesterol (LDL-chol), while high-density lipoprotein (HDL)-cholesterol (HDL-chol) is inversely correlated to mortality from CAD [14, 15]. In addition, TGs are strongly associated with incident CAD events in patients with low LDL-chol levels treated with statins [16].

FAs are organic compounds, which are components of lipids in cell membranes and plasma lipoproteins. Instead of simply acting as inert structural elements, FAs can be converted to bioactive molecules once released from the supramolecular complexes, having a significant impact on health and disease [17]. They play essential roles in metabolic homeostasis through their functional properties [18], including lipid metabolism, immune system modulation, endothelial function, platelet aggregation, and redox status. Changes in these functions can lead to several diseases associated with lipid dysregulation, including CAD [19]. FA chains may be saturated (SFAs) without double bonds, or unsaturated (UFAs), containing one (monounsaturated) or more (polyunsaturated) double bonds, the latter being of greater physiological importance due to their medicinal properties [20]. It is generally known that SFAs contribute to risk of CAD, while polyunsaturated FAs (PUFAs) significantly reduce the risk [2]. Erythrocyte FAs are a better long-term marker in comparison to plasma FA composition because of their slower turnover rate than plasma lipids [21]. Furthermore, alterations in FA content of circulating blood may have a profound effect on the lipid composition of erythrocyte membranes, which are in constant contact with plasma lipoproteins [22]. It is important to emphasize that oxygen-binding and oxygen-transport properties of hemoglobin depend on the morphological, physico-chemical and functional characteristics of erythrocyte membranes [23, 24]. Their lipid bilayer is one of the most important barriers of oxygen to hemoglobin, which depends on the individual phospholipids, the degree of oxidation, and the ratio of saturated and unsaturated fatty acids and is involved in the regulation of various functions through receptor systems [25]. Therefore, changes in the RBC membrane lipid composition such as length and degree of unsaturation of phospholipid fatty acids [26], which result in modifications in membrane fluidity, may also lead to increased risk of CAD and other clinical problems [27, 28].

In the present study, we aimed to compare the fatty acid profile in blood plasma and erythrocyte membrane, to calculate the ratio of saturated to unsaturated FAs in the blood plasma and erythrocyte membranes and lipid order of RBC membranes in patients with angiographically documented CAD and healthy volunteers (controls). These biochemical and biophysical parameters will be correlate with the serum levels of total and LDL cholesterol as well triglyceride levels to assess their possible impact on CAD in order to find novel prediction and treatment options.

2 Materials and Methods

2.1 Patients

Peripheral blood was collected into tubes containing acid citrate dextrose (ACD) solution, the latter being used to protect against coagulation. Erythrocytes isolated from peripheral blood of nonsmoking patients (5 men) diagnosed with CAD were used in this study.

All five patients have been followed a treatment to stabilize their CAD symptoms. The patients have been administered with Brilique/Clopidogrel/Clexane/Arixtra/Plavix, in combination with acetylsalicylic acid (antiplatelet agent), Prestarium/Valtensin Plus/Norvasc (against high blood pressure), Cardiket (vasodilator), Betaloc (beta-blocker), Rosvera/Crestor (lipid-lowering medications (statins)), Preductal (prevention of angina attacks). The peripheral blood from CAD patients were collected when patients were considered as CAD stable. This investigation was performed in cooperation with the St. Ekaterina University Hospital. Blood of healthy, nonsmoking volunteers ($n = 7$; 6 men and 1 woman) was used as a control.

Table 1. Characteristics of the control group and patients with CAD.

Parameter	Controls	CAD patients
Age (years)	45 ± 12	64 ± 16
BP_{sys} (mm/Hg)	119 ± 4	132 ± 8
BP_{dia} (mm/Hg)	72 ± 8	79 ± 15
Heart rate (beats/min)	75 ± 7	77 ± 9
Bodymass (kg)	76 ± 13	89 ± 11
BMI (kg/m^2)	25.4 ± 4.8	29.8 ± 5.6
Glucose (mmol/L)	5.6 ± 0.4	6.4 ± 0.9
hsCRP (mg/L)	1.9 ± 1.52	2.6 ± 0.5

Note: Values are means ± SD. BP_{sys}, systolic blood pressure; BP_{dia}, diastolic blood pressure; BMI, body mass index; hsCRP, high-sensitivity C-reactive protein

The characteristic features of patients and controls are shown in Table 1. Inclusion criteria were nonsmoking, no following particular diet patients and control volunteers with body mass index (BMI) between 25–30 kg/m^2. For CAD patients, myocardial infarction occurred more than 6 months prior to the time of screening. The serum levels of total, LDL and HDL cholesterol as well TG levels were measured for every individual from the patient and control group. The serum and plasma TGs, and cholesterol levels, differ not more than 3% as plasma levels are 3% lower compared to serum levels [29]. This difference in cholesterol or TG concentrations will usually not affect the decision as to whether or not a patient should be placed on, or continued on, a lipid-lowering drug regimen [29].

2.2 Blood Plasma and Erythrocyte Ghost Membranes Isolation

Human erythrocytes were isolated from fresh citrated blood [30]. All steps were performed at 4 °C. The blood was centrifuged for 10 min at 100 g. The resulting plasma was removed and used for FA determination. The remaining red blood cells were collected and washed several times with 5 volumes of phosphate buffered saline (150 mM NaCl, 5 mM sodium phosphate, pH 8.0) and hemolyzed in 5 mM sodium phosphate, pH 8.0 (40 volumes). The ghosts were collected by centrifugation for 20 min at 22 000 g (Beckman J2-HS), washed (x 4) with 5 mM phosphate buffer until "white ghosts" were obtained. The total protein concentration of the membranes was measured using the method of Bradford [31].

2.3 Determination of Lipid Packing in Erythrocyte Ghost Membranes

The polarity sensitive fluorescent probe Laurdan was used to assess membrane lipid packing by the change of its emission spectra with the polarity of the environment, which generally represents the hydration level of the bilayer [32]. Laurdan is incorporated into the membrane with the fluorescent naphthalene moiety at the level of the glycerol backbone and the lauric acid tail anchored in the hydrophobic core. Upon excitation, the dipole moment of Laurdan increases and could lead to a reorientation of water molecules in the vicinity of the probe. It has an emission maximum of 440 nm and 490 nm in liquid-ordered and liquid-disordered phase membranes, respectively [33, 34]. The spectral shift in emission between different states of membrane order is used to calculate the generalized polarization (GP), which is a relative quantitative measure for lipid packing [35, 36]:

$$GP = (I_{440} - I_{490})/(I_{440} + I_{490}),$$

where I_{440} and I_{490} are the emission intensities at 440 and 490 nm. Theoretically the GP values are from −1.0 (least ordered) to +1.0 (most ordered) [32].

Fluorescence measurements were performed with a FP-8300 spectrofluorometer (Jasco) equipped with a xenon arc lamp as a light source, using excitation at 355 nm. Samples were placed in quartz cuvette. The temperature in the cuvette holder was maintained at 37 °C using a water-circulating bath (Julabo). 200 μl of the erythrocyte ghost samples were mixed with 800 μl PBS. Laurdan (14 μl), dissolved in DMSO in stock concentration 1 mg/ml, was added to the erythrocyte ghost suspension and incubated at 37 °C for 1 h. The emission spectra were recorded from 400 to 600 nm. The bandwidths were 5 nm. Samples were run in triplicate and averaged. For analysis of the steady-state spectra was used OriginPro 9.0.

2.4 Fatty Acid Analysis of Erythrocyte Ghost Membrane and Blood Plasma

To determine the fatty acid composition of the total lipid extracts, FAs were hydrolyzed and methylated. Lipid extraction was performed with chloroform/methanol solvent, according to the method of Bligh and Dyer [37]. The obtained organic phase was concentrated and analyzed by thin layer chromatography (TLC). The lipids and FAs were separated on silica gel G 60 plates in a solvent system containing hexane/diethyl ether/acetic acid (90/30/1) for neutral lipids. The location of the separated fractions was visualized after revelation by iodine sprayed on the plates. The FAs spot was scraped and FAs were extracted with solvent. Methylation was performed using a 3 ml of solution methanol – acetyl chloride 50:4 (v/v) for 1 h at 75 °C [38].

The resulting FA methyl esters (FAMEs), dissolved in hexane, were separated by gas chromatography (Chromatec Crystal) equipped with a MS detector and a capillary BPX5 column (length of 60 m, inner diameter of 0.25 mm, film thickness of 0.25 μm). The instrument operated also with a split/splitless injector and a flame-ionization detector; the temperature of the detector and evaporator were 250 and 240 °C, respectively. Quantification was referenced to an internal standard of heptadecanoic methyl ester. The Supelco FAME mix 37 certified calibrator was used to identify the response factors of the various FAs.

2.5 Statistical Analysis

The statistical analysis was performed using OriginPro 9. The data was drawn from a normally distributed population. The mean values of each measured parameter were compared using Mann-Whitney Test. Statistically significant differences based on Tukey test linked to the control group: * - $p < 0.05$, *** - $p < 0.001$.

3 Results and Discussion

Patients with established CAD have a significantly higher risk of new cardiovascular events, including recurrent MI or death than healthy individuals [39]. It has been reported that plasma lipid profiling might have diagnostic and prognostic potential for the identification of patients at risk. Thus, the first step in our study was to compare the serum lipid spectrum by the content of TGs and cholesterol in LDLs and HDLs in patients with diagnosed and stabilized CAD condition and healthy controls. The obtained results from biochemical analysis of blood samples (Fig. 1) show that lipid content is in the range of normal values, i.e. the average levels are below the high borderline. Also, there are no significant differences in cholesterol levels between the two investigated groups. Nevertheless, TG levels are elevated in CAD patients compared to healthy volunteers.

Although an increase in LDL-chol is a risk factor for CAD patients, it cannot be considered as a disease marker. On one hand, high LDL-chol levels could be found in people without CAD for other reasons such as a positive family history of CVD. On the other hand, the use of statins (inhibitors of 3-hydroxy-3-methylglutaryl coenzyme A reductase, the enzyme that catalyses the rate-limiting step in cholesterol synthesis) in therapy for CVD treatment and prevention has induced impressive reductions in LDL-chol levels. However, only moderate effect on total mortality was achieved [40].

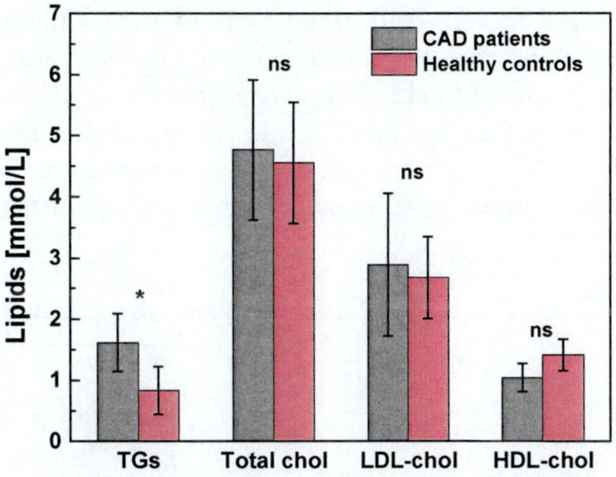

Fig. 1. Serum lipid concentrations mmol/L in patients with CAD (n = 5) and healthy controls (n = 7). Results are presented as means ± standard deviation (SD). TG levels are significantly higher in CAD patients compared with the controls, $p < 0.05$. There are no significant differences (ns) in cholesterol levels between the two investigated groups. Reference ranges are: TGs, 0.5–1.7; Total chol, 2.5–5.2; LDL-chol, 1.0–4.1; HDL-chol, 1.04–2.4.

The Prospective Cardiovascular Münster (PROCAM) study showed that fasting levels of TGs above 1.5 mmol/L were an independent risk factor for CAD events [41–44]. After all, the evidence for that remains to some extent unconvincing. In this regard several possible explanations have been proposed. The daily consumption of dietary fat is reflected in highly variable plasma TG levels [45]. Also, patients with high TG levels frequently exhibit additional risk factors that may affect their susceptibility to atherosclerosis. Moreover, increased plasma TG concentrations are strongly associated with low HDL-chol levels [46]. Our results confirm this statement, as the average HDL-chol content in patients is at the lower reference borderline (Fig. 1).

Basically, we did not detect significant changes in serum cholesterol levels in patients in comparison to controls. However, as far as LDL-chol is considered among the major risk factors for CAD, ISH (International Society of Hypertension) guidelines state that the levels should be less than 3.0 mmol/L. In case of having diabetes or stage 3–4 chronic kidney disease, or a history of heart attack, a heart stent, heart bypass surgery, or other heart or vascular condition, desirable levels are under 1.4 mmol/L. In people with the highest risk of heart attacks, with multiple previous cardiovascular events and/or revascularization procedures or generalized atherosclerosis, the recommended LDL-chol is below 1.0 mmol/L [47]. All mentioned above means that LDL-chol levels in CAD patients should not just be retained within the reference range but reduced according to the risk profile. Therefore, our results still could be indicative for increased risk of cardiovascular events in patients with stabilized CAD symptoms because of high LDL-chol levels ≈3.0 mmol/L (Fig. 1).

The quality and quantity of dietary fat intake reflect circulating lipid profiles and markers associated with CVD risk [48]. FAs are important lipid components and are assumed as essential nutrients for the human body. They are biologically active molecules with a key role in membrane architecture and are involved in various cellular processes such as cell signaling, cell-protein, cell-cell interactions, transcriptional and translational modulation, and energy storage [49]. As a second step in the present study, we determined and compared the FA levels in blood plasma and erythrocyte membranes in patients with CAD and controls. We used the obtained data to calculate ratios between saturated vs. unsaturated FAs (Fig. 2). The ratios indicate that there is a difference between FA content in plasma and RBC membranes in accordance with already published results [50–52].

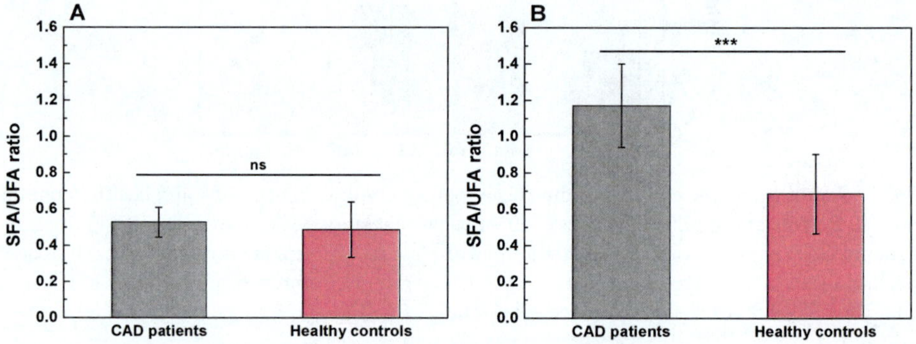

Fig. 2. SFA/UFA ratio in plasma (A) and erythrocyte ghosts (B) in patients with CAD ($n = 5$) and healthy controls ($n = 7$). Results are presented as means of three replicates for each subject ± standard deviation (SD). There are no significant differences (ns) in plasma SFA/UFA ratio between CAD patients and healthy controls, $p < 0.05$. The FA ratio in erythrocyte membranes is significantly higher in CAD patients in comparison to the controls, $p < 0.001$.

Studies have shown that SFAs increase serum levels of LDL and as well as total cholesterol by inhibiting the activity of LDL receptors [53]. Moreover, in serum phospholipids the proportions of omega-3 PUFAs, known to have a cardioprotective effect, are significantly lower in patients with CAD [54–56]. Nevertheless, in this study, no significant differences were observed in plasma SFA/UFA ratio between patients in the CAD group and healthy controls (Fig. 2A). However, in erythrocyte ghosts the ratio has almost two-fold higher value in CAD patients compared to control group (Fig. 2B). This implies that the total SFAs are elevated and/or UFAs are reduced in CAD cases.

Changes in the FA content of membranes could affect the properties of the lipid bilayer matrix, including its fluidity and consequently the lateral mobility and clustering of membrane proteins, their vertical orientation or conformation [57–60] and the affinity of receptors, thus causing cell senescence, autophagy, or apoptosis [61, 62]. The measurement of membrane fluidity reflects the order and dynamics of the phospholipid acyl chains in the bilayer implicated to particular membrane functions [63]. It is considered that changes in the structural-functional properties of erythrocyte membranes play a crucial role in the development of many cardiovascular diseases [64, 65]. For this reason, we probed the degree of hydration/lipid order at the glycerol level of lipids in erythrocyte ghosts by using Laurdan fluorescence spectroscopy. A characteristic feature of cholesterol-rich RBC membranes [66] is the high order degree. As shown in Fig. 3, Laurdan GP values for the control group are about 0.306, while for the patient group are lower −0.261. Hence, a significant decrease in lipid order (15%) of erythrocyte membranes isolated from treated with drugs CAD patients in comparison to healthy controls was observed.

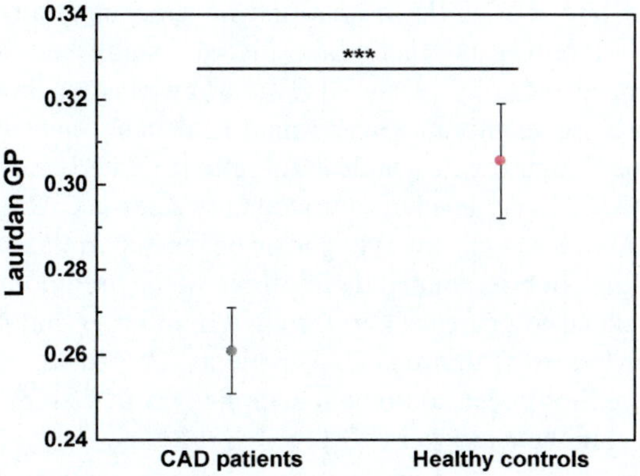

Fig. 3. Laurdan GP of erythrocyte membranes in treated patients with CAD (n = 5) and healthy controls (n = 7) calculated from spectra measured at 37 °C. GP values correspond to the lipid bilayer order. Results are presented as means of three replicates for each subject ± standard deviation (SD). The data show statistically significant Laurdan GP decrease of RBC membranes in CAD patients in comparison to controls, $p < 0.001$. This means a lower degree of lipid packing in the glycerol level of erythrocyte membranes in patients, i.e. more fluid bilayer.

The obtained results for treated CAD patients in this study contradict to those reported in the literature [67]. These authors observed significant differences in membrane rigidity of red blood cells between the healthy controls and the CAD patients: fluorescence anisotropy for TMA-DPH (trimethylamine-diphenylhexatriene) was 0.303 ± 0.014 vs. 0.351 ± 0.015 ($p < 0.001$) and fluorescence anisotropy for DPH (diphenylhexatriene) was 0.276 ± 0.004 vs. 0.297 ± 0.018 ($p < 0.001$). Our fluorescent probe Laurdan sense a membrane location between those of TMA-DPH (polar part of the outer leaflet) and DPH (hydrophobic lipid core). Such a difference does not appear to be the main reason for the observed discrepancy. We suggest that it is due to the significant increase in the concentration of total cholesterol in CAD patients (2.36 ± 0.20 mg_{chol}/mL_{PC} for control group versus 2.91 ± 0.21 mg_{chol}/mL_{PC} for CAD patients) in the experiments carried out by Pytel et al. [67]. An increased level of total cholesterol found in CAD patients may be responsible for the accumulation of cholesterol in the erythrocyte membranes. Moreover, membrane fluidity strongly depends on the composition of lipids. Higher cholesterol and saturated lipid levels may be responsible for the decrease in membrane fluidity. In our study neither total cholesterol nor LDL-cholesterol levels are significantly different for treated CAD patients compared to controls. However, higher saturated/unsaturated fatty acid ratio was found significant in erythrocyte ghost membranes from patients compared to the control group. Moreover, the TG level of CAD patients was also higher compared to the controls. Unexpectedly, although a higher level of saturated fatty acids was observed, lower erythrocyte membrane lipid order was found for treated CAD patients compared to the healthy individuals. The present results suggest that the serum TGs are able to control more significantly the lipid order of erythrocyte ghost membranes from patients than the saturated/unsaturated fatty acid ratio of phospholipids in accordance with the literature data reporting an inverse relationship between TG levels and the lipid anisotropy in model and erythrocyte membranes [68]. TGs alone as well as free FAs in blood plasma can be incorporated into RBC-TGs *in vivo* and *in vitro*, and can selectively modify lipid order [68, 69]. Moreover, in other studies, it was established that membrane lipid order can be influenced more by plasma TGs than by plasma cholesterol [70]. DPH anisotropy of erythrocyte membranes was found to depend inversely on plasma TGs [71, 72]. Lipids, in particular the PL/Cholesterol ratio is related inversely to TGs than to plasma cholesterol [73]. Higher level of saturated fatty acids in CAD patients is expected to exhibit higher lipid order in erythrocyte ghosts but such correlation was not observed in treated CAD patients. In brief, our study implies that the presence of higher TG levels in blood serum is related to decreased erythrocyte lipid order and may still serve as a potential risk for the treated with drugs CAD patients. Of course, other reasons for the decreased erythrocyte lipid order could be the applied statin therapy [74, 75] as well as vasodilators [76, 77] and beta blockers [78].

4 Conclusions

The findings in this research demonstrate no significant changes in cholesterol profile in treated CAD patients. However, the saturated/unsaturated fatty acid ratio in erythrocytes and triglycerides in blood serum were found to be increased in CAD patients compared to controls. In parallel, a significant difference was observed between CAD and the control group in the lipid organization of erythrocyte membrane. Lower membrane lipid order correlates inversely with higher TG level as well as statin, vasodilator and beta-blocker treatment according to literature data. We can conclude that the biochemical serum analysis may not always accurately reflect the status of chronically ill patients or elderly individuals, whereas the biochemical and biophysical structural and functional changes of RBC membrane at lipid level, i.e. FA composition, lipid order and others deserve more attention, since such information underlies a clearer understanding of the metabolic abnormalities and cellular disturbances in patients with CAD, both untreated and treated ones.

Acknowledgments. This study was fnancially supported from the National Science Fund of Bulgaria by Grant КП-06-H58/6-19.11.2021. The authors thank the Bulgarian Ministry of Education and Science for support: Scientifc Infrastructure on Cell Technologies in Biomedicine (SICTB) DO1-154/28/08/2018 and D01-392/2020 "National Center for Biomedical Photonics", part of Bulgarian National Roadmap for Scientifc Infrastructures 2020-2027. The authors also acknowledged COST Action 17 121: Correlated Multimodal Imaging in Life Sciences-COMULIS by grant KP-06-COST/10/2020.

References

1. Namazi, G., et al.: Investigation of membrane fatty acid profiles in erythrocytes of patients with stable coronary artery disease. J. Clin. Lipidol. **10**(4), 930–936 (2016)

2. Xuan, C., et al.: Serum fatty acids profile and association with early-onset coronary artery disease. Therapeutic Adv. Chronic Disease **12**, 20406223211033104 (2021)
3. Anitschkow, N., Chalatow, S.: Über experimentelle Cholesterinsteatose und ihre Bedeutung für die Entstehung einiger pathologischer Prozesse. Zentralbl. Allg. Pathol. **24**(1913), 1–9
4. Litwin, N., et al.: Assessment of red blood cell membrane fatty acid composition in relation to dietary intake in patients undergoing cardiac catheterization. J. Acad. Nutr. Diet. **114**(9, Supplement), A63 (2014)
5. Kannel, W.B., et al.: Factors of risk in the development of coronary heart disease—six-year follow-up experience. Ann. Intern. Med. **55**(1), 33–50 (1961)
6. Keys, A., et al.: The diet and 15-year death rate in the seven countries study. Am. J. Epidemiol. **124**(6), 903–915 (1986)
7. Stamler, J., Wentworth, D., Neaton, J.D.: Is relationship between serum cholesterol and risk of premature death from coronary heart disease continuous and graded?: Findings in 356 222 primary screenees of the multiple risk factor intervention trial (MRFIT). JAMA **256**(20), 2823–2828 (1986)
8. Hokanson, J.E., Austin, M.A.D.: Plasma triglyceride level is a risk factor for cardiovascular disease independent of high-density lipoprotein cholesterol level: a metaanalysis of population-based prospective studies. J. Cardiovasc. Risk **3**(2), 213–219 (1996)
9. Nordestgaard, B.G., et al.: Nonfasting triglycerides and risk of myocardial infarction, ischemic heart disease, and death in men and women. JAMA **298**(3), 299–308 (2007)
10. Baum, S.J., et al.: Fatty acids in cardiovascular health and disease: a comprehensive update. J. Clin. Lipidol. **6**(3), 216–234 (2012)
11. Ghosh, A., et al.: Role of free fatty acids in endothelial dysfunction. J. Biomed. Sci. **24**(1), 50 (2017)
12. Lecerf, J.-M.: Fatty acids and cardiovascular disease. Nutr. Rev. **67**(5), 273–283 (2009)
13. Paganelli, F., et al.: Altered erythrocyte n-3 fatty acids in Mediterranean patients with coronary artery disease. Int. J. Cardiol. **78**(1), 27–32 (2001)
14. Gordon, T., et al.: High density lipoprotein as a protective factor against coronary heart disease: the Framingham study. Am. J. Med. **62**(5), 707–714 (1977)
15. Kannel, W.B., Castelli, W.P., Gordon, T.: Cholesterol in the prediction of atherosclerotic disease. Ann. Intern. Med. **90**(1), 85–91 (1979)
16. Miller, M., et al.: Impact of triglyceride levels beyond low-density lipoprotein cholesterol after acute coronary syndrome in the PROVE IT-TIMI 22 trial. J. Am. Coll. Cardiol. **51**(7), 724–730 (2008)
17. Maulucci, G., et al.: Fatty acid-related modulations of membrane fluidity in cells: detection and implications. Free Radic. Res. **50**(sup1), S40–S50 (2016)
18. Fessler, M.B., Rudel, L.L., Brown, J.M.: Toll-like receptor signaling links dietary fatty acids to the metabolic syndrome. Curr. Opin. Lipidol. **20**(5), 379–385 (2009)
19. Shramko, V.S., et al.: The short overview on the relevance of fatty acids for human cardiovascular disorders. Biomolecules **10**(8) (2020)
20. Zárate, R., el Jaber-Vazdekis, N., Tejera, N., Pérez, J.A., Rodríguez, C.: Significance of long chain polyunsaturated fatty acids in human health. Clin. Transl. Med. **6**(1), 1–19 (2017). https://doi.org/10.1186/s40169-017-0153-6
21. Risé, P., et al.: Fatty acid composition of plasma, blood cells and whole blood: relevance for the assessment of the fatty acid status in humans. Prostaglandins Leukot. Essent. Fatty Acids **76**(6), 363–369 (2007)
22. Abu-Salah, K.M., Alothman, A.A., Lei, K.Y.: Lipid composition and fluidity of the erythrocyte membrane in copper-deficient rats. Br. J. Nutr. **68**(2), 435–443 (1992)
23. Jewell, S.A., Petrov, P.G., Winlove, C.P.: The effect of oxidative stress on the membrane dipole potential of human red blood cells. Biochim. et Biophys. Acta (BBA) – Biomembr. **1828**(4), 1250–1258 (2013)

24. Novitskiy, V.V., et al.: Molecular disorders of erythrocyte membrane due to pathology of different genesis are typical reaction of a body: contours of the problem. Bull. Sib. Med. **2**, 62–69 (2006)
25. Mohandas, N., Gallagher, P.G.: Red cell membrane: past, present, and future. Blood **112**(10), 3939–3948 (2008)
26. Schwarz, S.M., et al.: Lipid composition and membrane fluidity in the small intestine of the developing rabbit. Gastroenterology **86**(6), 1544–1551 (1984)
27. Dolegowska, B., et al.: Effect of hemodialysis on the content of fatty acids in monolayers of erythrocyte membranes in patients with chronic renal failure. Ren. Fail. **29**(4), 447–452 (2007)
28. Pilz, S., März, W.: Free fatty acids as a cardiovascular risk factor. Clin. Chem. Lab. Med. **46**(4), 429–434 (2008)
29. Cholesterol and triglyceride concentrations in serum/plasma pairs. Clin. Chem. **23**(1), 60–63 (1977)
30. Steck, T.L., Kant, J.A.: Preparation of Impermeable Ghosts and Inside-Out Vesicles from Human Erythrocyte Membranes, in Methods in Enzymology, pp. 172–180. Academic Press (1974)
31. Bradford, M.M.: A rapid and sensitive method for the quantitation of microgram quantities of protein utilizing the principle of protein-dye binding. Anal. Biochem. **72**(1), 248–254 (1976)
32. Parasassi, T., et al.: Laurdan and Prodan as polarity-sensitive fluorescent membrane probes. J. Fluoresc. **8**(4), 365–373 (1998)
33. Bagatolli, L.A., et al.: Giant Vesicles, Laurdan, and Two-Photon Fluorescence Microscopy: Evidence of Lipid Lateral Separation in Bilayers, in Methods in Enzymology, pp. 481–500. Academic Press (2003)
34. Dinic, J., et al.: Laurdan and di-4-ANEPPDHQ do not respond to membrane-inserted peptides and are good probes for lipid packing. Biochim. et Biophys. Acta (BBA) – Biomembr. **1808**(1), 298–306 (2011)
35. Jin, L., et al.: Cholesterol-enriched lipid domains can be visualized by di-4-ANEPPDHQ with linear and nonlinear optics. Biophys. J . **89**(1), L04–L06 (2005)
36. Parasassi, T., et al.: Quantitation of lipid phases in phospholipid vesicles by the generalized polarization of Laurdan fluorescence. Biophys. J . **60**(1), 179–189 (1991)
37. Bligh, E.G., Dyer, W.J.: A rapid method of total lipid extraction and purification. Can. J. Biochem. Physiol. **37**(8), 911–917 (1959)
38. Liu, K.-S.: Preparation of fatty acid methyl esters for gas-chromatographic analysis of lipids in biological materials. J. Am. Oil Chem. Soc. **71**(11), 1179–1187 (1994)
39. Jortveit, J., et al.: Unsatisfactory risk factor control and high rate of new cardiovascular events in patients with myocardial infarction and prior coronary artery disease. BMC Cardiovasc. Disord. **19**(1), 71 (2019)
40. Harchaoui, E.L.K., et al.: Triglycerides and cardiovascular risk. Curr. Cardiol. Rev. **5**(3), 216–222 (2009)
41. Assmann, G., et al.: The emergence of triglycerides as a significant independent risk factor in coronary artery disease. Eur. Heart J. **19**(Suppl M), M8-14 (1998)
42. Brunzell, J.D., et al.: Lipoprotein management in patients with cardiometabolic risk: consensus statement from the American diabetes association and the American college of cardiology foundation. Diabetes Care **31**(4), 811–822 (2008)
43. Brunzell, J.D., et al.: Lipoprotein management in patients with cardiometabolic risk: consensus conference report from the American diabetes association and the American college of cardiology foundation. J. Am. Coll. Cardiol. **51**(15), 1512–1524 (2008)
44. Maki, K.C., Galant, R., Davidson, M.H.: Non–high-density lipoprotein cholesterol: the forgotten therapeutic target. Am. J. Cardiol. **96**(9, Supplement 1), 59–64 (2005)

45. Mora, S., et al.: Fasting compared with nonfasting lipids and apolipoproteins for predicting incident cardiovascular events. Circulation **118**(10), 993–1001 (2008)
46. Ginsberg, H.N., et al.: Evolution of the lipid trial protocol of the action to control cardiovascular risk in diabetes (ACCORD) trial. Am. J. Cardiol. **99**(12, Supplement), S56-S67 (2007)
47. Filipiak, K.J.: LDL-cholesterol targets as Achilles' heel of 2020 ISH guidelines. Int. J. Cardiol. Hypertens. **8**, 100078 (2021)
48. Kim, O.Y., Lee, S.M., An, W.S.: Impact of blood or erythrocyte membrane fatty acids for disease risk prediction: focusing on cardiovascular disease and chronic kidney disease. Nutrients **10**(10), 1454 (2018)
49. Cascio, G., Schiera, G., Di Liegro, I.: Dietary fatty acids in metabolic syndrome, diabetes and cardiovascular diseases. Curr. Diabetes Rev. **8**(1), 2–17 (2012)
50. Duran, M., Wanders, R.J.A.: Plasmalogens and polyunsaturated fatty acids. In: Blau, N., Duran, M., Gibson, K. (eds.) Laboratory Guide to the Methods in Biochemical Genetics, pp. 207–220. Springer, Berlin, Heidelberg (2008). https://doi.org/10.1007/978-3-540-76698-8_11
51. Firl, N., et al.: Determination of the fatty acid profile of neutral lipids, free fatty acids and phospholipids in human plasma. Clin. Chem. Lab. Med. (CCLM) **51**(4), 799–810 (2013)
52. Zybko, W.C., Cluette-Brown, J.E., Laposata, M.: Improved sensitivity and reduced sample size in serum fatty acid ethyl ester analysis. Clin. Chem. **47**(6), 1120–1121 (2001)
53. Dietschy, J.M.: Dietary fatty acids and the regulation of plasma low density lipoprotein cholesterol concentrations. J. Nutr. **128**(2), 444S-448S (1998)
54. Kim, O.Y., et al.: Association of fatty acid composition in serum phospholipids with metabolic syndrome and arterial stiffness. Nutr. Metab. Cardiovasc. Dis. **23**(4), 366–374 (2013)
55. Leeson, C.P.M., et al.: Relationship between circulating n-3 fatty acid concentrations and endothelial function in early adulthood. Eur. Heart J. **23**(3), 216–222 (2002)
56. Mozaffarian, D., Wu, J.H.Y.: Omega-3 fatty acids and cardiovascular disease: effects on risk factors, molecular pathways, and clinical events. J. Am. Coll. Cardiol. **58**(20), 2047–2067 (2011)
57. Borochov, H., et al.: Modulation of erythrocyte membrane proteins by membrane cholesterol and lipid fluidity. Biochemistry **18**(2), 251–255 (1979)
58. Brûlet, P., McConnell, H.M.: Lateral hapten mobility and immunochemistry of model membranes. Proc. Natl. Acad. Sci. **73**(9), 2977–2981 (1976)
59. Gorbunov, N.V.: Effect of structural modification of membrane proteins on lipid-protein interactions in the human erythrocyte membranes. Bull. Exp. Biol. Med. **116**(5), 1364–1367 (1993)
60. Palsdottir, H., Hunte, C.: Lipids in membrane protein structures. Biochim. et Biophys. Acta (BBA) – Biomembr. **1666**(1), 2–18 (2004)
61. Janssen, C.I.F., Kiliaan, A.J.: Long-chain polyunsaturated fatty acids (LCPUFA) from genesis to senescence: the influence of LCPUFA on neural development, aging, and neurodegeneration. Prog. Lipid Res. **53**, 1–17 (2014)
62. Rambold, A.S., Cohen, S., Lippincott-Schwartz, J.: Fatty acid trafficking in starved cells: regulation by lipid droplet lipolysis, autophagy, and mitochondrial fusion dynamics. Dev. Cell **32**(6), 678–692 (2015)
63. Arsov, Z., Schara, M., Štrancar, J.: Quantifying the lateral lipid domain properties in erythrocyte ghost membranes using EPR-spectra decomposition. J. Magn. Reson. **157**(1), 52–60 (2002)
64. Vasiljeva, E.M.: Biochemical peculiarity of the blood red cells. The influence of the pathology. a review. Biomeditsinskaya Khimiya **51**(2), 118–126 (2005)
65. Yoshida, Y., et al.: Levels of lipid peroxidation in human plasma and erythrocytes: comparison between fatty acids and cholesterol. Lipids **42**(5), 439–449 (2007)

66. Yamaguchi, T., Ishimatu, T.: Effects of cholesterol on membrane stability of human erythrocytes. Biol. Pharm. Bull. **43**(10), 1604–1608 (2020)
67. Pytel, E., et al.: Increased oxidative stress and decreased membrane fluidity in erythrocytes of CAD patients. Biochem. Cell Biol. **91**(5), 315–318 (2013)
68. Zicha, J., Kuneš, J., Devynck, M.-A.: Abnormalities of membrane function and lipid metabolism in hypertension*: a review. Am. J. Hypertens. **12**(3), 315–331 (1999)
69. Song, Y., Jensen, M.D.: Red blood cell triglycerides—a unique pool that incorporates plasma-free fatty acids and relates to metabolic health. J. Lipid Res. **62**, 100131 (2021)
70. Muzulu, S.I., Bing, R.F., Norman, R.I.: Human erythrocyte membrane fluidity and calcium pump activity in primary combined hyperlipidaemia. Clin. Sci. **88**(3), 307–310 (1995)
71. Muller, S., et al.: Rheological properties and membrane fluidity of red blood cells and platelets in primary hyperlipoproteinemia. Atherosclerosis **83**(2), 231–237 (1990)
72. Miller, M.A., et al.: Comparison of calcium, magnesium-ATPase activity and membrane fluidity in patients with essential hypertension and in normotensive controls. J. Hypertens. **12**(8), 929–938 (1994)
73. Devynck, M.-A., et al.: Membrane microviscosity and plasma triacylglycerols in the rat. Clin. Sci. **94**(1), 79–85 (1998)
74. Broncel, M., et al.: Erythrocyte fluidity in patients with hyperlipidemia during statins therapy. Pol. Arch. Med. Wewn. **113**(6), 531–537 (2006)
75. Sheikh-Hasani, V., et al.: Atorvastatin treatment softens human red blood cells: an optical tweezers study. Biomed. Opt. Express **9**, 1256 (2018)
76. Zhao, Y., et al.: Nitric oxide in red blood cell adaptation to hypoxia. Acta Biochim. Biophys. Sin. **50**(7), 621–634 (2018)
77. Tsuda, K., et al.: Nitric oxide improves membrane fluidity of erythrocytes in essential hypertension: an electron paramagnetic resonance investigation. Biochem. Biophys. Res. Commun. **275**(3), 946–954 (2000)
78. Weitman, S.D., et al.: Propranolol-induced alterations in rat erythrocyte membrane fluidity and apparent phase-transition temperatures: a depth-dependent process. Biochem. Pharmacol. **38**(18), 2949–2955 (1989)

Stress Response of Gram-Positive and Gram-Negative Bacteria Induced by Metal and Non-metal Nanoparticles. In Search of Smart Antimicrobial Agents

Iliana Ivanova[1], Radostina Toshkovska[1], Lyubomira Yocheva[2], Dayana Benkova[3], Vesela Yordanova[3], Alexandrina Nesheva[3], Rusina Hazarosova[3], Galya Staneva[3](✉), and Aneliya Kostadinova[3](✉)

[1] Faculty of Biology, Sofia University "St. Kliment Ohridski", 8 Dragan Tzankov Blvd., 1164 Sofia, Bulgaria
[2] Faculty of Medicine, Sofia University "St. Kliment Ohridski", 1 Koziak Str., 1407 Sofia, Bulgaria
[3] Institute of Biophysics and Biomedical Engineering, Bulgarian Academy of Sciences, Acad. Georgi Bonchev Str., Bl. 21, 1113 Sofia, Bulgaria
gstaneva@obzor.bio21.bas.bg, aneliakk@yahoo.com

Abstract. The increasing resistance of pathogens to a number of antibiotics has been the subject of many research reports from the European Antibiotic Resistance Surveillance Network (Ears-Net) and The World Health Organization (The WHO). The aim of this work was to study the effects of nanomaterial dispersions as Selenium (Se), Gold (Au), Iron oxide (Fe_2O_3), Silicon dioxide (SiO_2) and Graphene oxide (GO) on bacteria like *Staphylococcus aureus, Staphylococcus epidermidis, Bacillus cereus* and two strains of *Escherichia coli.* Two classical methods were used to investigate the antibacterial effect of the nanoparticles (NPs): Spot and Well diffusion tests in agar medium. The tested nanoparticles were active against Gram-positive bacteria in concentrations between 3.0 and 1.5 mg/mL but they were not active against Gram-negative bacteria such as *E. coli.* Among tested nanomaterials, SeNPs express the strongest antimicrobial effect. Gold nanoparticles with Polyvinylpyrolidone (Au-PVP NPs) were more active against bacteria than pure AuNPs. Lower concentrations (1.0 mg/mL and 0.5 mg/mL) of Se, GO and the two types of Gold nanoparticles did not show activity against all test microorganisms. Fe_2O_3 NPs as well as SiO_2 NPs had no effect on any test bacteria in the mentioned concentrations.

In conclusion, the most cytotoxic for tested bacteria were SeNPs, followed by Au-PVP and AuNPs. GONPs also showed a certain cytotoxic effect, especially on *B. cereus* 1095.

Keywords: Antimicrobial effect · Nanoparticles · Gram-positive bacteria · Gram-negative bacteria

S. Sotirov et al. (Eds.): BioInfoMed 2022, LNNS 658, pp. 147–155, 2023.
https://doi.org/10.1007/978-3-031-31069-0_15

1 Introduction

The increasing resistance of pathogens to a number of antibiotics has been the subject of many research reports from the European Antibiotic Resistance Surveillance Network (Ears-Net) [1] and The World Health Organization (The WHO) [2]. The abuse and lack of effective control over antibiotic treatment, as well as the ability of bacteria to adapt and develop rapidly, threaten humanity with the possibility of re-entering an era of uncontrolled infections: the so-called era of resistant super-microbes - superbugs. Many infectious agents and environmental isolates do not respond to existing antimicrobials and the doses administered. They are common not only in hospital settings, but in the community, too. This puts small children, the elderly, patients with chronic diseases and immune disorders at high risk. The reasons for the increasing antibiotic resistance are varied: from self-medication to increased use of antibiotics in animals as a growth promoter. There is ample evidence of increasingly common infections of environmental isolates and commensals (opportunistic pathogens). The attention of the scientific community is focused on the study of various nanotechnology products as an alternative to antibiotics and more effective antimicrobial agents.

Nanotechnology involves the study, design and production of materials at the atomic and molecular level, the application of which in medicine is related to diagnosis and treatment. Due to their size, nanomaterials have the advantage of easy interaction with biomolecules, changing their precision and efficiency [3]. The application of nanotechnology in medical microbiology leads in two directions - one aimed at the production of bactericidal agents or new antibiotics [4], and the other focused on the synthesis and design of sensors for the detection of microorganisms or their metabolites [5, 6]. Metal medical devices made of stainless steel, Co (cobalt) and Ti (titanium) alloys have long been widely used as artificial hip joints, bone plates and dental implants due to their excellent mechanical properties and durability. Often as a side effect, the implantation of medical devices leads to the emergence of difficult-to-treat infections, as a consequence of the colonization of their surfaces by microorganisms forming a biofilm resistant to antimicrobial therapy. The promising infection-fighting strategy, related to medical devices, is based on nanocomposite materials that either repel microbes so they cannot attach to device surfaces or kill them on direct contact.

Iron oxide nanoparticles (Fe_2O_3 NPs) are physically and chemically stable, biocompatible, and environmentally safe [7]. Studies indicate that the potential of Iron nanoparticles to generate microbial toxicity is due to a series of interactions, including membrane depolarization with subsequent damage to cell integrity [8], production of reactive oxygen species (ROS) with lipid peroxidation and damage of DNA [9] and release of metal ions that affect cellular homeostasis and protein coordination.

Other nanoparticles (NPs) that are extremely stable and less toxic are Silicon dioxide, also known as Silica NPs (SiO_2 NPs). Mesoporous Silicon materials with pore sizes in the range between 2 and 50 nm have attracted attention due to their fine-tuned macroscopic shape, chemical functionality, and mesoporous structure. Silica NPs are divided into mesoporous and nonporous, in which the size of nanoparticles can be controlled by changing the surfactant composition during nanoparticle synthesis [10]. Silica nanoparticles are the safest non-toxic particles for DNA conjugation and drug delivery [11, 12].

Other nonmetallic material that attracts scientists' attention is Graphene. Graphene is composed of a close-packed single layer of carbon atoms forming a 2D honeycomb-like lattice. In monolayer Graphene, carbon atoms bond to surrounding carbon atoms by sp^2-hybridization, forming a benzene ring in which each atom carries one unpaired electron. The structure of Graphene is extremely stable. Graphene has the strongest and most rigid crystal structure among all known materials.

Graphene oxide (GO) is the oxidized analogue of Graphene, recognized as the only intermediate or precursor to obtain the reduced Graphene oxide on a large scale [13–15]. GO is gaining more attention because it is easily functionalized with a fluorescent probe and other compatible biomolecules [16, 17]. Its antibacterial activity has been proven against Gram-positive and Gram-negative bacteria - *Bacillus subtilis, Staphylococus epidermidis, Pseudomonas aeruginosa* and *Enterobacter aerogenes* [18].

In terms of antimicrobial activity, Gold nanoparticles (AuNPs) conjugated with small molecules such as antibiotics, drugs, vaccines and antibodies are more effective than single other nanoparticles and molecules. AuNPs are gaining increasing attention due to their optical and electrical properties, which strongly depend on their shape and size [19]. Some researchers observed toxic effects of AuNPs, which were attributed to the permeability of certain cells to small-diameter AuNPs, while others suggested that AuNPs were not toxic to the same organisms.

Selenium (Se) is an essential trace element in the human body with an antioxidant effect (about 40 μg Se/day), but at high doses Selenium can be toxic (400 μg Se/day) [20]. SeNPs are involved in defense systems against toxicants and play an important role in protection against oxidative stress [21, 22]. Elemental Selenium in the form of nanoparticles has shown antibacterial activity against *Staphylococcus aureus,* a major pathogen in hospital-acquired infections and medical devices, even some researchers claim that SeNPs are more effective with lower toxicity compared to silver nanoparticles [23–25].

The aim of this work was to study the response of bacteria to stress due to effects of nanoparticle materials as Selenium, Gold, Iron oxide, Silicon dioxide and Graphene oxide on bacteria like *Staphylococcus aureus, Staphylococcus epidermidis, Bacillus cereus* and two strains of *Escherichia coli.* All of them are the most often causing agents of nosocomial infections today.

2 Materials and Methods

2.1 Nanoparticles

All chemicals were of analytical grade and used directly without further purification (Table 1).

Table 1. Nanoparticles used in the study

Nanoparticles	Particle size (nm)	Distributor
Graphene oxide (GO)	100	Merck (Germany)
Silicon dioxide (SiO_2)	60–100	Nanoshel (India)
Iron oxide (Fe_2O_3)	50–100	Nanoshel (India)
Selenium (Se)	40	Merck (Germany)
Gold (Au)	20	Abcam (USA)
Gold-Polyvinylpyrolidone (Au-PVP)	10	Abcam (USA)

2.2 Microorganisms

The antimicrobial effect was tested on Gram-negative bacteria: two strains of *Escherichia coli* (ATCC 10536 and ATCC 25922), and Gram-positive ones: *Bacillus cereus* NBIMCC 1095, *Staphylococcus epidermidis* and *Staphylococcus aureus* ATCC 25923, all supplemented from the National Bank of Industrial Microorganisms and Cell Cultures (NBIMCC), Bulgaria. The test microorganisms were grown in nutrient broth (NB Conda, Spain) at 37 °C and 180 rpm for 24 h with 2 sub-cultivations. They were cultured in the most suitable condition to reach exponential phase of growth. The microbial density of the cultures 0.5 was determined according to McFarland standard and used for the tests.

2.3 Antimicrobial Activity

Two classical methods were used to investigate the antibacterial effect of the nanoparticles: Spot and Well diffusion tests in agar medium. The aliquots of 100 μL microbial suspension in exponential phase with optical density 0.5 MacFarland were randomly spread on solid medium (Nutrient agar, Conda, Spain) in both tests. At Spot test, 10 μL drops of the investigated nanomaterial were put on inoculated agar. The used concentrations were 3.0 mg/mL, 1.5 mg/mL, 1.0 mg/mL, and 0.75 mg/mL. At Well diffusion test, wells with diameter 6 mm were cut by stopper borer and were filled with 50 μL solutions of nanoparticles in concentrations 1.0 mg/mL, 0.5 mg/mL, 0.3 mg/mL, and 0.1 mg/mL. In both methods, the plates were left for 2 h at 4–6 °C to afford diffusion of the dispersions and after that cultivated for 24 h at 37 °C. The obtained sterile zones were measured with 0.5 mm accuracy. All experiments were performed in two replicates and results are presented as the mean value of two experiments.

3 Results and Discussion

The growth inhibitory effect of Selenium, Gold, Iron oxides, Silicon dioxide and Graphene oxide nanoparticles on test bacteria like *Staphylococcus aureus, Staphylococcus epidermidis, Bacillus cereus* and two strains of *Escherichia coli* was studied.

The results obtained after measuring the sterile zones formed on the inoculated plates with different microbial cultures are presented in Table 2.

As could be seen from Table 2 the tested nanoparticles were active against Gram-positive bacteria but they were not active against Gram-negative bacteria such as *E. coli. B. cereus* is the most sensitive species among the three tested Gram-positive species. Also *S. aureus* is more resistant than *S. epidermidis.* Among tested nanomaterials, SeNPs express the strongest antimicrobial effect. GONPs are active only against *B. cereus* (Table 2, Fig. 1).

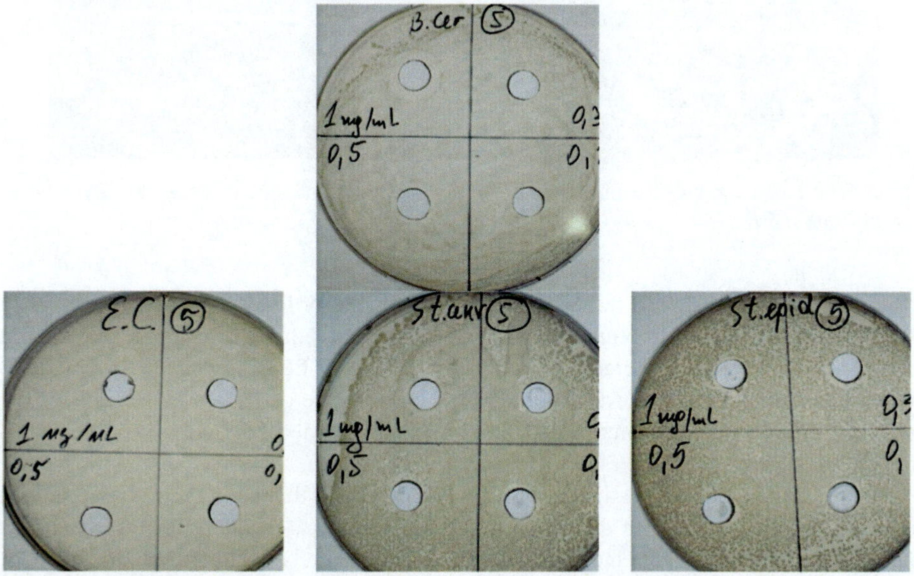

Fig. 1. Growth inhibitory effect of Graphene oxide on *Bacillus cereus* in concentrations of 1.0 and 0.5 mg/mL but absence of effect on other tested bacteria as *Escherichia coli, Staphylococcus aureus* and *Staphylococcus epidermidis.*

It is worth to mention that the Au-PVP NPs are more active against bacteria than pure AuNPs. These results are confirmed by Well diffusion method for concentrations of nanoparticles 1.0 mg/mL and 0.5 mg/mL. Lower concentrations (0.3 mg/mL and 0.1 mg/mL) of Selenium, Graphene oxide and two types of Gold nanoparticles do not show activity against all test microorganisms.

Results about Fe_2O_3 NPs are presented in Table 2, but they have no effect on any test bacteria in the mentioned concentrations, as well as Silica nanoparticles (Fig. 2).

NP concentration and size, charge and functionalization are the most important factors affecting their antimicrobial properties. The nanoparticles smaller than 10 nm can accumulate in the cell by penetration into the membranes, thereby making an effect on the nucleic acids and other macromolecules [26]. Another mechanism in which the NPs exhibit antimicrobial activity in the presence of oxygen is ROS production. By disrupting normal cellular functions of electron-transport chain, NPs trigger ROS formation such as OH^-, O_2^- and H_2O_2, and can cause death of bacteria [27]. On the basis of physicochemical characteristics and the photographs obtained by TEM (Transmission Electron Microscopy) (data not shown), we can arrange the obtained particles starting

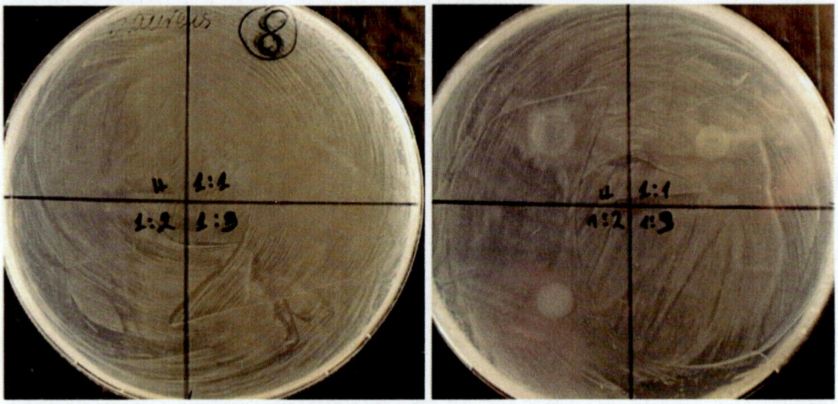

Fig. 2. The lack of any effect of different dilutions of Silica nanoparticles on *Staphylococcus aureus* (left) and *Bacillus cereus* (right).

from the smallest in size to the largest in the following order: Au-PVP < Au < Se < Fe_2O_3 < GO < SiO_2. Although the Au-PVP NPs possess the smallest size, they do not express the expected antibacterial effect. The presence of PVP enhance their antibacterial effect in comparison to Au, so our results confirm the dependence that always the presence of a metal functionalized with a precisely defined polymer would lead to a higher cytotoxicity [28].

SeNPs have an effect on the cells and these particles, thanks to their charge, attach to the cells and interact with them faster, and the additional available metal ion has, as mentioned above, an effect on the oxidative stress in the cell and the production of ROS. Selenium is an element that is known to have high antioxidant activity, but sometimes the presence of Selenium in higher concentrations increases its toxic effect [20]. Selenium in our research demonstrated the strongest antimicrobial activity, not only because of the small size of its particles, but also because of the highly toxic effect of Selenium in these concentrations. Selenium could be used as an antibacterial agent in complexes with other nanoparticles with antimicrobial effects to avoid antibiotic resistance in pathogens.

Our results obtained with Graphene oxide are in congruence with reported by [18]. The same could be valid for Gold nanoparticles; they could be used as drug delivery agents in medical practice.

In our work, the Iron oxide and Silica nanoparticles do not express antimicrobial activity but they could be used for drug delivery systems after testing their effect on tissues and organisms and proving the lack of toxic effect. There is a report about application of Silica nanoparticles as a capsulizing agent for peptide. As the Silica NPs are biodegradable they protect the protein from biodegradation and transport it to target place [29]. Nanocomposites of Silica and other nanoparticles could be less toxic to tissues and toxic to pathogenic bacteria [12, 30, 31].

Table 2. Growth inhibitory effect of Selenium, Graphene oxide, and two types of Gold nanoparticles on the test microorganisms in mm diameter of sterile zones.

Nanoparticles	Bacterial sterile zones in mm				
	E. coli 10536	*E. coli* 25922	*B. cereus* 1095	*S. aureus* 25923	*S. epidermidis*
Se 3 mg/mL	0	0	12 ± 0.5	11 ± 0.5	11.5 ± 0.5
Nondiluted	0	0	11.5 ± 0.5	10 ± 0.5	11.5 ± 0.5
1:1	0	0	10 ± 0.5	9.5 ± 0.5	10.5 ± 0.5
1:2	0	0	8.5 ± 0.5	5 ± 0.5	10.5 ± 0.5
1:3					
GO 3 mg/mL					
Nondiluted	0	0	8 ± 0.5	0	0
1:1	0	0	6 ± 0.5	0	0
1:2	0	0	0	0	0
1:3	0	0	0	0	0
Au 3 mg/mL					
Nondiluted	0	0	7 ± 0.5	5 ± 0.5	6 ± 0.5
1:1	0	0	4 ± 0.5	0	3 ± 0.5
1:2	0	0	0	0	0
1:3	0	0	0	0	0
Au-PVP 3 mg/mL					
Nondiluted	0	0	10 ± 0.5	7.0 ± 0.5	8 ± 0.5
1:1	0	0	9 ± 0.5	0	8 ± 0.5
1:2	0	0	5 ± 0.5	0	0
1:3	0	0	0	0	0
SiO_2 3 mg/mL					
Nondiluted	0	0	2 ± 0.5	0	0
1:1	0	0	1 ± 0.5	0	0
1:2	0	0	0	0	0
1:3	0	0	0	0	0
Fe_2O_3 3 mg/mL	0	0	0	0	0
Nondiluted	0	0	0	0	0
1:1	0	0	0	0	0
1:2	0	0	0	0	0
1:3					

4 Conclusions

The investigated nanoparticles have different cytotoxicity on the selected bacterial strains, due to their different size, chemical and physicochemical characteristics. On the basis of antibacterial studies, we can conclude that the most cytotoxic for tested bacteria are SeNPs, followed by Au-PVP NPs and AuNPs. GONPs also show a certain cytotoxic effect, especially on *B. cereus* 1095. Additionally, it can be assumed that the AuNPs and Fe_2O_3 NPs could be successfully used as a drug-delivery system for antibiotics, but in higher concentrations they would have better results as antibacterial agents.

Acknowledgments. This study was financially supported from the National Science Fund of Bulgaria by Grant KP-06-H58/6-19.11.2021. The authors thank the Bulgarian Ministry of Education and Science for support: Scientific Infrastructure on Cell Technologies in Biomedicine (SICTB) D01-154/28/08/2018 and "National Center for Biomedical Photonics" D01-392/2020, part of Bulgarian National Roadmap for Scientific Infrastructures 2020–2027.

References

1. European Antimicrobial Resistance Surveillance Network (EARS-Net). https://www.ecdc.europa.eu/en/about-us/partnerships-and-networks/disease-and-laboratory-networks/ears-net
2. WHO Report (2014). http://www.who.int/drugresistance/documents/surveillancereport/en
3. Cabral, R.M., Baptista, P.V.: The chemistry and biology of gold nanoparticle-mediated photothermal therapy: promises and challenges. Nano LIFE **03**(03), 1330001 (2013)
4. Thambirajoo, M., et al.: Potential of nanoparticles integrated with antibacterial properties in preventing biofilm and antibiotic resistance. Antibiotics **10**(11), 1338 (2021)
5. Ivanova, I., Tsacheva, I.: Microbial sensors based on nanostructures. Recent Pat. Nanomed. **5**(1), 59–65 (2015)
6. Aasi, A., et al.: A density functional theory study on the interaction of toluene with transition metal decorated carbon nanotubes: a promising platform for early detection of lung cancer from human breath. Nanotechnology **31**(41), 415707 (2020)
7. Lu, A.-H., et al.: Magnetic nanoparticles: synthesis, protection, functionalization, and application. Angew. Chem. Int. Ed. **46**(8), 1222–1244 (2007)
8. Pelgrift, R.Y., Friedman, A.J.: Nanotechnology as a therapeutic tool to combat microbial resistance. Adv. Drug Deliv. Rev. **65**(13–14), 1803–1815 (2013)
9. Gentile, F., et al.: DNA damage by lipid peroxidation products: implications in cancer, inflammation and autoimmunity. AIMS Genet. **04**(02), 103–137 (2017)
10. Tang, L., Cheng, J.: Nonporous silica nanoparticles for nanomedicine application. Nano Today **8**(3), 290–312 (2013)
11. Gangwar, R.K., et al.: Curcumin conjugated silica nanoparticles for improving bioavailability and its anticancer applications. J. Agric. Food Chem. **61**(40), 9632–9637 (2013)
12. Jeelani, P.G., Mulay, P., Venkat, R., Ramalingam, C.: Multifaceted application of silica nanoparticles. A review. SILICON **12**(6), 1337–1354 (2019). https://doi.org/10.1007/s12633-019-00229-y
13. Feicht, P., Eigler, S.: Defects in graphene oxide as structural motifs. ChemNanoMat **4**(3), 244–252 (2018)
14. Gutiérrez-Portocarrero, S., et al.: Study of structural defects on reduced graphite oxide generated by different reductants. Diam. Relat. Mater. **92**, 219–227 (2019)

15. Sun, L., et al.: A force-engineered lint roller for superclean graphene. Adv. Mater. **31**, 1902978 (2019)
16. Liu, Y., et al.: Investigation on fluorescence quenching of dyes by graphite oxide and graphene. Appl. Surf. Sci. **257**(13), 5513–5518 (2011)
17. Liu, Y., et al.: Biocompatible graphene oxide-based glucose biosensors. Langmuir **26**(9), 6158–6160 (2010)
18. Gupta, D.K., Rajaura, R.S., Sharma, K.: Synthesis and characterization of graphene oxide nanoparticles and their antibacterial activity. Int. J. Environ. Sci. Technol. **1**(1), 16–24 (2015)
19. Verissimo, T.V., et al.: In vitro cytotoxicity and phototoxicity of surface-modified gold nanoparticles associated with neutral red as a potential drug delivery system in phototherapy. Mater. Sci. Eng. C **65**, 199–204 (2016)
20. Shakibaie, M., et al.: Biosynthesis and recovery of selenium nanoparticles and the effects on matrix metalloproteinase-2 expression. Biotechnol. Appl. Biochem. **56**(1), 7–15 (2010)
21. Wang, H., et al.: Elemental selenium at nano size possesses lower toxicity without compromising the fundamental effect on selenoenzymes: comparison with selenomethionine in mice. Free Radic. Biol. Med. **42**(10), 1524–1533 (2007)
22. Iranifam, M., et al.: A novel selenium nanoparticles-enhanced chemiluminescence system for determination of dinitrobutylphenol. Talanta **107**, 263–269 (2013)
23. Wang, Q., Webster, T.J.: Nanostructured selenium for preventing biofilm formation on polycarbonate medical devices. J. Biomed. Mater. Res. A **100**(12), 3205–3210 (2012)
24. Tran, P.A., Webster, T.J.: Antimicrobial selenium nanoparticle coatings on polymeric medical devices. Nanotechnology **24**(15), 155101 (2013)
25. Boroumand, S., et al.: Selenium nanoparticles: synthesis, characterization and study of their cytotoxicity, antioxidant and antibacterial activity. Mater. Res. Express **6**(8), 0850d8 (2019)
26. Choi, O., et al.: The inhibitory effects of silver nanoparticles, silver ions, and silver chloride colloids on microbial growth. Water Res. **42**(12), 3066–3074 (2008)
27. Vargas-Reus, M.A., et al.: Antimicrobial activity of nanoparticulate metal oxides against peri-implantitis pathogens. Int. J. Antimicrob. Agents **40**(2), 135–139 (2012)
28. Suchak, N.M., et al.: Study on the concentration of gold nanoparticles for antibacterial activity. Adv. Nat. Sci. Nanosci. Nanotechnol. **12**(3), 035003 (2021)
29. Ge, H., et al.: Preparation of organic–inorganic hybrid silica nanoparticles with contact antibacterial properties and their application in UV-curable coatings. Prog. Org. Coat. **106**, 20–26 (2017)
30. Camporotondi, D., et al.: Antimicrobial properties of silica modified nanoparticles. Microb. Pathog. Strat. Combating Sci. Technol. Educ. **2**, 283–290 (2013)
31. Mosselhy, D., et al.: Silica-gentamicin nanohybrids: synthesis and antimicrobial action. Materials **9**(3), 170 (2016)

Check for updates

Reactivity of Recombinant and Native pLDH Antigens with Seven Commercially Available Rapid Diagnostic Test Kits for Malaria Diagnosis

Daniela Todorova-Balvay[1], R. Ravishankaran[2,3], C. R. Pillai[4], Xavier C. Ding[5], and P. K. Desai[1,2,3](✉)

[1] Span Diagnostics SARL, Parc Technologique des Rives de l'Oise, Rue les Rives de l'Oise, BP 50149, 60201 Venette, Compiegne Cedex, France
contact@span-diagnostics.com, pradipk.desai@gmail.com

[2] Span Bioproducts Pvt. Ltd., 9th Floor, Rajhans Bonista, Ghod-Dod Road, Surat 395001, India

[3] Surat Raktadan Kendra and Research Center, 1st Health Center, Udhana - Magdalla Rd, Khatodra Wadi, Surat 395002, India

[4] Indian Council of Medical Research-National Institute of Malaria Research, Sector 8, Dwarka, New Delhi 110077, India

[5] Foundation for Innovative New Diagnostics (FIND), Campus Biotech, 9 Chemin des Mines, 1202 Geneva, Switzerland

Abstract. The glycolytic enzyme, lactate dehydrogenase (LDH), is a widely used diagnostic marker for malaria caused by *Plasmodium* species. Each *Plasmodium* species has a unique genetic sequence for LDH; this can be used to construct species-specific antigens for validation of rapid diagnostic tests (RDTs) for malaria. This study aims to review seven RDTs widely used in India for the sensitivity and specificity of *Plasmodium falciparum* and non-*falciparum* species based on *Plasmodium* LDH pLDH test bands.

Keywords: Lactate dehydrogenase · *P. falciparum* · *P. vivax* · Plasmodium species · RDTs

1 Introduction

Malaria is a severe and fatal disease caused by five species: *Plasmodium falciparum (Pf)*, *P. vivax (Pv)*, *P. ovale (Po)*, *P. malariae (Pm)* and *P. knowlesi (Pk)* infections [1]. According to the World Health Organization (WHO), 229 million cases were recorded worldwide in 2019, where *Pf* infection is predominantly prevalent in Africa [2, 3]; *Pv* is found mostly in Asia, Latin America and some African regions [3, 4]; two genetically distinct, subspecies, *P. ovale curtisi (Poc)* and *P. ovale wallikeri (Pow)* are found in Africa, Asia and Western Pacific islands as mono/mixed infections [1, 4]; *Pk* in South east Asia [4]; *Pm* in South America, Asia, and Africa [2, 3]. *Pf* and *Pv* infections are

D. Todorova-Balvay and R. Ravishankaran—Contributed equally to this work.

S. Sotirov et al. (Eds.): BioInfoMed 2022, LNNS 658, pp. 156–166, 2023.
https://doi.org/10.1007/978-3-031-31069-0_16

the commonly reported in India [5], but mixed-*Pk*, *Pm* and *Po* infections have also been reported in various endemic regions [1].

WHO recommended that all the suspected malaria cases be tested using either rapid diagnostic tests (RDTs) or microscopy. Since, the percentage of malaria being tested increased from 38% (2010) to 85% (2018) [3]. Microscopy technique requires skilled manpower to differentiate the mono/mixed infections. This would be difficult in remote areas leading to higher mortality due to misdiagnosis or delayed treatment. RDT is simple device used to identify active infections with limited laboratory and financial resources. Malaria-RDTs used worldwide are based on the detection of histidine-rich protein-2 (HRP2), expressed only in *Pf* [6]. HRP2-based tests are not useful in detecting *Pf* parasites lacking the *pfhrp2/3* genes that prevail in many endemic countries [6, 7]. Alternately, WHO recommends RDTs which can detect pLDH that is expressed at high levels in the asexual stages of all active malarial parasites where its activity is correlated with the level of parasitaemia [2, 8]. Currently, better species-specific point of care RDTs are needed for detecting *Pf* and non-*Pf* infection as there is no prequalified non-HRP2 combination tests available in the market [2, 9].

Previous investigations on malarial RDTs analysed the performance with polymerase chain reaction and thick-film microscopy in endemic areas of India to compare the efficacy and utility of these tests in detecting *Pf/Pv* infections [10–13]. Study on the epidemiology and geographic distribution of malaria in India emphasized the importance of including *Pm, Po* and *Pk* infections in routine screening alongside *Pf* and *Pv* species [1]. Although malaria-RDTs targeting *Pf* and non-*Pf* infections are available in India, no exclusive RDT studies were performed to assess their potential use in detecting *Pk*, *Pm* and *Po* infections. To fill this gap, evaluation of widely use RDTs in detecting *Pf*/*Pv* along with non-*Pf*/*Pv* is the need of the hour which also contributes to the improvisation of the RDTs.

This study aims to evaluate the LDH antigen-based detection RDTs which was widely used in India to identify Pf/Pv infections as well as highly neglected non-Pf/Pv malarial antigens. To our knowledge, this is the first study in India to assess the performance of existing RDTs for the detection of LDH antigens such as Pm, Poc, Pow, and Pk, along with Pf and Pv antigens.

2 Materials and Methods

2.1 Samples Used in This Study

Six species-specific, commercially available recombinant pLDH (rPf, rPv, rPm, rPk, rPoc and rPow) antigens of the five human-infecting *Plasmodium* species were developed by SPAN Diagnostics Sarl., France, in collaboration with the Foundation for Innovative New Diagnostics (FIND), Switzerland.

Both infected and healthy samples were obtained from Surat Raktadan Kendra and Research Center (SRKRC), India. The institutional ethical committee approved this study (No. SRKRC/RP/02/2021) and the participants' samples were taken after obtaining their informed consent. Infected samples were examined microscopically for parasitaemia.

Clinical isolates of *Pf/Pv* parasites were cultivated in the laboratory of Span Bioproducts Pvt. Ltd, India.

2.2 Cultivation of Native *P. falciparum* and *P. vivax*

Pf/Pv clinical samples were used for cultivation after confirming a parasitaemia level of approximately 1% as reported previously [14, 15]. Parasitaemia maturation were examined under light microscopy (Labomed Vision 2000) after staining by Jaswant Singh Bhattacharji (JSB) using Malaria Stain kit (JSB I and II) from Biolab Diagnostics (catalogue no.HE-807).

2.3 Testing Protocol Followed for Seven RDTs

Details of the seven RDTs used in this study are shown in Table 1.

Table 1. Details of the rapid diagnostic tests (RDTs) used in this study.

Diagnostic kit	Referral name in the paper	Manufacturer	Detection claimed	Coated capture antibody	Time	Reported relative sensitivity	Reported relative specificity	Reference (lot no.)	Date of manufacture/expiry (month, year)
Para-HIT® Total+	**ParaHIT**	ARKRAY Healthcare Pvt. Ltd., India	*P. falciparum* *P. vivax* *P. ovale* *P malariae*	**Pf**: anti-HRP2 **Pan**: anti-pLDH	20 min	100%	99.5%	55IC206-25 (4000022766)	November 2019/November 2021
FalciVax™	**FalciVax**	Zephyr Biomedicals (a division of Tulip Diagnostics (P) Ltd., India	*P. falciparum* *P. vivax*	**Pf**: anti-HRP2 **Pv**: anti-Pv pLDH	20 min	100%	100%	503150025 (81278)	March 2019/February 2021
TRUSTline Malaria Pf/Pv Ag Rapid Test	**TRUSTline**	Athenese-Dx Pvt. Ltd., India	*P. falciparum* *P. vivax*	**Pf**: anti-HRP2 **Pv**: anti-Pv pLDH	30 min	95.45%	97-98%	AR0112C (F1511191)	November 2019/November 2020
SD BIOLINE Malaria Ag Pf/Pan	**SD-BIOLINE**	Standard Diagnostics, INC, Korea*	*P. falciparum* *P. vivax* *P. ovale* *P malariae*	**Pf**: anti-HRP2 **Pan**: anti-pLDH	15–30 min	95.5 (Pf), 99.7% (non-Pf)	99.5%	05FK60 (05EDE025B)	May 2019/May 2021
MERISCREEN Malaria Pf/Pv Ag test	**MERISCREEN**	MERIL Diagnostics Pvt. Ltd, India	*P. falciparum* *P. vivax*	**Pf**: anti-HRP2 **Pv**: anti-Pv pLDH	20–30 min	>95%	>95%	RPDMFV-01 (MI111908)	November 2019/October 2021
First Response® Malaria Ag Pf/Pv card test	**First-Response**	Premier Medical Corporation Private Limited, India	*P. falciparum* *P. vivax*	**Pf**: anti-HRP2 **Pv**: anti-Pv pLDH	20 min	100%	100%	PI19FRC25 (76E0320S)	May 2020/April 2022
ADVANTAGE MALARIA PAN+P.F CARD test kit	**ADVANTAGE MALARIA**	J. Mitra & Co. Pvt. Ltd., India	*P. falciparum* *P. vivax* *P. ovale* *P malariae*	**Pf**: anti-HRP2 **Pan**: anti-pLDH	20 min	100%	100%	IR231050 (APF031019)	October 2019/September 2021

*now Abbott, Korea

Specimen, whole blood; sample volume, 5 μL; format, cassette

Evaluation studies were carried out using three different concentrations (0.5, 1 and 2 μg/mL) of the six recombinant antigens which were prepared in the healthy individual blood samples. Antigen data at a concentration of 2 μg/mL was displayed in Table 2, however all three concentrations showed identical results. Whole blood (5 μL) which was used as a diluent served as a negative control. Next, RDT test bands were analysed using cultivated parasites, both low and high parasitaemia of *Pf* (0.5% and 4%) and *Pv* (0.5% and 2%) respectively and 56 clinical blood samples, of which 27 samples were from healthy individual and 29 samples were from infected individuals where 15 samples had shown about ≤0.5% parsitaemia & 14 samples about 1–2% parasitaemia in microscopic analysis.3 Results.

2.4 Evaluation of RDTs Using Six Recombinant pLDH Antigens

We evaluated all the results with respect to the pLDH antibody used in the test band (Pan-malaria or Pv specific antibody) of the respective RDTs. The Pf test bands of all seven RDTs were coated with anti-HRP2 antibodies, and none of them showed any cross-reactivity with the six recombinant antigens (Table 2).

Table 2. Evaluation of the pan-malaria/*Plasmodium vivax* test bands of seven commercially available RDTs, using six novel pLDH recombinant antigens.

Recombinant antigen	**ParaHIT**	**FalciVax**	**TRUSTline Malaria**	**SD BIOLINE**	**MERISCREEN**	**First Response**	**ADVANTAGE MALARIA**
	Pan-pLDH Pf, Pv, Pm, Po	**Pv-pLDH** Pv-specific	**Pv-pLDH** Pv-specific	**PAN-pLDH** Pf, Pv, Pm, Po	**Pv-pLDH** Pv-specific	**Pv-pLDH** Pv-specific	**PAN-pLDH** Pf, Pv, Pm, Po
rPf	+	–	+	+	–	+	+
rPv	+	+	+	+	+	+	+
rPm	–	–	–	–	–	–	–
rPk	–	–	–	–	–	–	–
rPoc	–	–	+	–	–	+	–
rPow	+	–	+	+	–	+	–
Negative control	–	–	–	–	–	–	–

(+) Positive ± Weak positive (–) Negative

All kits used anti-HRP2 antibodies for their Pf test bands; none of the recombinant antigens showed any cross-reactivity with the Pf test band of any RDT.

2.5 Cultivation of Pf and Pv Parasites

Different growth stages of *Pf* are shown in Fig. 1b, 1c and 1d. The parasitaemia level of the cultures was monitored every 24 h using JSB staining, and the cultures were propagated accordingly. Different stages of *Pv* cultures are shown in Fig. 1f, 1g and 1h. The low and high parasitaemia samples collected from both cultures were used to evaluate the RDTs.

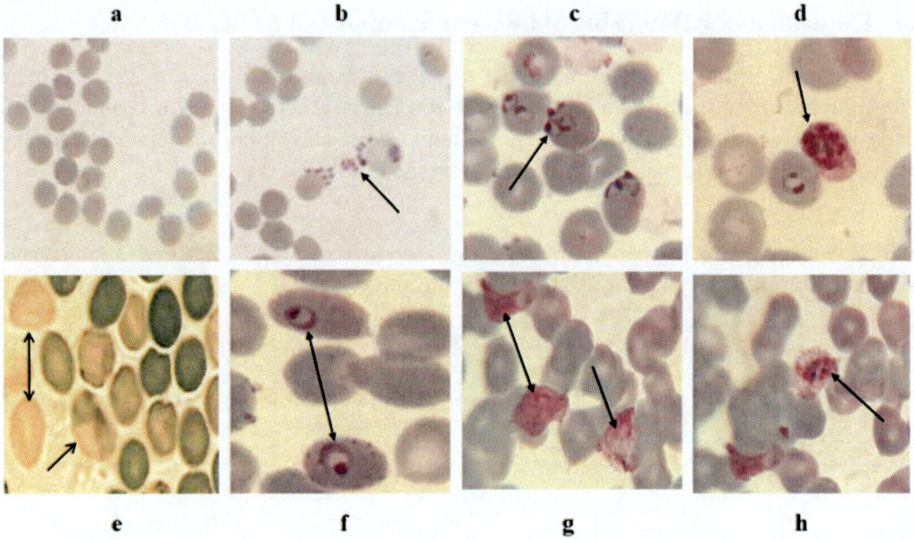

Fig. 1. Photomicrographs of stained blood smears of *in vitro* cultures of *P. falciparum and P. vivax* (a) uninfected erythrocytes, (b) erythrocyte infection by merozoites, (c) trophozoites (ring forms), and (d) developing schizont stages of *P. falciparum* parasites; (e) uninfected reticulocytes, (f) early trophozoites (ring forms), (g) late trophozoites, and (h) developing schizont stages of *P. vivax* parasites.

2.6 Evaluation of RDTs Using Native pLDH Antigens

Pf Culture Samples

The pan-pLDH test band of ParaHIT and SD-BIOLINE exhibited positive reactivity with *Pf* culture samples at both the 0.5% and 4% levels of parasitaemia, while the ADVANTAGE-MALARIA test showed a stronger positive reaction at both of these parasitaemia levels. Pv test bands in the FalciVax, TRUSTline and MERISCREEN RDTs displayed no reactivity at either parasitaemia level (0.5% or 4%). A weak-positive response was observed using the First-Response kit with the high (4%) parasitaemia sample which might be possibly due to the prozone effect, but not with the low parasitemia sample. All seven kits showed positive reactivity on the Pf (HRP2) test band, especially the TRUSTline and First-Response kits, which showed strong reactivity even at the low parasitaemia level. These results are shown in Table 3.

Table 3. Assessment of seven commercially available RDTs using cultivated *P. falciparum* and *P. vivax* parasites.

Culti-vated parasites	Par-asi-taemia per-centage	ParaHIT		FalciVax		TRUSTline		SD BIOLINE		MERISCREEN		First Response		ADVANTAGE MALARIA	
		HRP2	Pan pLDH	HRP2	Pv-pLDH	HRP2	Pv-pLDH	HRP2	Pan pLDH	HRP2	Pv-pLDH	HRP2	Pv-pLDH	HRP2	Pan pLDH
P. falciparum	0.5%	±	+	+	-	++	-	±	+	±	-	++	-	+	++
	4%	+	+	++	-	++	-	+	+	+	-	++	±	+	++
P. vivax	0.5%	-	-	-	+	-	±	-	-	-	±	-	++	-	±
	2%	-	+	-	+	-	+	-	+	-	+	-	++	-	+

(++) Strongly positive (+) Positive ± Weakly positive (-) Negative

Considered ~ 200,000 parasites/μL blood as 4% parasite density; 100,000 /μL blood as 2% parasite density and; 25,000 parasites/μL blood as 0.5% parasite density.

Pv Culture Samples

A signal was observed on the pan-pLDH test band when using the ADVANTAGE-MALARIA test with *P. vivax* culture samples at both the 0.5% and 2% levels of parasitaemia, while reactivity was only recorded at the 2% parasitaemia level when using ParaHIT and SD- BIOLINE. The four RDTs with Pv-specific test bands detected native *Pv* antigens, but the sensitivity varied among the different kits. The First-Response and FalciVax tests exhibited a positive response to both levels of parasitaemia, with the signal intensity (colour density of the test band) particularly strong with the First-Response test kit. With the two other remaining kits, a good signal intensity was observed, mainly with samples at 2% parasitaemia. All kits used anti-HRP2 antibodies in their Pf test bands, and none of the kits showed any non-specific reactivity against *Pv* culture samples (Table 3).

Clinical Samples

In infected clinical samples, all seven WHO-approved RDTs demonstrated identical results for (Pan/Pv - pLDH) test bands, none of the seven tests shown cross-reactivity with samples from healthy individuals, as claimed by RDT manufacturers (Table 4), and the results was equivalent to those obtained with Pf and Pv culture % parasitemia samples (Table 3).

Table 4. Results obtained in the pan-malaria/*P. vivax* test bands of seven commercially available RDTs when tested using clinical blood samples.

Clinical sample	Parasitaemia percentage	ParaHIT	FalciVax	TRUSTline	SD BIOLINE	MERISCREEN	First Response	ADVANTAGE MALARIA
		Pan-pLDH	Pv-pLDH	Pv-pLDH	Pan-pLDH	Pv-pLDH	Pv-pLDH	Pan-pLDH
P. falciparum	≤0.5%	+	–	–	+	–	–	++
	1–2%	+	–	–	+	–	–	++
P. vivax	≤0.5%	–	+	±	–	±	++	±
	1–2%	+	+	+	+	+	++	+
Healthy individual		–	–	–	–	–	–	–

(+) Positive ± Weak positive (–) Negative

Totally 56 samples were tested, out of which 27 samples were from healthy individual and 29 samples were from infected individuals where 15 samples had shown about ≤0.5% parsitaemia & 14 samples about 1–2% parasitaemia. All the tested samples exhibited same reactive pattern results in the seven RDTs. All kits used anti-HRP2 antibodies for their Pf test bands; only *P. falciparum* clinical samples were detected; none of the *P. vivax* samples showed any cross-reactivity in the Pf test bands. These results are not included in this table.

3 Discussion

Seven RDTs were able to identify and discriminate species-specific pLDH antigens based on their Pan/Pv-specific test bands. Other studies have compared PCR/microscopy with RDTs [10, 13] for mono and mixed infections. But this is the first of its kind of study in India that has compared only RDTs to facilitate quick testing.

Pan test band in the ADVANTAGE-MALARIA detected native, clinical and r*Pf* antigens of *falciparum* more effectively than ParaHIT and SD-BIOLINE kits. The pan-PLDH band intensity of ADVANTAGE-MALARIA was found stronger with native *Pf* in comparison with that of native *Pv*-LDH antigen. We also found that the pan-pLDH test band in the paraHIT, SD-BIOLINE, and ADVANTAGE-MALARIA kits detected both native and clinical *falciparum* samples with greater signal intensity than the Pf-specific HRP2 test band.

First-Response kit had shown remarkably high sensitivity in detecting the native, clinical and rPv antigens of *vivax*-LDH, followed by FalciVax test kit. Pan-pLDH test band in the ParaHIT, SD-BIOLINE and ADVANTAGE-MALARIA kits and Pv-LDH test band of MERISCREEN and TRUSTline detects *Pv* infection only in severe conditions (2% parasitaemia).

Poc and *Pow* exhibit similar morphology but have different genotyping and clinical features. The latency period of *Poc* has been shown to be twice that of *Pow* [17]. Based on observations made during a study of imported cases of malaria in Spain (2005–2011), the clinical features of *Pow* infection were reported to produce more severe thrombocytopaenia due to its relatively higher pathogenicity compared with that of *Poc*

[18]. Currently, molecular techniques are the only means for clinicians to differentiate subspecies where microscopy and RDT analysis have been found to be unreliable. When testing recombinant antigens, we noticed that the pan test band of the ParaHIT and SD-BIOLINE tests was able to identify *Pow* infection, but not *Poc*. With respect to the detection of *Po*, our results were similar to those previously reported by Tang et al. [19], i.e. that SD-BIOLINE can detect only *Pow* infection (with 18% sensitivity) and not *Poc*. While the Pv test band of TRUSTline and First-Response detected the recombinant antigens of both *Poc and Pow*. This suggests that these kits could potentially be used in areas where there is an outbreak of malaria caused by *Po*.

Although TRUSTline and First-Response are claimed to be Pv-specific card tests, a positive response was observed in the Pv test bands when using the recombinant rPf, rPoc and rPow, as well as when using the rPv antigens. It is possible that the positive responses observed using the TRUSTline and First-Response card tests using the recombinant pLDH antigens might be due to interactions with epitopes that are common to *Pf*, *Pv*, and *Poc and Pow* [20–22]. To verify this cross-reactivity with Pf LDH antigen, both kits were tested using the *Pf* cultivated (4%) and clinical samples (2%). While examining the First-Response Pv test band, we observed a faint positive signal (±) when using the 4% parasitemia culture sample but no band at 2% parasitemia level whereas no cross-reactivity in the TRUSTline kit. Faint positive signal in First-Response card tests might be possibly due to the prozone effect which can occur at high parasitemia conditions. Similar positive results were reported in six out of nine Pv specific test bands in the *Pf* blood samples infected with high parasitaemia (>2%) [16].

None of the seven RDTs (Table 2) recognised *Pm* or *Pk* recombinant pLDH antigens. Considering the overall non-*Pf/Pv* infections in India, *Pm* is more prevalent than *Po* and *Pk* as mixed infection with *Pf* [1]. *Pk* is a more serious malarial parasite than *Pf*, due to its shorter erythrocytic schizogony period of 24 h [23]. The potential for capillary sequestration of *Pk* -infected erythrocytes may even lead to cerebral malaria as well, although no cases have been recorded until now, except for one report of the post-mortem brain histopathology of a fatal case [24]. The transmission of *Pk* infection is possible in the eastern part of India, either ecologically (via a favourable forest ecosystem) or geographically (by travellers from South-East Asia) [25–27]. Recent finding of *Pf*, *Pv* and *Pk* mixed infection spotted in the histopathological analysis of severe malaria patients associated with acute kidney injury by All India Institute if Medical Sciences, India clearly exposed that *Pk* infection has emerged in India [28]. This, in turn, raises the question of our readiness to diagnose and handle the threat of *Pk* in India/Asian continent if an outbreak occurs in the near future as there is no RDT currently available that can specifically detect *Pk*. Even globally available malaria-RDTs show suboptimal performance in detecting *Pk, Pm, Po* infections [29].

Based on these constraints of the existing kits, we strongly recommend that RDTs be developed that can specifically detect *Pk*, *Pm* and *Po* (at subspecies level). The panel of six recombinant antigens, corresponding to the five *Plasmodium* spp., that was used in our study can facilitate the development of hexavalent Pan-LDH or species-specific RDTs which can support the key interventions of India's "Elimination Programme to achieve 'Malaria free nation' by 2030" [5]. In addition, recombinant pLDHs are useful for evaluating the performance of RDTs in situations when there is no or very limited chance of acquiring clinical samples from patients to identify *Plasmodium* spp. microscopically. Finally, it could be possible to perform species-specific, early diagnosis of imported malaria cases using RDTs for travellers who have migrated from endemic countries. A final important use for these recombinants as a control material, according to SOPs published by WHO [30]. This is the preliminary study done with limitation to seven WHO prequalified (2020) Pv-specific and Pan-malaria LDH kits which were highly preferred in the literature as they have shown significant sensitivity and specificity. Expansion of this study using the other kits will help the Indian and International manufacturers to develop/improve the species-specific LDH combination tests as recommended by WHO [9]. This type of evaluation study will help to improve the manufactured in India and apply in health facilities with reference to epidemiology as well as to prevent the indigenous transmission of malaria infection throughout the country.

4 Conclusions

Overall, our study found that the First-Response kit has the advantage of high sensitivity in detecting *Pv* infection but with less specificity. Other than the FalciVax and MERISCREEN tests, no kit could differentiate *Plasmodium* infections at the species level. We anticipate that this evaluation of malaria-RDTs commercially available on the Indian market, using native antigens and six species-specific recombinant pLDH antigens, will help manufacturers in two keyways. First, to improve and validate their products with respect to their specificity in differentiating the five *Plasmodium* species to authenticate their claims made during the commercialisation process. Second, due to the usage of the antimalarial medicine Hydroxychloroquine as a preventative measure against Covid-19 during the period of 2020–21, very few malaria cases were recorded. In such cases of low or no malaria sample availability, validation of the produced RDTs can be performed with the six recombinant antigens of SPAN Diagnostics Sarl., France as well as the Pf and Pv native antigens from cultured parasites as an alternate. Altogether, improvement in the RDT efficacy could reduce antimalarial usage, drug resistance and, the fatality due to misdiagnosis.

Acknowledgements. We acknowledge the European union and the Fonds Europeen de Dévelloppement Regional (FEDER), as well as the Banque publique d'Investissements (BPI), the region Hauts-de-France, for the co-financing the establishment of the lyophilisation platform at Span Diagnostics SARL, France. We also thank Dr.Walid Yakoub and Mrs. Sandra Dirson of Span Diagnostics Sarl, France, for preparing the recombinant antigens used in this study. We acknowledge the help of SRKRC in providing the clinical samples used in this study. We thank Dr Ramya Ramachandran, Scientist at Span Bioproducts Pvt. Ltd., for compiling the data and writing the original draft of the manuscript; Shukla Maitri and Thakur Angira, Research Assistants at Span Bioproducts Pvt. Ltd., for their involvement in culturing *P. falciparum* and *P. vivax* and for performing tests of the RDTs. We thank assoc. Prof. Dr. Milen Todorov, University "Asen Zlatarov", Bourgas, Bulgaria for the informatic expertise.

References

1. Chaturvedi, R., Deora, N., Bhandari, D., Parvez, S., Sinha, A., Sharma, A.: Trends of neglected Plasmodium species infection in humans over the past century in India. One Health **1**(11), 100190 (2021)
2. World Health Organization. World Malaria Report. https://www.who.int/publications/i/item/world-malaria-report-2020
3. World Health Organization. World Malaria Report. https://www.who.int/publications/i/item/world-malaria-report-2019
4. Center for Disease Control and Prevention. Malaria: Disease. https://www.cdc.gov/dpdx/malaria/index.html
5. Directorate of National Vector Borne Disease Control Programme. National Framework for Malaria Elimination in India 2016–2030. Directorate General of Health Services Ministry of Health and Family Welfare, Govt of India: New Delhi, India. 2016:1-43
6. Mukkala, A.N., Kwan, J., Lau, R., Harris, D., Kain, D., Boggild, A.K.: An update on malaria rapid diagnostic tests. Curr. Infect. Dis. Rep. **20**(12), 1–8 (2018)
7. Verma, A.K., Bharti, P.K., Das, A.: HRP-2 deletion: a hole in the ship of malaria elimination. Lancet Infect. Dis. **18**(8), 826–827 (2018)
8. Talman, A.M., et al.: Evaluation of the intra-and inter-specific genetic variability of Plasmodium lactate dehydrogenase. Malar. J. **6**(1), 1–6 (2007)
9. WHO list of prequalified in vitro diagnostic products (2021). https://extranet.who.int/pqweb/sites/default/files/documents/210115_List_PQ_IVDs.pdf. Accessed 15 Jan 2021
10. Singh, N., et al.: Field and laboratory comparative evaluation of rapid malaria diagnostic tests versus traditional and molecular techniques in India. Malar. J. **9**(1), 1–3 (2010)
11. Singh, N., et al.: Comparative evaluation of bivalent malaria rapid diagnostic tests versus traditional methods in field with special reference to heat stability testing in central India. PLoS ONE **8**(3), e58080 (2013)
12. Sahu, S.S., Gunasekaran, K., Jambulingam, P.: Field performance of malaria rapid diagnostic test for the detection of Plasmodium falciparum infection in Odisha State, India. Indian J. Med. Res. **142**(Suppl 1), S52 (2015)
13. Joseph, N., Uchila, A.K.: Validation of malaria antigen detecting rapid diagnostic test kit: a study from highly endemic area in coastal India. J. Clin. Diagn. Res. **12**(9) (2018)
14. Trager, W., Jensen, J.B.: Human malaria parasites in continuous culture. Science **193**(4254), 673–675 (1976)
15. Devi, C.U., Pillai, C.R., Subbarao, S.K., Dwivedi, S.C.: Short term in vitro cultivation of erythrocytic stages of Plasmodium vivax. J. Parasit. Dis. **24**(1), 61–66 (2000)
16. Maltha, J., Gillet, P., Cnops, L., Van den Ende, J., van Esbroeck, M., Jacobs, J.: Malaria rapid diagnostic tests: plasmodium falciparum infections with high parasite densities may generate false positive Plasmodium vivax pLDH lines. Mala. J. **9**(1), 1–7 (2010)
17. Nolder, D., et al.: An observational study of malaria in British travellers: plasmodium ovale wallikeri and Plasmodium ovale curtisi differ significantly in the duration of latency. BMJ Open **3**(5) (2013)
18. Rojo-Marcos, G., et al.: Comparison of imported Plasmodium ovale curtisi and P. ovale wallikeri infections among patients in Spain, 2005–2011. Emerg. Infect. Dis. **20**(3), 409 (2014)
19. Tang, J., et al.: Assessment of false negative rates of lactate dehydrogenase-based malaria rapid diagnostic tests for Plasmodium ovale detection. PLoS Negl. Trop. Dis. **13**(3), e0007254 (2019)
20. Piper, R.C., Buchanan, I., Choi, Y.H., Makler, M.T.: Opportunities for improving pLDH-based malaria diagnostic tests. Malar. J. **10**(1), 1–4 (2011)

21. Hurdayal, R., Achilonu, I., Choveaux, D., Coetzer, T.H., Goldring, J.D.: Anti-peptide antibodies differentiate between plasmodial lactate dehydrogenases. Peptides **31**(4), 525–532 (2010)
22. McCutchan, T.F., Piper, R.C., Makler, M.T.: Use of malaria rapid diagnostic test to identify Plasmodium knowlesi infection. Emerg. Infect. Dis. **14**(11), 1750 (2008)
23. Singh, B., Daneshvar, C.: Human infections and detection of Plasmodium knowlesi. Clin. Microbiol. Rev. **26**(2), 165–184 (2013)
24. Cox-Singh, J., et al.: Severe malaria-a case of fatal Plasmodium knowlesi infection with post-mortem findings: a case report. Malar. J. **9**(1), 1–7 (2010)
25. Subbarao, S.K.: Centenary celebrations article: Plasmodium knowlesi: from macaque monkeys to humans in South-east Asia and the risk of its spread in India. J. Parasit. Dis. **35**(2), 87–93 (2011)
26. Tyagi, R.K., Das, M.K., Singh, S.S., Sharma, Y.D.: Discordance in drug resistance-associated mutation patterns in marker genes of Plasmodium falciparum and Plasmodium knowlesi during coinfections. J. Antimicrob. Chemother. **68**(5), 1081–1088 (2013)
27. Lee, K.S., Vythilingam, I.: Plasmodium knowlesi: Emergent Human Malaria in Southeast Asia. In: Lim, Y., Vythilingam, I. (eds.) Parasites and their vectors, pp. 5–31. Springer, Vienna (2013). https://doi.org/10.1007/978-3-7091-1553-4_2
28. Kaur, C., et al.: Renal detection of Plasmodium falciparum, Plasmodium vivax and Plasmodium knowlesi in malaria associated acute kidney injury: a retrospective case–control study. BMC. Res. Notes **13**(1), 1–7 (2020)
29. Yerlikaya, S., Campillo, A., Gonzalez, I.J.: A systematic review: performance of RDTs for the detection of Plasmodium knowlesi, Plasmodium malariae, and Plasmodium ovale monoinfections in human blood. J. Infect. Dis. (2018)
30. World Health Organization, Guidance on control materials for antigen detecting malaria RDTs: tools for preparation and validation (2019). https://apps.who.int/iris/bitstream/handle/10665/326092/WHO-CDS-GMP-2019.08-eng.pdf

Phytochemical Composition and Therapeutic Potential of *Bistorta major* Gray: A Review

Yordan Nikolaev Georgiev(✉), Manol Hristov Ognyanov, and Petko Nedyalkov Denev

Institute of Organic Chemistry with Centre of Phytochemistry, Bulgarian Academy of Sciences, 139 Ruski Blvd., 4000 Plovdiv, Bulgaria
{yordan.georgiev,manol.ognyanov,petko.denev}@orgchm.bas.bg

Abstract. *B. major* is an edible medicinal plant, which aqueous and hydroethanolic rhizome extracts are used in folk medicine for the relief of diarrhea and more gastrointestinal disorders, hemorrhages, inflammations, respiratory and other infections, wounds, etc. The progress on the bistort extract research enables the discovery of new bioactive compounds, evaluation of their therapeutic potential and safety for the treatment of modern socially significant diseases. Therefore, the present review aimed to analyze the therapeutic potential of *B. major* extracts with an emphasis on their active molecules, and the toxicological risk of bistort use. The antioxidant phenolics are among the most investigated phytochemicals in the plant, as two new flavonoids with anti-inflammatory properties have been discovered. The herb is a source of chlorogenic, gallic acids, catechins, procyanidins, and derivatives of caffeic acid, quercetin, kaempferol, luteolin and apigenin. Some triterpenoids, phenolic acids, flavan-3-ols, flavonols, tannins, and fatty acids are among the elucidated physiologically active compounds in the extracts. They are responsible for their antibacterial, antioxidant, hemostatic, immunostimulatory, anti-inflammatory, hepatoprotective, gastroprotective, and anticancer effects. Particularly, 5-glutinen-3-one and tannic acid are promising anti-rheumatic and hepatoprotective agents, respectively. However, the role of polysaccharides in the bioactivity of the aqueous extracts is not studied. It seems that by decreasing the polarity of the extragent it can increase the possible toxic effects of the extract on the basis of toxicological studies. Considering the biological and pharmacological investigations with bistort extracts, their biomedical potential deserves to be tested in malignant, infectious, chronic inflammatory, liver, gastrointestinal, cardiovascular diseases and diabetes.

Keywords: *Bistorta major* · Bistort · Phytochemistry · Phenolics · Biological Activity · Biomedicine · Nutrition

List of Abbreviations

AqE(s)	Aqueous extract(s)
AqRE(s)	Aqueous rhizome extract(s)
BM	*Bistorta major*
BMR	BM rhizomes

S. Sotirov et al. (Eds.): BioInfoMed 2022, LNNS 658, pp. 167–191, 2023.
https://doi.org/10.1007/978-3-031-31069-0_17

BR	Bistort rhizomes
B.W.	Body weight
GAE	Gallic acid equivalents
IC_{50},	Half-maximal inhibitory concentration
I.P.	Intraperitoneal injection
HAE(s)	Hydroalcoholic extract(s)
HARE(s)	Hydroalcoholic rhizome extract(s)
MBC	Minimal bactericidal concentration
MIC	Minimal inhibitory concentration
P.O.	*per os.*

1 Introduction

BM Gray (some synonyms: *B. officinalis* Delarbre, *Polygonum bistorta* L. and *Persicaria bistorta* L. Samp.) is a perennial herbaceous medicinal plant with a rhizomatous root pattern, unbranched stem, narrow triangular leaves and bright or pale pink flowers. BM belongs to the Polygonaceae family and it is found in moist and marshy meadows, mainly in the mountain belt [1, 2]. The species is native to Europe and Asia, but it has been introduced to other continents, such as North America, and it is not threatened or protected by special laws [3]. Some of its common names are bistort, snakeroot/snakeweed and pudding grass. The plant is important for honey bees, as they collect nectar and black-brown pollen from its flowers [4]. It has been found that bistort essential oil has a protective effect against some bee pathogens [5]. It has also been reported that a methanolic extract of BM leaves reveals growth-inhibitory activity against several phytopathogenic *Xanthomonas* pathovars [6]. The allelopathic activity of bistort fresh leaves on the germination of lettuce has confirmed that the plant is not poisonous, and it should be classified as a medicinal herb [7]. Therefore, the cultivation of BM can be considered part of the strategies for the conservation of bee populations worldwide and to support the development of ecological farming.

In the past, BMR were used in the tanning of leather and painting of wool. The whole plant is consumed as a vegetable in some regions of Europe (England) and Asia (China) to make salads, soups, stews, puddings, and it is used as an additive in the bread making [3, 8]. The BMR are used in European ethnomedicine, and especially in Bulgarian folk medicine. The herb is one of the most potent natural astringent and it helps in diarrhea, irritable bowel syndrome, external and internal bleedings, hemorrhagic colitis, ulcerative colitis, peptic ulcers, various other inflammations, runny nose, sore throat, pharyngitis, respiratory infections, cystitis, wounds, etc. [1, 8, 9]. It is included in the European, Russian and Chinese pharmacopoeias [10, 11]. The rhizomes of the plant are mainly extracted with hot water and hydroethanolic solutions in folk medicine. An infusion of 2 teaspoons of the herb is prepared with 250 mL boiling water, and after 20 min it is drained and consumed [1]. The AqRE of bistort is applied as a hemostatic, anti-inflammatory, diuretic and laxative agent [1]. Furthermore, the therapeutic potential of the aerial parts of the herb has not been properly examined. Although the AqRE of bistort has been traditionally prepared for a long time, most of its active constituents, aside phenolics, are still unknown. Nowadays, bistort extracts are also obtained by 100% methanol, acetone and other organic solvents, because they specifically isolate compounds with more potent cytotoxic activity than hot water. It is in progress elucidation of their selectivity on microbes, viruses, and abnormal cells, and the possible toxicity against normal mammal cells.

Perhaps, some of the most intriguing therapeutic effects of BMR extracts are their anti-inflammatory properties, which have to be explained by some secondary metabolites. The rhizomes are part of traditional herbal mixtures for the treatment of colitis and reproductive problems in Bulgaria [12, 13]. Awei and Fei-Liu-Ping herbal formulations, containing BR, are used in the clinical practice in China for adjunctive therapy in the treatment of malignant diseases and hyperlipidemia, respectively [14, 15]. Furthermore, the Chinese traditional formula Yangxue Jiedu Soup also contains BM. It has been revealed that this decoction exhibits protective activity against the inflammatory psoriasis-like lesions in model male BALB/c mice [16]. The molecular mechanism of the immunomodulatory action of the decoction is explained by inhibition of the secretion of exosome HSP70, which stops TLR4 activation, and inhibits the phosphorylation of NF-κB signaling pathway, and further inflammation. Another Chinese formulation Zeng-Sheng-Ping, which also contains BM, has been applied in the treatment of precancerous lesions in the esophagus in a clinical trial [17]. Bistort is found in the Mongolian herbal drug Deva-5, which is traditionally applied for the treatment of acute infectious diseases. This formulation has expressed *in vivo* anti-inflammatory effects against acute lung injury in rats reducing TNF-α, IL-1β, and IL-6 secretions, and mononuclear cell infiltration in the bronchioles and alveoli [18]. Except for Deva-5 [19], in most of the studies with herbal mixtures, the individual biological role of BM and its own metabolites need to be clarified to find their contribution in the main or supplementary therapeutic effects observed in folk medicine.

As a result of the traditional use and pharmacological studies of BR, the commercial interest in its use in cosmetics and preventive medicine is increasing. For instance, Perlaura™ (BASF Care Creations) is a clinically tested product, containing BMR extract, which is added to cosmetic products to help skin repair. There is a patent for the topical use of the AqRE of BM stating that it can stimulate the expression of perlecan and dystroglycan in the skin for the treatment of rosacea, telangiectasias, chapping and/or disorders of the buccal and ocular mucous membranes [20].

The phytochemical composition and pharmacological effects of BM have been discussed in reviews focused on the species [8, 21] and including on other species in the genus or other family members [22–27]. This supports that the plant has documented and scientifically proven health-promoting effects. However, the structure-activity relationship between bistort compounds and their bioactivity still needs to be examined. Furthermore, the toxicological aspects of BM use have to be carefully reviewed in terms of its application in functional and dietary nutrition. Therefore, the present review aimed to analyze the therapeutic potential of BM extracts with an emphasis on their active compounds, and the toxicological risk of bistort use.

2 Phytochemical Composition

In Table 1 are presented published results from the phytochemical characterization of bistort extracts. It has been found that the BMR are rich in starch and tannins, but they contain more specifically phenolic acids, flavonol glycosides, flavan-3-ols, flavones, coumarins, as well as terpenoids, sterols and fatty acids [1, 28–37]. For instance, Demiray et al. have obtained 70% acetone, 70% methanol and AqRE of bistort with total phenolic contents of 4.2, 2.2 and 1.9 mg GAE/g dried plant material, respectively [38]. The predominant identified phenolic compounds in the AqE are coumaric acid, protocatechuic acid, and naringenin, which are determined to be 229.9, 217.8, and 129.7 μg/g dried plant material. Interestingly, the authors have not detected catechin, which should be expected in the AqEs and HAEs. On the other hand, Khushtar et al. have reported that the total phenolic content in a 70% ethanolic extract from the rhizomes is about 4.4 mg GAE/g extract [39]. Although the plant material in both studies was different, it would be interesting to compare the total amount of phenolics extracted with all solvents, if the yields of the extracts had been presented. It is evident that acetone is a good extragent for polyphenols, but it is used only in several of the reviewed studies on BM. In another study, it has been estimated that the major elucidated phenolic compounds in 80% ethanolic extract of BR are chlorogenic acid (33.9 mg/g) and catechin (14.9 mg/g) [30]. Furthermore, the hydroxycoumarins scopoletin and umbelliferone have also been isolated from the BMR [32].

Identification and quantification of as much as possible phenolic and other metabolites in the bioactive BR extracts are of great importance for understanding their molecular mechanisms of action and safety. In some studies only qualitative data or a simple identification of several phenolics is performed, but hypotheses are built for the structure-activity relationships [37, 40]. This illustrates that more in-depth phytochemical studies on BM extracts are still necessary. For example, Pawłowska et al. have identified by UHPLC-DAD-MS^3 and NMR 28 tannin-related phenolics in an AqRE, as they have shown that it is rich in chlorogenic acid and gallic acid derivatives, such as catechins and procyanidins [29]. This confirms that tannins and their building blocks are important bioactive molecules extracted with hot water in the bistort extracts. Similarly, Wang et al. have elucidated 31 phenolic compounds in a 70% methanolic extract of BR, using anion-exchange solid phase extraction and further HPLC-QTOF MS analysis [41]. The detected compounds are classified as benzoyl and caffeoyl derivatives, but lignans, sesquiterpenes, and fatty acids are also separated and not further analyzed. It would be interesting, if the authors had presented quantitative data for the identified amount of phenols in comparison with the total phenolic content in the sample. There are some differences in the identified phenolic profile of the 70% methanolic extract of BR analyzed by Demiray et al. and Wang et al. [38, 41]. Furthermore, it has been estimated that the contents of gallic acid, chlorogenic acid, and catechin, extracted with 30% methanol, in 18 BMR samples vary between 0.24–0.88, 0.60–1.52, and 0.33–1.70%, respectively [42]. This confirms that not only the extraction conditions and analytical detection technique, but also the environmental and geographic conditions, and the degree of maturity, are essential for preparation of extracts rich in valuable phytonutrients.

Table 1. Phytochemical composition of *Bistorta major* extracts.

Extraction	Compounds	Ref.
acetone:methanol:water (3:1:1, v/v/v) extract from aerial parts	4-*O*-caffeoylquinic acid, 3-*O*-caffeoylquinic acid, isoorientin, luteolin 6-C-(2″-*O*-rhamnoglucoside, quercetin 5-*O*-glucoside, quercetin *O*-rhamnoglucoside, isovitexin, apigenin 6-*C*-(2″-*O*-rhamnoglucoside), rutin, isoquercitrin, quercetin *O*-galloyl-*O*-rhamnoglucoside, quercetin 3-*O*-glucoside-7-*O*-rhamnoside, kaempferol 3-*O*-glucoside-7-*O*-rhamnoside, quercetin 3-*O*-(6″-*O*-malonyl)-glucoside, avicularin, astragalin, quercitrin, 3,5-*O*-dicaffeoylquinic acid, quercetin 3-*O*-(5″-*O*-malonyl)-arabinofuranoside, kaempferol 3-*O*-(6″-*O*-malonyl)-glucoside, quercetin 3-*O*-(4″-*O*-malonyl)-rhamnoside, kaempferol 3-*O*-(5″-*O*-malonyl)-arabinofuranoside, and others	[2]
hydrodistillation from aerial parts	3-methylbut-3-en-1-ol, 2-phenylacetaldehyde, linalool, *β*-cyclogeraniol, isovaleric acid, myrcene, limonene, eucalyptol, geraniol, eugenol, *α*-terpineol, trans-*β*-caryophyllene, dodecanoic acid, methyl dodecanoate, hexadecanoic acid, tetradecanoic acid, tricosane, henicosane, nonanal, and others	[5]
AqRE	gallic acid, chlorogenic acid, 6-*O*-galloylglucose isomers, 1,6-*O*,*O*-digalloyl glucose, catechin, epicatechin, procyanidin B3, procyanidin B1, procyanidin B7, galloyl salindroside, (−)-epicatechin gallate-(4,8)-(+)-catechin, (−)-epigallocatechin gallate-(4,8)-(+)-catechin, and others	[29]
80% ethanolic extract from rhizomes	quinic acid, gallic acid, chlorogenic acid, caffeic acid, protocatechuic acid, catechin, epicatechin, quercetin	[30]
fractionated 95% ethanolic extract from rhizomes	*β*-sitosterol, dihydromyricetin, quercetin, kaempferol, epicatechin, chlorogenic acid, rutin, gallic acid, protocatechuic acid, ellagic acid	[31]
fractionated methanolic extract from rhizomes	catechol, 4-hydroxybenzaldehyde, umbelliferone, scopoletin, pyrogallol	[32]

(*continued*)

Table 1. *(continued)*

Extraction	Compounds	Ref.
fractionated 70% ethanolic extract from rhizomes	5-glutinen-3-one, friedelanol, β-sitosterol	[33]
fractionated 90% methanolic extract from rhizomes	3β-Acetoxy-dammara-20, 24-diene, arborinone, adianenone, arborinol, isoarborinol, 6,7-methylenedioxycoumarin, 6-hydroxystigmast-4-en-3-one	[34]
fractionated chloroform extract from rhizomes	24(*E*)-ethylidenecycloartanone, 24(*E*)-ethylidenecycloartan-3α-ol, cycloartane-3,24-dione, 24-methylenecycloartanone, γ-sitosterol, β-sitosterol, β-sitosterone, friedelin, 3β-friedelinol	[35]
40% methanolic extract from rhizomes	gallic acid, protocatechuic acid, *p*-hydroxy benzoic acid, chlorogenic acid, pyrogallol, hydroquinone, vanillic acid, syringic acid, catechol, syringol, 4-methyl catechol, myristic acid, palmitic acid, linoleic acid	[37]
70% acetone, 70% methanolic and AqEs from rhizomes	protocatechuic acid, gallic acid, ferulic acid, coumaric acid, caffeic acid, chlorogenic acid, quercetin, naringenin	[38]
70% methanolic extract from rhizomes	gallic acid, galloylglucose isomers, protocatechuic acid, *p*-hydroxybenzoic acid, galloyl salidroside isomers, caffeic acid, 4-*O*-galloylarbutin isomer, galloyl salidroside isomers, digalloylglucose, catechin-5-*O*-(6-*O*-galloyl-β-glycopyranoside) isomers, lanceoloside A isomer, dodegranoside A isomer, russelianoside A isomer, terpineol galloylglycopyranoside, and others	[41]
hydrodistillation from the flowers	α- and β-pinene, limonene, 1-undecene, nonanal, lavandulol, 2E,4E-decadienal, tetradecane, farnesane, tetradecanoic acid, nonadecane, hexadecanoic acid, heneicosane, docosane, tricosane, tetracosane, hexacosane	[46]
fractionated chloroform extract from rhizomes	24,31-epoxy-24-ethylcycloartan-3α-ol	[47]
fractionated 95% ethanolic extract from rhizomes	bistortaside A	[48]
ethanolic extract from rhizomes	2,3',4',4,6-pentahydroxy flavone and 2,5',6-trihydroxy-4,2'-dimethoxy flavone	[49]

In general, the phytochemical composition of the plant rhizomes is more studied in comparison with the aerial parts, whose application in nutrition and medicine is still underestimated. Interestingly, Voronkova et al. have determined that bistort dried leaves, inflorescences and rhizomes contain 39.5, 23.3, and 29.8% tannins and 1.8, 3.0, and 10.5% catechins, respectively [28]. Additionally, it has been demonstrated that the fresh leaves can be a source of dietary fiber, but their composition and biological activity have to be studied [43]. The fresh leaves also contain vit. C and A, but their mineral content also deserves attention [44]. Apart from that, Yoshitama et al. have determined that BM fresh leaves, collected in June, contain 40.1 μg/g anthocyanidins, as they have identified only the presence of cyanidin-3-*O*-glucoside and delphinidin [45]. Interestingly, Klimczak et al. have determined 34 phenolic compounds in the aerial parts of BM, such as different phenolic acids (caffeic acid derivatives), flavan-3-ols (catechin, procyanidins), flavonols (quercetin and kaempferol derivatives) and flavones (luteolin and apigenin derivatives) [2]. They have found that flavonoids are the largest group of the detected compounds among the phenolics, however, they have not presented quantitative data for them.

Another important source of anti-inflammatory and antimicrobial compounds is the essential oil of the plant. It has been reported that the BM aerial parts contain between 0.004 and 0.010% essential oil on a fresh material basis [5]. They have determined that the major constituents in the vegetative, flowering and fruiting phases are 3-methylbut-3-en-1-ol (11.0 μg/g), linalool (5.3 μg/g), and dodecanoic acid (20.8 μg/g) and its methyl ester (17.2 μg/g). Additionally, 44 different metabolites have been elucidated in the essential oil of the flowers of *P. bistorta* subsp. *Carneum*, which is rich in hydrocarbons (54.5% of area), followed by carboxylic acids (16.5%), terpenoids (14.0%), esters (6.1%), and aldehydes (4.1%) [46]. Intisar et al. have obtained between 0.11–0.29% essential oil from the BMR, collected from three different regions, and they have identified 77 metabolites [36]. Some of the main determined fatty acids have been oleic acid (4.3–8.9%), oleic acid methyl ester (0.3–8.6%), palmitic acid (4.8–6.6%), linoleic acid (0.6–4.2%), and linoleic acid methyl ester (0.2–4.0%), but the essential oils have been also rich in furfurals. Similarly, linoleic, myristic and palmitic acids have been isolated from a 40% methanolic BR extract. Further information is needed for the composition of polyunsaturated fatty acids in the whole plant, because they can be involved in its anti-inflammatory properties [37].

Several new compounds have also been detected in BM, such as the triterpenoids 24(E)-ethylidenecycloartanone, 24(E)-ethylidenecycloarta-3a-ol and 24,31-epoxy-24-ethylcycloartan-3α-ol, the tannin-like compound bistortazide A, and two malonylated flavonoids quercetin and kaempferol 3-*O*-(5″-*O*-malonyl)-α-L-arabinofuranosides [2, 35, 47, 48]. The lack of commercial standards for some compounds strongly restricts the quantitative chromatographic analyses, however, more classes of bioactive compounds should be isolated and characterized from BM. The phytochemical analyses of the plant should not be focused on phenolic compounds only, because they cannot be responsible for all of the therapeutic effects known from ethnomedicine.

3 Pharmacological Activities

The BMR and aerial parts are used for the preparation of extracts with hot water and organic solvents, which exhibit diverse biological activities with potential therapeutic effects. In this section, the biological activities of the AqE are firstly discussed and then of those obtained with organic extragents, because some of the active constituents in the different extracts are expected to be non-identical. For the similar effects of the AqEs and other extracts comparative analyses are performed. In general, the recent biological and pharmacological studies confirm and expand the known from folk medicine beneficial effects of the plant extracts.

3.1 AqE

The anti-inflammatory activity of the AqEs of rhizomes and aerial parts has been revealed by their suppressive effects on the production of reactive oxygen species and myeloperoxidase release by fMLP-treated neutrophils, and the TNF-α secretion of LPS-treated human neutrophils [2, 29]. The AqRE has expressed a growth-inhibitory activity against several strains of *Staphylococcus aureus*, *S. epidermidis* and low activity against *Helicobacter pylori* [29]. The inhibited staphylococci strains are known skin pathogens. This fact supports the usefulness of BM extracts for treating wounds in folk medicine. Similarly, Sharma & Kaushik have confirmed that BM leaf decoction inhibits the growth of two, out of 6, clinical isolates of *S. aureus* [50].

Furthermore, the *in vivo* anticancer activity of an AqRE has been demonstrated in an animal model with a liver carcinoma [40]. The administration of the extract (232 mg/kg, *per os*) has resulted in the suppression of the endoplasmic reticulum functions in the tumor cells by activation of the reactive oxygen species production. This is a consequence of the increased autophagosome formation and accumulation in Hep3B cells, provoked by the extract, followed by inhibition of the next autolysosome formation via downregulation of mTOR activity. It is interesting that the BM extract can inhibit mTOR, which regulates the protein synthesis in the cells, and participates in the signal transduction of the insulin receptor, but it is also a known target for antitumour therapies [51].Therefore, there are plenty of interesting molecules in the BM extracts, which could target against the tumour and diabetes progression. Additionally, the extract decreases cyclin and cyclin-dependent kinase in the tumour cells to block the cell cycle progression. Finally, the proteasome activity is inhibited by the extract and leads to the activation of the intrinsic and extrinsic apoptotic caspase pathways. On the basis of the results with animals, Liu et al. have calculated that a dose of 1.4 g extract/day for an adult of 60 kg b.w. could be applied, as a dietary supplement, in a clinical study for patients with hepatocellular carcinoma [40].

An AqRE (100 mg/kg, p.o.) has shown a protective activity against liver and kidney injury caused by drugs and toxicants in male albino rats [52–54]. The extract decreases the levels of liver transaminases, alkaline phosphatase, albumin, bilirubin, urea, creatinine, triglycerides, cholesterol and normalizes the levels of glucose-6-phosphatase, glutathione, and some antioxidant enzymes. It reduces the DNA damage in the hepatocytes and reverses the histopathological alterations. The hepatoprotective activity of the AqE is in a logical agreement with its cytotoxic activity against hepatoma cells

[40]. The restorative effects of the AqE on DNA, lipid peroxidation, glutathione levels, glutathione peroxidase, glutathione-S-transferase, superoxide dismutase and catalase should be partly addressed to its antioxidant properties [55, 56]. However, more *in vivo* studies have to be performed to understand on a molecular level the antioxidant protective action of the extract on some inflammatory signaling pathways.

Additionally, the AqRE has exhibited analgesic activity in mice and it has been effective against coccidiosis in broilers [57, 58]. The AqREs have been successfully used for preparation of nanoparticles for the purpose of *in vitro* treatment of human breast cancer (MCF-7) or ecological disposal of wastewater from rose oil distilleries [55, 59]. Therefore, BM AqEs can be successfully incorporated in new functional formulations for application in nanomedicine and ecology. On the other hand, Oyuntsetseg et al. have shown that an AqE from the aerial parts of BM does not exhibit antiviral activity against influenza A virus H3N8, but it stimulates the growth of the host mammalian cells [19]. This means that the extract could express a cytoprotective activity, in case of virus infections, which deserves to be studied in more detail.

3.2 HAE and Other Extracts

Despite the fact that the AqEs are easier for administration because they are linked with the safe traditional herbal use, the share of biological studies with organic extracts is always bigger. This could be explained by the potency of their biological effects. For example, a methanolic extract of BMR has demonstrated better antibacterial activity than an AqE against 14 different microorganisms [60]. The (H)AEs of BR have shown *in vitro* antitumour, antibacterial, antifungal, antioxidant and cytoprotective activities [37, 61–63]. Jovanović et al. have reported that an 80% ethanolic extract of BM, rich in phenolic compounds, expresses moderate antibacterial activity against S. *aureus* ATCC 25923 and weak activity against *Pseudomonas aeruginosa* PAO1 with MIC of 156 (MBC $= 312$ µg/mL) and 1000 µg/mL, respectively [30]. It is not active on *Escherichia coli*, *Shigella flexneri*, *Listeria monocytogenes* or *Enterococcus faecalis* even at a dose of 5 mg/mL. For comparison purposes, Pawłowska et al. have obtained an AqRE with a MIC < 50 µg/mL and MBC $= 4000$ µg/mL against the same strain of *S. aureus* [29]. Liu et al. have obtained with 96% ethanol a rhizome extract with MIC values of 100 and 200 µg/mL against *S. aureus* ATCC6530 and *P. aeruginosa* ATCC9027, which has been also active on *E. coli* [61]. These examples illustrate that the quality of the herbal material, extraction conditions, cell concentration, viability, and origin of the microbial isolates, are very important for an adequate estimation of the antimicrobial activity of natural extracts. In some studies with BM only liquid extracts are used and inhibition zones are measured, without determination of the MIC and MBC, which is not enough for proper evaluation of the antimicrobial activity [60, 62]. The antimicrobial activity of a HARE of BM has been suggested to be linked to its ability to induce production of interferons in living cells [63]. AqE and 60% ethanolic extract of BM stems have exhibited *in vitro* anti-mutagenic activity, which could be explained by their antioxidant activities [64]. The *in vitro* antibacterial, antifungal and antitumour effects of the HAEs can be explained, but not restricted to, their antioxidant properties [38, 41, 62, 78]. However, the performed ABTS, DPPH and FRAP radical scavenging assays with the bistort extracts are not reliable enough to carry their effects in *in vivo* systems.

On the other hand, the HAE has also shown *in vivo* anti-diarrheal and anti-ulcer effects [39, 65]. The gastroprotective action of a dose of 1000 mg/kg b.w./day extract has been comparable to that of 20 mg/kg b.w./day ranitidine [39]. This is a direct scientific evidence for the documented in folk medicine traditional use of BMR to relieve (ulcerative) colitis, diarrhea, irritable bowel syndrome and peptic ulcers. However, the 70% ethanolic extract, at a moderate concentration of 100 μg/mL, has not inhibited the *H. pylori* cell viability and DNA fragmentation of AGS tumour cells, which suggests that the gastroprotective properties of the extract should not be correlated with its cytotoxic effects [66]. Furthermore, the MBC of an AqRE against *H. pylori* has been estimated to be 32 mg/mL [29]. Therefore, the beneficial effect is due to the anti-inflammatory and antioxidant properties of the extract.

The *in vivo* immunostimulatory effect of HARE has been demonstrated by increase of thymus and spleen weights, proliferation of T cells and cytotoxic activity of NK cells in mice [67]. Interestingly, the anti-inflammatory activity of HARE can be partly explained by their suppression of IL-8 production in *H. pylori*-infected AGS cells and xanthine oxidase-inhibitory activity *in vitro* [66, 68]. It has been confirmed *in vivo* by restoration of a carrageenan-induced paw oedema in rats, as the half maximal effective concentration has been calculated to be about 158.5 mg/kg b.w. [69]. Apart from that, Salehi et al. have reported that the HARE of BM lowers blood glucose levels in streptozotocin-induced diabetic mice within a two week treatment [70]. This study is in agreement with the results of Kumar et al. for the AqE of the herb, which can positively regulate carbohydrate metabolism [53]. Interestingly, a butanolic extract of BMR has expressed *in vivo* cardioprotective activity in rats with ischemia reperfusion through activation of some antioxidant enzymes, decreasing lipid peroxidation in the myocardial tissue and lowering the serum levels of lactate dehydrogenase and creatine phosphokinase [71]. Furthermore, a butanolic extract of the rhizomes has expressed *in vivo* protective antioxidant and vasodilator activities against retinal ischemia in experimental rats [72]. Both studies suggest that the herb contains bioactive agents that could reverse these ischemia conditions by exhibiting antioxidant and other activities.

Chloroform and hexane extracts of BR have revealed promising cytotoxic activities with IC_{50} values below 100 μg/mL against P388, HepG2, HL60, LL2 and MCF-7 tumour cells *in vitro* [73]. The AqREs, rich in phenolics, have also exhibited *in vitro* cytotoxic activity against HepG2 and MCF-7 cells [40, 54, 55]. An AqRE of BM, at a concentration of 694.20 μg/mL, has inhibited the cell viability of HepG2 cells to 53.88% [54]. This confirms that organic extracts can express more potent cytotoxic effects than the aqueous ones. Not only phenolics, found in AqEs and HAEs, but also some hydrophobic metabolites in BM seem promising antitumour agents. However, it is quite important to study the effect of the samples and on normal cell lines to evaluate if the cytotoxic effect is selective or not. It is quite interesting to investigate the molecular mechanisms of cytotoxic action of the hydrophobic compounds, but the insolubility in water restricts their application in amphiphilic formulations only.

In Table 2 are summarized the reviewed biological and pharmacological effects of the bistort extracts with experimental models, dosages of administration and positive controls or reference drugs used.

Table 2. Biological activities of *Bistorta major* extracts.

Extraction	Experimental model	Administration	Biological activity	Control	Ref.
fractionated acetone:methanol:water (3:1:1, v/v/v) extract from aerial parts	fMLP-stimulated human neutrophils	isolated flavonoids (1–25 μM)	*Ex vivo* anti-inflammatory activity through inhibition of oxidative burst and myeloperoxidase release from neutrophils	quercetin (1–25 μM)	[2]
acetone-methanol-water (3:1:1, v/v/v) extraction; AqE from rhizomes	LPS-stimulated human neutrophils	12.5, 25, and 50 μg/mL (AqE); 6.25, 12.5, and 25 μM (pure compounds)	*Ex vivo* anti-inflammatory activity via decrease of TNF-α	urolithin A (12.5, 25, and 50 μM)	[29]
acetone-methanol-water (3:1:1, v/v/v) extraction; AqE from rhizomes	*S. aureus* ATCC6538, ATCC25923, ATCC43300, MRSA 13318, *S. epidermidis* 13199, *S.* coagulase-negative 16248, *Helicobacter pylori* ATCC43504	94 to 0.004 mg/mL (extract) 500 to 0.25 μg/mL (pure compounds)	*In vitro* antibacterial activity of AqE with MIC = 125–30 μg/mL	ampicillin	[29]
80% ethanolic extract from rhizomes	*E. coli* ATCC 8739, *S. flexneri* ATCC 9199, *S. enteritidis* ATCC 13076, *S. aureus* ATCC 25923, *L. monocytogenes* ATCC 19111, *E. faecalis* ATCC 29212, *P. aeruginosa* PAO1, *P. aeruginosa* PA14	10, 50, and 100 μg/mL and higher for 24 h	*In vitro* antibacterial activity with IC_{50} = 156 and 1000 μg/mL *In vitro* antibiofilm activity	-	[30]
fractionated 70% ethanol extract from rhizomes	Carrageenan-induced paw oedema in male Sprague Dawley rats	400 and 80 (5-glutinen-3-one and friedelano) mg/kg b.w	*In vivo* anti-inflammatory activity via reduction of maximal and total paw oedema	2.5 mg/kg b.w. indomethacin (p.o.)	[33]
40% methanolic extract from rhizomes	human hepatocellular carcinoma cell line (HCCLM3)	200–800 μg/mL for 24 h	*In vitro* cytotoxic activity with IC_{50} = 86.5–126.8 μg/mL	-	[37]
70% acetone, 70% ethanol and AqEs from rhizomes	ABTS radical scavenging assay	Different concentrations	Low *in vitro* antioxidant activity	vit. C	[38]

(*continued*)

Table 2. *(continued)*

Extraction	Experimental model	Administration	Biological activity	Control	Ref.
70% ethanolic extract from rhizomes	indomethacin-induced gastric ulcers in Sprague Dawley rats	500–1000 mg/kg/day (p.o.) for 10 days	*In vivo* gastroprotective activity through alleviation of muco-oxidative stress	20 mg/kg/day ranitidine	[39]
AqRE	female, non-obese diabetic/severe combined immunodeficiency (NOD-SCID) mice inoculated with Hep3B and HepG2 cells	232 mg/kg (p.o.), 5 times per week, for 28–35 days	*In vitro* antitumour activity through induction of cell death, inhibition of cell adhesion and motility, activation of proteasome degradation, and inhibition of protein synthesis in Hep3B and HepG2 cells *In vivo* antitumour activity	-	[40]
phosphate buffer saline extract from leaves	multi-drug resistant *S. aureus* N1, N3, N6, S3, E4, P1	6 mg/well	*In vitro* antibacterial activity against *S. aureus* N1 and E4	vancomycin (1.2 mg/well)	[50]
AqRE	CCl_4 (1.5 mL/kg b.w., i.p.)-induced hepatotoxicity in albino male Sprague Dawley rats	100 mg/kg b.w. (p.o.) tannic acid (25 mg/kg b.w., p.o.)	*In vivo* hepatoprotective activity	silymarin (50 mg/kg, p.o.)	[52]
AqRE	Albino male rats Sprague Dawley, treated with paracetamol (2 g/kg, p.o.) or CCl_4 (1.5 ml/kg, i.p.)	100 mg/kg b.w. (p.o.) tannic acid (25 mg/kg b.w., p.o.) for 5 days	*In vivo* hepatorenal protective activity, normalizing activities of transaminases, alkaline phosphatase, adenosine triphosphatase, glucose-6-phosphatase, glutathione levels and reduction of lipid peroxidation	silymarin (50 mg/kg, p.o.)	[53]

(continued)

Table 2. (*continued*)

Extraction	Experimental model	Administration	Biological activity	Control	Ref.
AqRE	CCl_4 (1.5 mL/kg, i.p.)-induced hepatotoxicity in albino male Sprague Dawley rats HepG2 cancer cells	100 mg/kg b.w. for 5 days 10.85–694.20 µg/mL	*In vivo* hepatoprotective activity via restoring the activities of glutathione peroxidase, glutathione reductase, glutathione-S-transferase, glucose-6-phosphate dehydrogenase *In vitro* anticancer activity	silymarin (50 mg/kg p.o.), paclitaxel (IC_{50} = 27.82 µg/ml), tamoxifen (IC_{50} = 33.21 µg/ml)	[54]
nanoparticles of AqRE with ZnO	MCF-7 human breast cancer cell line ABTS radical scavenging assay	15.6, 31.2, 62.5, and 125 µg/mL for 72 h	*In vitro* cytotoxic activity with IC_{50} = 20 µg/mL for 72 h *In vitro* antioxidant activity with IC_{50} = 40 µg/mL	BHA (antioxidant test)	[55]
AqRE	Male Sprague Dawley rats DPPH radical scavenging assay	100 mg/kg b.w. 10–50 µg/mL	*In vivo* antipyretic activity *In vitro* antioxidant activity	paracetamol (33 mg/kg b.w.), vit. C (10 µg/mL)	[56]
AqRE	healthy adult Kunming mice	0.10 and 0.15 mg/g b.w. (i.p.)	*In vivo* analgesic effect not via opiate receptor	morphine, amidazofen (0.10 mg/g b.w.)	[57]
AqRE	Coccidial oocysts of *Eimeria* spp. Infection in broiler chicks	2 mL (twice a day) for 5 days	*In vivo* anticoccidial activity	Darvisul liquid	[58]
AqE and methanolic extract from leaves	*Klebsiella pneumoniae* MTCC109, *P. aeruginosa* MTCC1688, *S. paratyphi* B, *Shigella boydii, S. sonnei* MTCC 2957, *S. aureus* MTCC737, *Streptococcus faecalis* MTCC459	liquid extract (50 µL)	*In vitro* antibacterial activity	gentamicin, streptomycin (10 µg/disc)	[60]
96% ethanolic extracts from rhizomes	*Bacillus subtilis* ATCC6633, *S. aureus* ATCC6530, *E. coli* ATCC1229, *P. aeruginosa* ATCC9027	200, 100, 50, and 25 µg/mL	*In vitro* antibacterial activity with MIC = 100–200 µg/mL	-	[61]

(*continued*)

Table 2. (*continued*)

Extraction	Experimental model	Administration	Biological activity	Control	Ref.
methanolic and ethanolic extracts from rhizomes	*Aspergillus niger*, *Alternaria solani;* DPPH radical scavenging assay	various concentrations	*In vitro* antifungal activity *In vitro* antioxidant activity with IC_{50} = 49.20 μg/mL (methanolic extract) and 61.14 μg/mL (ethanolic extract)	vit. C (IC_{50} = 40.81 μg/mL)	[62]
80% ethanolic extract from rhizomes and fruits	$ECHO_9$ virus in monkey kidney cells (GMK)	0.1–1.0 mL liquid extracts for 24 h	*In vitro* antiviral with cytoprotective effects	-	[63]
AqE and 60% ethanolic extract from stems	*S. typhimurium* TA98 treated with 3-amino-1,4-dimethyl-5H-pyrido[4,3-b]indole	0.1 and 0.5 mL liquid extract for 24–48 h	*In vitro* antimutagenic activity	-	[64]
80% ethanolic extract from rhizomes	Castor oil-induced diarrhea in BALB/c mice	300–1.000 mg/kg b.w. (p.o.)	*In vivo* anti-diarrheal activity mediated through K + -channels activation and a weak Ca^{2+} antagonist effect	loperamide (10 mg/kg b.w., p.o.)	[65]
70% ethanolic extract from rhizomes	*H. pylori* 193C and human gastric cancer cell line AGS	100 μg/mL for 4 h	*In vitro* anti-inflammatory activity via inhibition of IL-8 production in *H. pylori*-infected AGS cells	curcumin (40 μM)	[66]
ethanolic extract from rhizome	Raw264.7 cells and mice	no full access	*In vitro* immunostimulatory activity via activation phagocytosis, proliferation of T cells and enhanced cytotoxicity of NK cells. Increased spleen and thymus weights, and IL-2 production in mice	-	[67]
fractionated methanolic extract from rhizomes and leaves; AqRE	xanthine oxidase (bovine milk) assay	400 μg/mL	*In vitro* xanthine oxidase-inhibitory activity with IC_{50} = 33.4 μg/mL of residual 50% methanolic extract from rhizomes	allopurinol (1.02 μg/mL)	[68]

(*continued*)

Table 2. *(continued)*

Extraction	Experimental model	Administration	Biological activity	Control	Ref.
70% ethanolic extract from rhizomes	carrageenan-induced rat paw oedema in male Sprague-Dawley rats and adjuvant-induced arthritis in male Wistar rats	100 and 200 mg/kg b.w. (p.o.) for 14 and 19 days	*In vivo* anti-inflammatory activity against acute and chronic phases of the adjuvant-induced rat paw swelling and oedema with a half maximal effective concentration of 158.5 mg/kg b.w	indomethacin (2.5 mg/kg b.w.)	[69]
HARE	streptozotocin-induced diabetes in mice	150 μg/kg b.w. for 15 days	*In vivo* antidiabetic activity via alleviation of blood glucose levels	Glibenclamide (0.5 mg/kg b.w.)	[70]
n-butanolic extract from rhizomes	male Wistar rats with ischemia reperfusion	60 and 120 mg/kg b.w. (p.o.)	*In vivo* cardioprotective activity via elevating the superoxide dismutase and decreasing malondialdehyde in the myocardial tissue, lowering serum lactate dehydrogenase, and creatine phosphokinase activities, preventing lipid peroxidation	-	[71]
n-butanolic extract from rhizomes	male Sprague Dawley with retinal ischemia/reperfusion	0.3 mg and 1.0 /kg b.w. (in sublingual vein)	*In vivo* protective antioxidant and vasodilator effects via elevating the activities of T-NOS and eNOS, and decreasing the activity of iNOS; increased production of NO in the blood vessels	-	[72]
fractionated chloroform extract from rhizomes	P338, HepG2, J82, HL60, MCF-7, LL2, WEHI164 cancer cell lines	Various concentrations for 3 days	*In vitro* cytotoxic activity with IC_{50} values under 100 μg/mL	6-mercaptopurine, doxorubicin ($IC_{50} \leq$ 10 μg/mL)	[73]
70% ethanolic extract from aerial parts	mouse fibroblast NIH3T3 cells luminol - H_2O_2 antioxidant activity assay	1–100 μg/mL for 48 h	*In vitro* cytotoxic activity with IC_{50} = 12.88 μg GAE/mL sample Antioxidant activity with IC_{50} = 0.57 μg GAE/mL extract	gallic acid (IC_{50} = 0.85 μg/mL), rutin (IC_{50} = 2.54 μg/mL)	[74]

4 Structure-Activity Relationship

Some triterpenoids, phenolic acids, flavan-3-ols, flavonols, tannins, and fatty acids are among the elucidated biologically active compounds, isolated from bistort extracts. For example, 5-glutinen-3-one and friedelanol have been the active constituents of a 70% ethanolic rhizome extract responsible for the reduction of carrageenan-induced rat paw oedema [33]. In fact, 5-glutinen-3-one has been more potent than friedelanol, however, the activity of the former has been much lower than that of indomethacin. The authors have explained that by the acidic character of indomethacin, which at lower pH is more lipophilic to pass the cell membrane of the inflamed cells.

In the previous section, it was demonstrated that a BM AqRE alleviated the hepatorenal damage in albino rats, treated with CCl_4 and paracetamol [53]. Furthermore, tannic acid, found in the extract, has expressed similar effects at a dose four times lower. The hepatoprotective activity of tannic acid has been more prominent and its protective effects have been comparable to that of silymarin. Interestingly, tannic acid has been able to alleviate the activity of glucose-6-phosphatase in the intoxicated rats. This is a key enzyme from carbohydrate metabolism, found in the liver and kidney, which is responsible for the release of glucose in the blood. Kumar et al. have demonstrated that tannic acid is one of the active phytochemicals of the AqRE of BM responsible for its *in vivo* antipyretic and choleretic activities in experimental rats [56].

Intisar et al. have suggested that the *in vitro* anticancer activity of fractionated 40% methanolic extract of BMR against hepatocellular carcinoma HCCLM3 cells is attributed to gallic acid, protocatechuic acid, chlorogenic acid, pyrogallol, and fatty acids detected in higher amounts in the extract [37]. In another study with BM, gallic, chlorogenic and protocatechuic acids have inhibited the cell adhesion and migration of hepatocellular carcinoma Hep3B and HepG2 cells [40]. However, these compounds have not exhibited alone or in a combination significant cytotoxic effects against the tumour cells. Therefore, there are some other active constituents (incl. Action in synergism) in the AqRE to make its antitumour activity more potent than the mentioned derived compounds.

On the other hand, at least partly responsible for the gastroprotective effects of the HAE should be phenolic antioxidants, which have alleviated the oxidative stress in the stomachs of rats [39]. A phenolic fraction, isolated from the aerial parts of bistort, has also shown immunomodulating effects [75]. Particularly, Klimczak et al. have isolated two new flavonoids quercetin- and kaempferol 3-*O*-(5″-*O*-malonyl)-α-L-arabinofuranosides, from the aerial parts of BM, which have suppressed the oxidative burst in fMLP-stimulated human neutrophils and myeloperoxidase release [2]. These flavonoids should be considered new anti-inflammatory compounds, however, further studies have to be performed for elucidation of the molecular mechanisms of anti-inflammation. It will be useful to test their effects on NF-*k*B, JAK-STAT and MAPK signaling pathways in case of inflammation.

Apart from that, Liu et al. have demonstrated that tannins in the rhizome extract are at least partly involved in its antibacterial activity [61]. Later, Pawłowska et al. have shown that flavan-3-ols and galloyl glucose compounds, isolated from BMR, contribute to the *in vitro* antibacterial and anti-inflammatory activities of the herbal extracts and most probably in the traditional topical use of the plant [29]. For example, the MBC of (−)-epicatechin-(4,8)-(−)-epicatechin gallate-(4,8)-(−)-epicatechin, isolated from an

AqRE, against *H. pylori* ATCC43504 and *S. aureus* ATCC25923 have been determined to be 64 and 32 times lower than that of the initial extract. Similarly, Jovanović et al. have reported that gallic acid and quercetin-3-*O*-glucoside (50 μg/mL), detected in an 80% ethanolic extract of BM, have been involved in the antibiofilm activity against *P. aeruginosa* PAO1, expressed by the extract [30]. Furthermore, the scientists have found that the anti-quorum sensing activity of the same extract has been expressed through the inhibition of *las* signaling pathway (LasR receptor) with quercetin-3-*O*-glucoside and catechin identified as the active compounds. Additionally, two coagulant flavonols 2,3',4',4,6-pentahydroxy flavone and 2,5',6-trihydroxy-4,2'-dimethoxy flavones have been isolated from HARE of the herb [49]. Not only phenolics, but also fatty acids are involved in the antibacterial activity of BM preparations. It has been revealed that lauric acid is among the potent antibacterial substances in the essential oil of the plant, but its methyl ester has lower activity [5].

5 Toxicological Aspects

It was demonstrated that BM extracts exhibit plenty of health-promoting effects in the previous sections. However, a major concern for their use is how safe they are, because in most of the reviewed studies toxicological data are missing. There are some *in vitro* studies with normal cell lines and *in vivo* experiments with animal models to evaluate the possible toxicological risk of application of BM extracts as dietary supplements in functional and therapeutic nutrition. In general, the high content of tannins in BM extracts can cause constipation in people with sensitive gastrointestinal tract, as it is known from traditional medicine. Particularly, it is not recommended to use BM during pregnancy and lactation, or in case of thrombophlebitis, angiocholitis and pancreatitis. It has been reported that a dose up to 5 g/kg b.w. of a BMR 80% ethanolic extract administered (p.o.) is safe in BALB/c mice and it does not lead to acute toxicity and mortality [65]. Additionally, the Chinese herbal formulation Zeng-Sheng-Ping, containing BR, has expressed chronic liver injury in golden hamsters, however, the possible hepatotoxic compounds are furanoids detected in two other herbs from the formulation, but not in bistort [76]. On the other hand, Pîrvu et al. have determined that a 70% ethanolic extract of BM aerial parts can be toxic to cell adhesion, spreading and proliferation of mouse fibroblast NIH3T3 cells at a dose above 25 μg total phenols (GAE)/mL sample [74]. It has also been revealed that an AqE is toxic at a dose higher than 2% for Madin-Darby canine kidney cells, but below 2% its toxicity is not significant [19]. Pillai has determined that the median lethal dose of crude rhizome chloroform extract and derived hexane and chloroform subfractions to be 142.82, 200 and 200.17 mg/kg/b.w. in male and female Swiss albino mice [77]. A HARE has exhibited cytotoxic activity against normal human lung fibroblasts MRC-5 cells with an IC_{50} value of 1.5 mg/mL. The determined 50% lethal concentration and half maximal effective concentration for the HARE against zebrafish embryos have been 207.7 and 184.6 μg/mL, respectively after 5 days of treatment [30]. The extract does not have a negative effect on the development and survival rate of the embryos up to 125 μg/mL, and it does not express any cardiotoxic activity at doses up to 200 μg/mL. However, it leads to teratogenic malformations in the treated embryos at a dose of 250 μg/mL. In conclusion, the dosage of bistort extracts is vital for their safety use, as it seems that the AqEs are safer. Therefore, before administration of the extracts toxicological risk assessment is highly recommended to be performed, which should be strictly used for determination of the doses of application.

6 Future Perspectives

Bioactive compounds of BM are mainly obtained by conventional extraction techniques with hot water and organic solvents, but only ultrasound-, microwave-assisted and anion-exchange solid phase extraction techniques have been applied from the non-conventional methods [29, 41, 62, 68]. It will be also interesting to use some green extraction methods with subcritical water, supercritical CO_2, ionic liquids, deep eutectic solvents under changed pressure, voltage power, in pulsed electric fields or with the help of enzyme (pre)treatments. This could increase the extraction yield, chemical diversity, bioactivity of the extracts, and the technological efficiency of the process for a possible industrial scaling up.

Tannic acid is one of the major active phytochemicals in the BM plant. From a biochemical point of view, its diverse biological activities can be linked to the high content of hydroxyl groups that interact with free and membrane-bound biomolecules in the living cells. In the current review, it was demonstrated that tannic acid and tannins express astringent, antimicrobial, antioxidant, anti-inflammatory, hemostatic, antidiarrheal, hepatorenal protective and antipyretic activities [52–54, 56, 61, 65, 78]. Tannic acid can also exhibit antiviral, wound-healing, and neuroprotective effects in Alzheimer's disease, and it is included in biomaterials as a crosslinking agent and physiologically active constituent [79]. However, the active hepatoprotective constituents of the AqRE of BM cannot be restricted to tannic acid only. Therefore, other tannins and lignans from BM should be also of a scientific interest. At the same time, the phytochemical research on BM should be extended to more classes of physiologically active organic compounds, including macromolecules, to explain and successfully exploit its potential to protect human health.

For example, the water-extractable polysaccharides deserve special attention, which could probably contribute to the observed anticancer, immunomodulatory properties of the AqRE with pronounced anti-inflammatory and immunostimulatory effects. It is important to note that there is a lack of information, in the available literature, regarding the chemical diversity and biological activity of the heteropolysaccharides from BR. Interestingly, Lv et al. have already isolated from the rhizomes of *Polygonum multiflorum* two Glc-rich polysaccharides, which exhibit antioxidant activity, and reduce lipid peroxidation and formation of advanced glycation end products *in vitro* [80]. However, no further information for the structural features of the polysaccharides and structure-activity relationship is found. Additionally, Lai & Li have isolated a water-soluble polysaccharide from *P. perfoliatum*, which suppresses the growth of lung adenocarcinoma A549 cells via upregulation of caspase-3, caspase-9, and Bax, and downregulation of Bcl-2 [81]. However, they have not studied the chemical features of the polymer, which is important for further investigation of its receptor recognition on tumour cells. At the same time, more phytochemical studies on the aerial part of BM are needed to evaluate its nutritional and medicinal uses. Again, for the aerial part of the herb, information is lacking regarding the type and biological activity of the contained polysaccharides, and chemical studies are mostly limited to phenolic compounds, essential oils and vitamins [2, 5, 44]. The rhizomes and aerial parts of BM could contain bioactive polysaccharides with immunomodulatory and prebiotic potential. Our group has already isolated and characterized such kinds of pectic polymers from the roots and leaves of *Geranium sanguineum*, which is another widely used medicinal plant [82]. In fact, the immunostimulating properties of pectins, in the absence of an allergen, and the manifestation of immunosuppressive activity in its presence, i.e. anti-inflammatory effects, have been exciting scientists for a long time. This is an up-to-date topic that is in the process of rapid development at the present moment at the level of clarifying the receptors to which polysaccharides bind on a competitive or non-competitive basis with the allergen. Our *in vitro* and *ex vivo* studies have shown that arabinogalactan regions of pectins are responsible for suppressing the endotoxin shock via decrease of NO production (incl. iNOS expression) in murine macrophages and PMA- and OZP-induced oxidative stress in human whole blood phagocytes and derived neutrophils [83, 84]. Therefore, the application of the AqEs and HAEs of BM, after the discovery of their active ingredients, in various biomedical formulations (incl. in nanomaterials), may be promising to support the treatment of different socially significant diseases.

7 Summary

The phytochemical composition, biological and pharmacological activities of BM have already been reviewed. However, an updated structure-activity relationship, and evaluation of the biomedical application, and safety of bistort extracts in the treatment of modern diseases, is needed. In general, the plant rhizomes are more extensively studied than the aerial parts, whose potential for functional and dietary nutrition is somehow underestimated. Furthermore, phytochemical studies are focused chiefly on phenolics, then on the lipid-soluble metabolites, however, the contribution of the water-soluble bioactive polysaccharides and other large molecules in the manifestation of bioactivity of the AqE is unclear.

The bistort AqEs and HAEs are characterized by antioxidant, antibacterial, antifungal, anti-inflammatory, cytoprotective, antitumour, immunostimulating, gastroprotective, hepatoprotective, cardioprotective, analgesic, antipyretic, and antidiabetic activities. The anti-inflammatory activity of the extracts and their flavonoids is confirmed by the inhibitory activity on xanthine oxidase, suppression of the oxidative burst, and secretion of inflammatory cytokines of inflamed human neutrophils and infected with *H. pylori* gastric tumour cells. Furthermore, it was shown that 5-glutinen-3-one and friedelanol, isolated from 70% ethanolic rhizome extract, are promising antirheumatic agents *in vivo*. Apart from that, some of the antibacterial constituents of the extracts are catechins, gallic acid, galloyl glucose compounds, quercetin-3-*O*-glucoside and tannins, which affect the biofilm formation and quorum sensing via *las* signaling pathway. Interestingly, the essential oil of the herb is a rich source of antibacterial fatty acids active against phyto- and human pathogens. Tannic acid is one of the active constituents responsible for the gastroprotective, hepatoprotective and antipyretic activities of the AqRE. It was demonstrated that the AqRE, HARE and hydrophobic rhizome extracts are effective against human leukaemia, hepatocellular, gastric, breast, lung, urothelial carcinomas and fibrosarcoma. However, the most promising results are observed in hepatocellular carcinoma. The AqRE expresses antitumour activity against hepatocellular carcinoma via imbalance of autophagosome and autolysosome formation, downregulation of mTOR activity, which increases the oxidative stress and blocks the cell cycle progression and triggers apoptosis. Gallic, chlorogenic and protocatechuic acids are among the cytotoxic compounds in the AqEs and HAEs, which inhibit the cell adhesion and migration of tumour cells. On the other hand, some toxicological studies reveal that the HARE is safe up to 5 g/kg b.w. (p.o.) in BALB/c mice. The median lethal dose of a crude rhizome chloroform extract is calculated to be 142.82 mg/kg/b.w. (p.o.) in male and female Swiss albino mice. Therefore, it was suggested that by decreasing the polarity of the extragent can increase the possible toxic effects of the final extract. In conclusion, BM extracts demonstrate promising therapeutic effects in animal models of malignant, infectious, chronic inflammatory, liver, gastrointestinal, cardiovascular diseases and diabetes. Therefore, bistort extracts and their active compounds can be further tested in biomedical studies, as part of the supplementary therapy for the treatment of some modern socially significant diseases.

References

1. Staneva, D., Panova, D., Rainova, L., Asenov, I.: 5.4. Bilki, izpolzuvani pri narusheno kryvosysirvane i anemia. In: Bachvarova, M. (ed.), Bilkite vyv Vseki Dom. 1st edn. Medicina i Fizkiltura, pp. 161–162. Sofia, Bulgaria (1982). (In Bulgarian)
2. Klimczak, U., Woźniak, M., Tomczyk, M., Granica, S.: Chemical composition of edible aerial parts of meadow bistort (*Persicaria bistorta* (L.) Samp.). Food Chem. **230**, 281–290 (2017). https://doi.org/10.1016/j.foodchem.2017.02.128
3. Tilford, G.L.: Edible and Medicinal Plants of the West. 11th edn. Mountain Press Publishing Press, pp. 18–19. Missoula, Montana, USA (2009)
4. Somme, L., et al.: Pollen and nectar quality drive the major and minor floral choices of bumble bees. Apidologie **46**(1), 92–106 (2014). https://doi.org/10.1007/s13592-014-0307-0
5. Cecotti, R., Carpana, E., Falchero, L., Paoletti, R., Tava, A.: Determination of the volatile fraction of *Polygonum bistorta* L. at different growing stages and evaluation of its antimicrobial activity against two major honeybee (*Apis mellifera*) pathogens. Chem. Biodivers. **9**(2), 359–369 (2012). https://doi.org/10.1002/cbdv.201100326
6. Babu, S., Satish, S., Mohana, D.C., Raghavendra, M.P., Raveesha, K.A.: Anti-bacterial evaluation and phytochemical analysis of some Iranian medicinal plants against plant pathogenic *Xanthomonas* pathovars. J. Agric. Technol. **3**(2), 307–316 (2007)
7. Mardani, H., Kazantseva, E., Onipchenko, V., Fujii, Y.: Evaluation of allelopathic activity of 178 Caucasian plant species. Int. J. Basic Appl. Sci. **5**(1), 75–81 (2016). https://doi.org/10.14419/ijbas.v5i1.5631
8. Zhang, Z.-F., Shen, Q., Cheung, H.-Y.: Qualitative evaluation and bioactivity assessment of *Polygonum bistorta*. Herb. Med. Nutraceuticals **19**(4), 164–167 (2012)
9. Nekratova, N.A., Shurupova, M.N.: Resources of medicinal plants in the Kuznetsky Alatau. Int. J. Environ. Stud. **71**(5), 656–666 (2014). https://doi.org/10.1080/00207233.2014.944790
10. Shikov, A.N., Pozharitskaya, O.N., Makarov, V.G., Wagner, H., Verpoorte, R., Heinrich, M.: Medicinal plants of the Russian pharmacopoeia; their history and applications. J. Ethnopharmacol. **154**(3), 481–536 (2014). https://doi.org/10.1016/j.jep.2014.04.007

11. Song, J., et al.: Authentication of the family Polygonaceae in Chinese pharmacopoeia by DNA barcoding technique. J. Ethnopharmacol. **124**(3), 434–439 (2009). https://doi.org/10.1016/j.jep.2009.05.042
12. Staneva, D., Panova, D., Rainova, L., Asenov, I.: 5.1.9. Bilkovi chaeve za lechenie na zabolyavaniya na hranosmilatelnata sistema. In: Bachvarova, M. (ed.), Bilkite vyv Vseki Dom, 1st edn. Medicina i Fizkiltura, p. 95. Sofia, Bulgaria (1982). (In Bulgarian)
13. Neshev, G., Landzhev, I.: Bilkovi retsepti pri bolesti na polovite organi. In: Tanev, K. (ed.), Spravochnik za Bilkite. 1st edn. D-r Petyr Beron, p. 207. Sofia, Bulgaria (1989)
14. Han, R., Xue, J., Hua, E., Zhou, L.: The high-performance liquid chromatographic fingerprints study of Awei capsules. Pharmacogn. Mag. **9**(33), 67–71 (2013). https://doi.org/10.4103/0973-1296.108143
15. Zheng, H., et al.: Chinese patent medicine Fei-Liu-Ping ointment as an adjunctive treatment for non-small cell lung cancer: protocol for a systematic review. BMJ Open **7**(1), e015045 (2017). https://doi.org/10.1136/bmjopen-2016-015045
16. Lv, J., et al.: Protective effect of Yangxue Jiedu Soup against psoriasis-like lesions by regulating TLR4/NF-κB signaling pathway mediated by secretion of exosome HSP70. Biomed. Pharmacother. **147**, 112604 (2022). https://doi.org/10.1016/j.biopha.2021.112604
17. Ding, Z., Gao, F., Lin, P.: Long-term effect of treating patients with precancerous lesions of the esophagus. Zhonghua Zhongliu Zazhi **21**, 275–277 (1999)
18. Oyuntsetseg, N., Chimedragchaa, C., Nomin-Erdene, J., Tsend-Ayush, D.: Effect of De-Mon syrup on lipopolysaccharide-induced acute lung injury in rats. Int. J. Appl. Res. Med. Plants **106** (2018). https://doi.org/10.29011/IJARMP-106.100006
19. Oyuntsetseg, N., et al.: Evaluation of direct antiviral activity of the Deva-5 herb formulation and extracts of five Asian plants against influenza a virus H3N8. BMC Complement. Altern. Med. **14**, 235 (2014). https://doi.org/10.1186/1472-6882-14-235
20. Andre-Frei, V., Bechetoille, N., Pain, S., Rousselle, P.: Cosmetic or dermatological use of a *Polygonum bistorta* extract. United States Patent Application Publication US 2016/0051462 A1, 25 February 2016. https://patents.google.com/patent/US20160051462A1/en
21. Mehar, A., Tabarak, H.I.M.: A review on *Polygonum bistorta* L. with reference to its pharmacology & phytochemistry. Glob. J. Res. Med. Plants Indig. Med. **2**(9), 669–674 (2019)
22. Voronkova, M.S., Vysochina, G.I.: *Bistorta* Scop. genus (Polygonaceae): Chemical composition and biological activity. Chem. Sustain. Dev. **22**, 207–212 (2014)
23. Dong, X., et al.: Pharmacological and other bioactivities of the genus *Polygonum* - a review. Trop. J. Pharm. Res. **13**(10), 1749–1759 (2014). https://doi.org/10.4314/tjpr.v13i10.27
24. Narasimhulu, G., Reddy, K.K., Mohamed, J.: The genus *Polygonum* (Polygonaceae): an ethnopharmacological and phytochemical perspectives – review. Int. J. Pharm. Pharm. Sci. **6**(2), 21–45 (2014)
25. Hao, D.C., Gu, X.-J., Xiao, P.G.: Chapter 12. Phytochemical and Biological Research of Polygoneae Medicinal Resources. Medicinal Plants: Chemistry, Biology and Omics. 1st edn., pp. 465–529. Woodhead Publishing, Cambridge, UK (2015). http://dx.doi.org/https://doi.org/10.1016/B978-0-08-100085-4.00012-8
26. Shen, B.-B., et al.: Analysis of the phytochemistry and bioactivity of the genus *Polygonum* of Polygonaceae. Digit. Chin. Med. **1**(1), 19–36 (2018). https://doi.org/10.1016/S2589-3777(19)30005-9
27. Seimandi, G., et al.: An update on phytochemicals and pharmacological activities of the genus *Persicaria* and *Polygonum*. Molecules **26**(19), 5956 (2021). https://doi.org/10.3390/molecules26195956
28. Voronkova, M.S., Vysochina, G.I., Kukushkina, T.A.: Phenolic compound content in individual plants of three species of the *Bistorta* (L.) Scop. Genus. Proc. Univ. Appl. Chem. Biotechnol. **9**(3), 529–535 (2019). (In Russian). https://doi.org/10.21285/2227-2925-2019-9-3-529-535

29. Pawłowska, K.A., Hałasa, R., Dudek, M.K., Majdan, M., Jankowska, K., Granica, S.: Antibacterial and anti-inflammatory activity of bistort (*Bistorta officinalis*) aqueous extract and its major components. Justification of the usage of the medicinal plant material as a traditional topical agent. J. Ethnopharmacol. **260**, 113077 (2020). https://doi.org/10.1016/j.jep.2020.113077
30. Jovanović, M., et al.: Anti-virulence potential and *in vivo* toxicity of *Persicaria maculosa* and *Bistorta officinalis* extracts. Molecules **25**(8), 1811 (2020). https://doi.org/10.3390/molecules25081811
31. Wang, H., Liu, B., Du, X., Zeng. J.: Identification of the chemical constituents of the ethyl acetate extraction portion of *Polygonum bistorta*/ 中国药房. China Pharm. **12**, 3494–3497 (2017)
32. Choi, S.Y., Kwon, Y.S., Kim, C.M.: Chemical constituents from *Polygonum bistorta* rhizomes. Korean J. Pharmacogn. **31**(4), 426–429 (2000)
33. Duwiejua, M., Zeitlin, I.J., Gray, A.I., Waterman, P.G.: The anti-inflammatory compounds of *Polygonum bistorta*: isolation and characterisation. Planta Med. **65**(4), 371–374 (1999). https://doi.org/10.1055/s-2006-960791
34. Sun, X.-B., Zhao, P.-H., Xu, Y.-J., Sun, L.-M., Cao, M.-A., Yuan, C.-Y.: Chemical constituents from the roots of *Polygonum bistorta*. Chem. Nat. Compd. **43**, 563–566 (2007). https://doi.org/10.1007/s10600-007-0193-z
35. Manoharan, K.P., Benny, T.K.H., Yang, D.: Cycloartane type triterpenoids from the rhizomes of *Polygonum bistorta*. Phytochemistry **66**, 2304–2308 (2005). https://doi.org/10.1016/j.phytochem.2005.07.008
36. Intisar, A., Zhang, L., Luo, H., Zhang, R., Wu, Z., Zhang, W.: Difference in essential oil composition of rhizome of *Polygonum bistorta* L. from different Asian regions and evaluation of its antibacterial activity. J. Essent. Oil Bear. Plants **15**(6), 964–971 (2012). https://doi.org/10.1080/0972060X.2012.10662600
37. Intisar, A., Zhang, L., Luo, H., Kiazolu, B.J., Zhang, R., Zhang, W.: Anticancer constituents and cytotoxic activity of methanol-water extract of *Polygonum bistorta* L. Afr. J. Tradit. Complement. Altern. Med. **10**(1), 53–59 (2013). https://doi.org/10.4314/ajtcam.v10i1.9
38. Demiray, S., Pintado, M.E., Castro, P.M.L.: Evaluation of phenolic profiles and antioxidant activities of Turkish medicinal plants: *Tilia argentea*, *Crataegi folium* leaves and *Polygonum bistorta* roots. Int. Sch. Sci. Res. Innov. **3**(6), 74–79 (2009). https://doi.org/10.5281/zenodo.1075066
39. Khushtar, M., Ahmad, A., Rahman, A.: Gastroprotective effect of hydro-alcoholic extract of *Polygonum bistorta* Linn root in indomethacin-induced gastric ulcers in Sprague Dawley rats. Indian J. Pharm. Educ. Res. **52**(4), 618–625 (2018). https://doi.org/10.5530/ijper.52.4.72
40. Liu, Y.-H., et al.: Aqueous extract of *Polygonum bistorta* modulates proteostasis by ROS-induced ER stress in human hepatoma cells. Sci. Rep. **7**, 41437 (2017). https://doi.org/10.1038/srep41437
41. Wang, S.-T., Yang, H., Gao, W., Li, H.-J., Li., P.: Trace enrichment and characterization of polyphenols in Bistort Rhizoma using weak anion-exchange solid phase extraction and high performance liquid chromatography-quadrupole time-of-flight mass spectrometry. J. Pharm. Biomed. Anal. **119**, 91–98 (2016). https://doi.org/10.1016/j.jpba.2015.11.033
42. Liu, X.-Q., Du, L.-L., Li, W.-W., Guan, H.-F., Liu, F.: Simultaneous qualitative and quantitative analysis of commercial *Bistorta* rhizome and its differentiation from closely related herbs using TLC and HPLC-DAD fingerprinting. Chem. Pharm. Bull. **56**(1), 75–78 (2008). https://doi.org/10.1248/cpb.56.75
43. Duke, J.A., Ayensu. E.S.: Medicinal Plants of China. Medicinal Plants of the World, no. 4. Reference Publications, Inc., Algonac, Michigan, USA (1985)
44. Moerman. D.E.: Native American Ethnobotany. 1st edn. Timber Press, Portland, Oregon, USA (1998)

45. Yoshitama, K., Nishino, H., Ozawa, H., Sakatani, M., Okabe, Y., Ishikura, N.: Distribution pattern of anthocyanidins and anthocyanins in Polyponaceous plant. Bot. Mag. Tokyo **100**, 143–149 (1987)
46. Iskender, N.Y., Güleç, C.A., Yücel, M., Sinek, K., Yayli, N.: Analysis of the essential oil from the flower of *Polygonum bistorta* L. subsp. *carneum* (Koch). Asian J. Chem. **23**(5), 1940–1942 (2011)
47. Pillai, M.K., Huat, B.T.K., Yang, D.: A New Cycloartane-type triterpenoid from *Polygonum bistorta*. Chem. Nat. Compd. **55**(6), 1085–1089 (2019). https://doi.org/10.1007/s10600-019-02900-7
48. Liu, X.-Q., Hua, H.-M., Liu, J., Chen, F.-K., Wu, L.J.: A new tannin-related compound from the rhizome of *Polygonum bistorta* L. J. Asian Nat. Prod. Res. **8**(4), 299–302 (2006). https://doi.org/10.1080/10286020500034956
49. Partovi, T., Zabihi, M.: Coagulant compounds from rhizomes of *Polygonum bistorta* (Linn.). Glob. Adv. Res. J. Eng. Technol. Innov. **1**(6), 127–130 (2012)
50. Sharma, N., Kaushik, S.: Antimicrobial and phytochemical studies on twelve medicinal plants against multidrug resistant *Staphylococcus aureus*. Int. J. Adv. Res. Biotechnol. Nanobiotechnol. **1**(I), 39–46 (2019)
51. Zou, Z., Tao, T., Li, H., Zhu, X.: mTOR signaling pathway and mTOR inhibitors in cancer: progress and challenges. Cell Biosci. **10**, 31 (2020). https://doi.org/10.1186/s13578-020-00396-1
52. Mittal, D.K., Joshi, D.: Evaluation of DNA damage by comet and MTT-assay in liver treated with *Polygonum bistorta* and tannic acid. Int. J. Pharm. Bio. Sci. **3**(1), 375–382 (2012)
53. Kumar, M.D., Deepmala, J., Sangeeta, S.: Hepatoprotective effects of *Polygonum bistorta* and its active principles on Albino rats intoxicated with carbon tetrachloride and paracetamol. Open Access Sci. Rep. **1**(4), 226 (2012). https://doi.org/10.4172/scientificreports.226
54. Mittal, D.K., Jena, A.K., Joshi, D.: Mechanism of *Polygonum bistorta* and *Zingiber roseum* against toxicity. J. Drug Metab. Toxicol. **5**(1), 161 (2014). https://doi.org/10.4172/2157-7609.1000161
55. Safavi, E., Homayouni-Tabrizi, M., Karimi, E., Mohammad, G.R.K.S.: Biosynthesis of zinc oxide nanoparticles using anjbar (root of *Persicaria bistorta*) extract and their cytotoxic effects on human breast cancer cell line (MCF-7). IET Nanobiotechnol. **13**(7), 736–741 (2019). https://doi.org/10.1049/iet-nbt.2018.5060
56. Kumar, M.D., Deepmala, J., Sangeeta, S.: Antioxidant, antipyretic and choleretic activities of crude extract and active compound of *Polygonum Bistorta* (Linn.) in albino rats. Int. J. Pharm. Biol. Sci. **2**(1), 25–31 (2012)
57. Zeng, Z.-Y., Wang, M., Ye, H.-Y., Zhou, L., Zhou, Q., Zeng, J.: Analgesic effect of *Polygonum bistorta* L. water extract. Chin. J. Clin. Rehabil. **10**, 199–201 (2006)
58. Khan, M.A., Younas, M., Khan, I., Abbas, R.Z., Ali, M.: Comparative efficacy of some herbal and homeopathic preparations against coccidiosis in broilers. Int. J. Agric. Biol. **10**(3), 358–360 (2008)
59. Azadi F., Karimi-Jashni, A., Zerafat, M.M.: Green synthesis and optimization of nano-magnetite using *Persicaria bistorta* root extract and its application for rosewater distillation wastewater treatment. Ecotoxicol. Environ. Saf. **165**, 467–475 (2018). https://doi.org/10.1016/j.ecoenv.2018.09.032
60. Satish, S., Mohana, D.C., Raghavendra, M.P., Babu, S., Raveesha, K.A.: Antibacterial evaluation of some Iranian medicinal plants against some human pathogenic bacteria. Asian J. Microbiol. Biotechnol. Environ. Exp. Sci. **11**(4), 735–738 (2009)
61. Liu, Y., Nielsen, M., Staerk, D., Jäger, A.K.: High-resolution bacterial growth inhibition profiling combined with HPLC–HRMS–SPE–NMR for identification of antibacterial constituents in Chinese plants used to treat snakebites. J. Ethnopharmacol. **155**, 1276–1283 (2014). https://doi.org/10.1016/j.jep.2014.07.019

62. Munir, N., Ijaz, W., Altaf, I., Naz, S.: Evaluation of antifungal and antioxidant potential of two medicinal plants: *Aconitum heterophyllum* and *Polygonum bistorta.* Asian Pac. J. Trop. Biomed. **4**(Suppl 2), S639–S643 (2014). https://doi.org/10.12980/APJTB.4.201414B182
63. Smolarz, H.D., Skwarek, T.: The investigations into the interferon-like activity of *Polygonum* L. genus. Acta Poloniae Pharm. Drug Res. **56**(6), 459–462 (1999). https://doi.org/10.21285/2227-2925-2019-9-3-529-535
64. Nikawa, M., Wu, A.-F., Sato, T., Nagase, H., Kito, H.: Effects of Chinese medicinal plant extracts on mutagenicity of trp-p-1. Nat. Med. **49**(3), 329–331 (1995)
65. Ali, M.Z., Janbaz, K.H., Mehmood, M.H., Gilani, A.-H.: Antidiarrheal and antispasmodic activities of *Polygonum bistorta* rhizomes are mediated predominantly through K+ channels activation. Bangladesh J. Pharmacol. **10**(3), 627–634 (2015). https://doi.org/10.3329/bjp.v10i3.23714
66. Zaidi, S.F., et al.: Anti-inflammatory and cytoprotective effects of selected Pakistani medicinal plants in *Helicobacter pylori*-infected gastric epithelial cells. J. Ethnopharmacol. **141**, 403–410 (2012). https://doi.org/10.1016/j.jep.2012.03.001
67. Li, K., Luan, X.: Immunoregulatory function of the ethanol extracts from Rhizoma Bistortae. Med. Plant **4**(1), 75–78, 82 (2013)
68. Orbán-Gyapai, O., Lajter, I., Hohmann, J., Jakab, G., Vasas, A.: Xanthine oxidase-inhibitory activity of extracts prepared from Polygonaceae species. Phytother. Res. **29**(3), 459–465 (2015). https://doi.org/10.1002/ptr.5275
69. Duwiejua, M., Zeitlin, I.J., Waterman, P.G., Gray, A.I.: An*ti-inflammatory activity of Polygonum bistorta, Guaiacum officinale and Hamamelis virginiana i*n rats. J. Pharm. Pharmacol. **46**(4), 286–290 (1994). https://doi.org/10.1111/j.2042-7158.1994.tb03795.x
70. Salehi, E., Sharifabad, M.M., Karimi, M., Zavareh, M.S.: The effect of hydro alcoholic extract of Bistort (*Polygonum bistorta* L) root on blood glucose concentration of streptozotocin induced diabetic mice. Q. Sci. J. Appl. Biol. **35**(1), 156–166 (2022). Ser. 70. 22051/JAB.2021.34309.1398
71. Ye, H.-Y., et al.: Protective effect of *Polygonum bistora* L. *n*-butyl alcohol extract on myocardial ischemia/reperfusion injury in rats in a dosage dependent manner. LiShiZhen Med. Mater. Med. Res. **17**(6), 907–909 (2006)
72. Li, L., Huang, Z., Xiao, H., Chen, X., Zeng, J.: Effect of PBNA on the NO content and NOS activity in ischemia/reperfusion injury in the rat retina. In: Anderson, R., Hollyfield, J., LaVail, M. (eds.) Retinal Degenerative Diseases. Advances in Experimental Medicine and Biology, vol. 664, pp. 501–507. Springer, New York, NY (2010). https://doi.org/10.1007/978-1-4419-1399-9_57
73. Manoharan, K.P., Yang, D., Hsu, A., Huat, B.T.: Evaluation of *Polygonum bistorta* for anticancer potential using selected cancer cell lines. Med. Chem. **3**(2), 121–126 (2007). https://doi.org/10.2174/157340607780059495
74. Pîrvu, L., et al.: *Polygonum bistorta* L. Herba et flores. Polyphenols profile, antioxidant properties and cytotoxic effect on murine fibroblast cell line NH3T3. Farmacia **65**(4), 571–576 (2017)
75. Pershukova, A.M., Makarova, N.V., Kryukova, L.I.: Izvestiya Akademiya Nauk Kazakhskoi S. S. R. Seriya biologicheskaya **1**, 78 (1991)
76. Wu, Y., et al.: NMR-based metabolomic approach to understanding Zeng-Sheng-Ping-induced hepatotoxicity, and identifying possible toxic constituents by LC-MS profiles. J. Pharm. Biomed. Anal. **217**, 114833 (2022). https://doi.org/10.1016/j.jpba.2022.114833
77. Pillai, M.K.: Evaluation of the extracts from rhizomes of *Polygonum bistorta* for the median lethal dosages in Swiss albino mice. Iran. J. Toxicol. **15**(3), 135–142 (2021). https://doi.org/10.32598/ijt.15.3.716.1

78. Wang, S.-T., et al.: Phenol profiles and antioxidant capacities of Bistort Rhizoma (*Polygonum bistorta* L.) extracts. RSC Adv. **6**, 27320–27328 (2016). https://doi.org/10.1039/C6RA00687F
79. Baldwin, A., Booth, B.W.: Biomedical applications of tannic acid. J. Biomater. Appl. **36**(8), 1503–1523 (2022). https://doi.org/10.1177/08853282211058099
80. Lv, L., Cheng, Y., Zheng, T., Li, X., Zhai, R.: Purification, antioxidant activity and antiglycation of polysaccharides from *Polygonum multiflorum* Thunb. Carbohyd. Polym. **99**, 765–773 (2014). https://doi.org/10.1016/j.carbpol.2013.09.007
81. Lai, X.-X., Li, Y.-P.: Antitumor effect and mechanism of action of polysaccharides extracted from *Polygonum perfoliatum* L whole plant in human lung carcinoma A549 cell line. Trop. J. Pharm. Res. **15**(6), 1243–1249 (2016). https://doi.org/10.4314/tjpr.v15i6.17
82. Georgiev, Y.N., et al.: Structural characterization of polysaccharides from *Geranium sanguineum* L. and their immunomodulatory effects in response to inflammatory agents. J. Nopharmacol. **294**, 115390 (2022). https://doi.org/10.1016/j.jep.2022.115390
83. Georgiev, Y.N., et al.: The common lavender (*Lavandula angustifolia* Mill.) pectic polysaccharides modulate phagocytic leukocytes and intestinal Peyer's patch cells. Carbohydr. Polym. **174**, 948–959 (2017). https://doi.org/10.1016/j.carbpol.2017.07.011
84. Georgiev, Y.N., et al.: *Tilia tomentosa* pectins exhibit dual mode of action on phagocytes as β-glucuronic acid monomers are abundant in their rhamnogalacturonans I. Carbohyd. Polym. **175**, 178–191 (2017). https://doi.org/10.1016/j.carbpol.2017.07.073

Methods of Treatment of Congenital Deformities of the Musculoskeletal System: Talipes Equinovarus

Maria Dragomirova(✉) and Marina Yaneva

Faculty of Medicine, University "Prof.d-R Assen Zlatarov" - Burgas, Burgas, Bulgaria
maria5k@abv.bg

Abstract. Talipes equinovarus, also called clubfoot, is characterized by plantar flexion, inward tilting of the heel (from the midline of the leg), and adduction of the forefoot (medial deviation away from the leg's vertical axis). Idiopathic congenital talipes equinovarus is the most common pediatric deformity and occurs in 1 in every 1000 live births. Even though it has been widely researched, the etiology of the condition remains poorly understood and is often described as being based on different factors. Genetic and environmental factors seem to have a major role in the development of this disease. If left untreated, it can result in long-term disability, deformity and pain. Interventions can be conservative (such as splinting or stretching) or surgical. Prenatal screening is key as treatment usually begins days after birth. Typical management of talipes equinovarus is non-operative and utilizes serial manipulation and casting, followed by bracing with Ponseti method. For cases unresponsive to the conservative methods, a variety of surgical approaches is used to address the deformity and sustain a full correction of the musculoskeletal system. A key point for full recovery is the utilization of intensive physiotherapy in the posttreatment period. We follow the case of a male patient, who was diagnosed with talipes equinovarus prenatally and observe his non – surgical and surgical treatment over the course of 28 months, which resulted in full recovery.

Keywords: talipes equinovarus · ponseti method · clubfoot

1 Introduction

1.1 Definition

Talipes equinovarus, known also as clubfoot, is a group of congenital deformities of the lower limb that are usually present at birth. They are characterized by a complex malalignment of the foot, involving soft and bony structures in the rear, middle and forefoot. The shortened tendons determine the external manifestation of the deformity: the child's foot is turned inward, and the heel points up. The frequency of the disease is twice as common (2:1) in men than in women. In 50% of cases, the deformity is bilateral. There are data in the literature that the disease occurs mostly in first-born

S. Sotirov et al. (Eds.): BioInfoMed 2022, LNNS 658, pp. 192–196, 2023.
https://doi.org/10.1007/978-3-031-31069-0_18

boys. For parents with no family medical history of the disorder, who have other healthy children, the chance of having a newborn with this condition is 1 in 1000.

The name *Talipes Equinovarus* originates from: [1]

"Tali" means ankle.

"Pes" means leg.

"Equinus" means leg, which is pointed down.

"Varus" means deviated to the midline.

1.2 Etiology

The etiology of congenital *talipes equinovarus* is still unclear. Suspected causes are intrauterine compression of the foot due to lack of amniotic fluid, delay in the development of the hock bone and intrauterine polio, but in most cases, the disease is idiopathic. It affects 1 in 1000 newborns and, according to multiple sources, is more common in first-born boys. It is possible that the disease will develop in the next pregnancy. There is no data on the role of maternal age, dietary habits such as vegetarianism or stress on fetal development (Table 1).

Table 1. Risk factors for developing talipes equinovarus

Family history of clubfoot
Active smoking
Use of drugs
Other congenital deformities – spina bifida [3]

1.3 Methods

We present the clinical case of patient X, following his development in the prenatal, postnatal and postoperative period. After a routine prenatal examination at 19 weeks' gestation, the mother is referred for 3D ultrasound and the diagnosis of bilateral talipes equinovarus is confirmed. The parents are informed about the specifics of the condition, after which they consult with two more specialists. A postnatal treatment plan is prepared in order to correct the deformity and restore normal musculoskeletal function. It includes a conservative and surgical approach. The conservative one is manual reposi-tioning and placement in a plaster cast every 14 days for four months, and the surgical one – Achilles tendon lengthening, which is performed surgically at the age of 5 months in order to release the tension in the structure of the musculoskeletal system of the foot.

2 Results

2.1 Prenatal Period

Patient X's mother is 31 years old and this is her first pregnancy. There is no history of concomitant diseases before and during pregnancy. The conception occurred naturally without medication or hormone therapy.

2.2 Postnatal Period

Despite the diagnosis, the birth is normal. 10 days later, patient X begins treatment with the Ponseti method, which continues for four months. The method is conservative and involves manual adjustment of the feet in the correct position, after which the lower limbs are fixated with a plaster cast [2]. Every 14 days, repositioning is carried out, and one day before that, the parents themselves remove the cast and conduct gymnastics for the child. The goal is, through gradual and consistent adjustment, to minimize the deformity and for the feet to regain their normal structure as much as possible, and placement in a plaster cast prevents the return of the musculoskeletal system to the original deformed position.

2.3 Surgical Intervention

At 5 months of age, Achilles tendon lengthening surgery is conducted under full anaesthesia. This is a surgical procedure in which a small incision is made in the Achilles tendon, allowing it to stretch and lengthen, releasing tension in the adjacent structures of the foot. Duration is 1 hour. Then the patient is placed in a cast again for one month, but without repositioning of the feet and without changing the cast.

2.4 Ponseti Method

After performing the surgical intervention, the improvement of patient X's condition begins. At the age of 7 months, the child begins to wear specialized Ponseti shoes, which fixate the feet again, but allow significant movement of the entire lower limb. The parents start taking the child to rehab once a week, but soon after, due to the COVID-19 pandemic, they are forced to conduct the specific exercises themselves in a home environment without specialized help. At the age of 11 months, the first progress is evident: the patient slowly begins to crawl, but the left leg does not participate actively enough and lags behind. At the age of 14 months, the child takes its first independent steps, but still lacks stability. Rehabilitation continues, but the child begins to wear the specialized shoes only at night. In the present, at 28 months of age, the function of the musculoskeletal system is fully recovered, but the Ponseti shoes are worn at night in order to fixate the position of the feet as well as possible, while the child is still in the period of growth and development.

If the condition is not treated at early age, it leads to adaptive and degenerative changes in the structure of the foot. It cannot move up and down as it normally would, and this can cause the child to walk on the side of the foot.

2.5 Palliative Care

Palliative care is specialized medical care for people living with a serious illness. This type of care is focused on providing relief from the symptoms and stress of the illness. The goal is to improve quality of life for both the patient and the family. The medical staff and other participants in the process of healing must provide it in order to optimize the results of the treatment. In the case of patient X, only a small part of palliative care is applied.

We recommend the following for patients with *congenital talipes equinovarus:*

1. A geneticist consultation for the mother and child. Several recent genetics studies have identified the PITX1-TBX4 transcriptional pathway as being important in clubfoot etiology.
2. A strict physiotherapeutic plan for treatment to be applied after permanent removal of the casts or surgical intervention.
3. After each cast replacement, the foot's blood supply to be checked by pressing and releasing the toes. In one case, the patient's cast was too tight and the foot started turning blue after he was already under parents' custody.
4. The child must undergo a check-up exam every six months after the end of treatment to exclude the possibility of relapse of the condition.

3 Discussion

The conservative treatment can be executed by Ponseti or Kite's method. The latter consists of a series of manipulations and castings followed by night splinting with the feet held in dorsiflexion and slight abduction. The difference between the two is that Kite's aims to gradually correct each deformity by itself, starting with the mediotarsal adduction, followed by correction of the internal rotation calcanopedal block and the calcaneal varus, and finally the equinus, while the Ponseti method aims to to correct the defining deformities in the same time, whereas the equinus will be corrected lastly.

A number of researches [4–6] evaluate Ponseti vs Kite, using the Dimeglio's scoring system prior and after treatment to compare the effectiveness. Dimeglio's classification is applied to assess the severity of the condition. It takes into account 4 major and 4 minor criteria [7]. The major criteria target the reducibility of the 4 malpositions, as follows: equinus, varus, adduction of the forefoot, and suppination. The grading is from 0 (reducible, almost normal) to 4 (nonreducible, very severe). Minor criteria are: posterior groove, medial groove, cavus, and affected muscle function, which are all awarded 1 point each of present and 0 points of absent. Thereby a score between 0 and 20, and 0 points represents a normal foot; also, an increasing score means a greater severity. The Dimeglio scoring system distinguishes 4 categories: benign (stage I), moderate (stage II), severe (stage III), and very severe (stage IV).

It is established that the Ponseti method is superior to Kite's due to higher correction rate in shorter treatment time and statistically there are less relapses, and in benign and moderate stages, surgical intervention is usually not required. Thus, during the last 20 years, this method replaced Kite's, which was the primary solution in the past century. However, a risk of Ponseti method is over-correction and stiff scar healing. [8]

The vast majority of idiopathic clubfeet are also treated with a percutaneous Achilles tendon tenotomy once the conservative approach is finished. For the resistant cases, a posterior release operation is necessary to release the extensive soft tissue. The surgical treatment is applied in the cases when the serial casting did not result in full correction or, on the contrary, resulted in overcorrection. Due to it is minimally invasive approach and fast recovery in the matter of days, the intervention is considered safe for infants and toddlers. It is also the first choice of treatment in cases of clubfoot, which have been neglected for years and patients seek correction at an older age.

4 Conclusion

There are probably more than one various causes and factors that act in combination to lead to this diagnosis. Medical treatment and palliative cares for a newborn with *talipes equinovarus* begin as early as 1 week of age and continue until about 5 years of age or more depending on the extent of the disease. Prenatal screening is key to diagnosis, and postnatal treatment and postoperative rehabilitation combined with adequate palliative care provide a favourable prognosis for full recovery.

References

1. Africa Clubfoot Training Project. Chapter 2 Africa Clubfoot Training Basic & Advanced Clubfoot Treatment Provider Courses Participant Manual. University of Oxford: Africa Clubfoot Training Project (2017)
2. Dietz, F.R., Noonan, K.: Treatment of clubfoot using the Ponseti method. JBJS Essent. Surg. Tech. 10;6(3), e28 (2016). https://doi.org/10.2106/JBJS.ST.14.00112
3. Mammen, L., Benson, C.B.: Outcome of fetuses with clubfeet diagnosed by prenatal sonography. J. Ultrasound Med. **23**, 497–500 (2004). https://doi.org/10.7863/jum.2004.23.4.497
4. Derzsi, Z., Nagy, Ö., Gozar, H., Gurzu, S., Pop, T. S.: Kite versus ponseti method in the treatment of 235 feet with idiopathic clubfoot **94** (33), e1379 (2015)
5. Sud, A., Tiwari, A., Sharma, D., Kapoor, S.: Ponseti's vs Kite's method in the treatment of clubfoot-a prospective randomised study. Int. Orthop. **32**(3), 409–413 (2008). https://doi.org/10.1007/s00264-007-0332-y
6. Sanghvi, A.V., Mittal, V.K.: Conservative management of idiopathic clubfoot: Kite versus Ponseti method. J. Orthop. Surg. **17**(1), 67–71 (2009)
7. Dimeglio, A., Bensahel, H., Souchet, P., et al.: Classification of clubfoot. J. Pediatr Orthop. B **4**, 129–136 (1995)
8. Delbrück, H., Schaltenbrand, M., Schröder, S., et al.: Clubfoot treatment through the ages: the Ponseti method in comparison to other conservative approaches and operative procedures. Orthopade **42**, 427–433 (2013)

Rehabilitation Approach After Arthroscopic Rotator Cuff Repair

Gergana Angelova-Popova(✉)

Department of Health and Pharmaceutical Care, Medical College, University "Prof. D-R Asen Zlatarov", Burgas, Bulgaria
giggi_popa@abv.bg

Abstract. Rotator cuff tears are often the cause of incapacitating shoulder pain, reduced shoulder function, and compromised joint mechanics with clinical manifestations of shoulder stiffness, weakness, instability and limitation of daily activities. The arthroscopic approach is increasingly used due to potential benefits such as minimal invasiveness, less postoperative pain while allowing secure fixation and early return of range of motion. Although arthroscopic rotator cuff repair appears to be a relatively mild procedure, postsurgical rehabilitation is critical in terms of long-term recovery of the reconstructed tendons. The main goal of rehabilitation after rotator cuff tear is to reduce the stress on the operated tissues and improve their healing, while preventing muscle stiffness and atrophy.

Keywords: rotator cuff tears · physical therapy · rehabilitation · shoulder

1 Introduction

Rotator cuff tears are often the cause of incapacitating shoulder pain, reduced shoulder function, and compromised joint mechanics with clinical manifestations of shoulder stiffness, weakness, instability and limitation of daily activities [37, 39].

Etiological factors usually include traumatic injury or, more commonly, age-related degenerative changes in the tendons, prolonged repetitive, stressful activities, subacromial impingement, which lead to partial and later complete tears. Asymptomatic partial and complete tears of the rotator cuff occur in 4% of people under 40 years of age and in more than 50% of people over 60 years of age [3]. Moreover, the rotator cuff damages are one of the most common complications after proximal humeral fractures, treated with or without surgical intervention accociated with poor functional outcomes [14–17]. Surgical repair via open or arthroscopic techniques is associated with improved function and patient satisfaction. The arthroscopic approach is increasingly used due to potential benefits such as minimal invasiveness, less postoperative pain [38] while allowing secure fixation and early return of range of motion [25]. Although arthroscopic rotator cuff repair appears to be a relatively mild procedure, postsurgical rehabilitation is critical in terms of long-term recovery of the reconstructed tendons [30]. Postoperative rehabilitation goals are to preserve the repaired muscle and tendon and recover the biomechanics and functionality of the shoulder by balancing the scapulothoracic and

S. Sotirov et al. (Eds.): BioInfoMed 2022, LNNS 658, pp. 197–204, 2023.
https://doi.org/10.1007/978-3-031-31069-0_19

glenohumeral force couples [19]. The main goal of rehabilitation after rotator cuff tear is to reduce the stress on the operated tissues and improve their healing, while preventing muscle stiffness and atrophy [40].

The aim of the present study is to specify the kinesitherapeutic techniques in different phases of rehabilitation and to present our own experience in patients with arthroscopic rotator cuff repair. As a result, two main approaches have been proposed in the literature regarding the optimal rehabilitation algorithm in the early postoperative phase after rotator cuff surgery. One program is accelerated, allowing early passive motion in an attempt to reduce postoperative stiffness. However, there is a risk that this rehabilitation regimen will lead to an increased rate of re-rupture. The second, more traditional program involves a more conservative rehabilitation sequence with immobilization for 4 - 6 weeks after surgery to protect tendon integrity. An analysis of multiple case-control studies revealed that a delayed kinesitherapy approach after arthroscopic rotator cuff rupture resulted in a lower recurrence rate, a higher healing rate, and an accelerated, early protocol resulted in a faster recovery of range of motion, but also has a higher recurrence rate than re-rupture [2]. For proper rehabilitation, it is important to know all the necessary details about the surgery, such as the size of the tear, the exact tendons that are affected and the quality of the tissues and repair technique used, in order to establish a suitable protocol for the patient. Rotator cuff repair is followed by a program consisted of four commonly used and accepted phases, beginning with the aim to protect mainly the repair in the immediate postoperative phase and gradually return to preoperative activity level.

PHASE I: Passive ROM (0 to 4–6 Weeks)

Protection of the repaired tissue should be the focus of the initial phase of postoperative rehabilitation, to ensure an adequate balance between assisting tendon healing and preventing the development of postoperative adhesions. This is the healing phase. The strength of the repaired tendon is determined and depends solely on the strength of the sutures and the anchors used to grip the bone. For this reason, no active movements of the shoulder are recommended until 6 weeks, as they may disrupt the attachment of the tendon to the bone. By week 4, the strength of the restored tendon is about 20% of normal.

The goals of this initial phase are to preserve the integrity of the repaired tendon without inducing stress (protection of the surgical repair), gradually increase passive range of motion, reduce swelling, inflammation and minimize pain, prevent muscle inhibition.It is recommended immobilization with an abduction pillow brace at 30°-45° abduction and neutral position with respect to rotation for 4–6 weeks. It is widely described that early passive motion may be beneficial in helping the process of tendon to bone healing. After the 3rd week, passive exercises are first included, performed carefully and controlled within a safe range. Passive ROM exercises should be performed gently. Pendulum exercises can also be performed, cryotherapy can be helpful regarding postoperative pain and inflammation. It is extremely important to make a strict differentiated judgment and to approach analytically to the affected muscles, so that when performing passive movements, stretching and unnecessary stress is not caused to the affected muscles, especially for rotators.Passive external and internal rotation in the scapular plane up to 20°-30° the first 3 weeks) and up to 45° by the end of the 6th week, with the arm positioned at 20 to 45 degrees in the scapular plane, as it has been shown that rotation with the arm in adduction increases stress in the repaired tissues [11]. Similarly, external rotation beyond 20 degrees should be avoided if the operative intervention affects the m. subscapularis.

At this stage, no self-assisted exercises (with a stick) and active-assisted exercises for the shoulder joint and no active movements for the shoulder girdle are applied until week 3. Retraction of the scapula is avoided when restoring the m. teres minor.

PHASE 2: ACTIVE RANGE OF MOTION (4–6 TO 12 WEEKS)

After 4 weeks, tendon recovery is approximately 20% of normal strength, which is sufficient to allow assisted and active movements to be performed. They include: flexion using the strong arm with the elbow flexed (short lever), flexion with a cane, external rotation with a cane, elevation to 90° from the side lying. By 6 weeks postoperatively, the restorative and inflammatory processes in the healing phase progress to a stage of collagen remodeling. The healing of the tendons to the bones is progressively increased and is able to withstand the applied muscle forces generated during light active movements such as simply raising the arm forward. This low-intensity active muscle loading helps establish the orientation of collagen fibers during their maturation stage, ultimately resulting in increased (tensile strength) of the repaired construct [27]. Appropriate at this time are the use of pulleys, sticks, actively assisted and auto-assisted range of motion. At this stage, submaximal isometric contractions can be initiated for external and internal rotation. The scapulothoracic articulation should also continue to be a focus of therapy as it will allow for the optimal neuromuscular control necessary to achieve maximal volume without pain. Exercises include depression, retraction of the scapula.

During this phase, the patient switches to isotonic and light closed-circuit stability exercises. Until the 8th week, passive stretching of the affected structures is not included, but only auto-assisted by the patient himself until a slight stretch, without pain. After the 8th - 10th week, can be applied stretching to the operated structures.

PHASE III strengthening phase (10–12 to 16–18 Weeks)

It should begin approximately 10 to 12 weeks postoperatively depending on the size of the tear and surgical fixation. In this phase, the histological remodeling phase is complete and the healing of tendons to bone is strong enough to allow a strengthening, loading program. The main goals of this phase are to achieve full passive pain-free ROM, optimize neuromuscular control and improve endurance.

Only when both glenohumeral and scapulothoracic kinematics are restored or range of motion is increased are resistance exercises allowed to increase muscle load. It is important to avoid any overhead activities that result in subacromial contact. Such contact will only exacerbate the impingement or stress the reconstruction.

The initial exercise for internal and external rotators should be performed with the arm below shoulder level. This strengthening phase begins with elastic resistance exercises, concentrating on high repetitions with moderate resistance. The purpose of these exercises is to build muscle endurance. After 10 weeks, stretching can also be applied to internal rotators to increase external rotation at 90 abd., hand behind the back, hand behind the back-internal rotation, stretching for internal rotators.

After 12 weeks, full pain-free range of motion should be achieved. After 12 weeks, resistance training from an elastic band for rotators, for serratus anterior - “bear hug”, push-ups, T, Y, W - exercises against resistance, PNF diagonals with an elastic band are included. Exercises to restore motor control such as rhythmic stabilization, closed kinetic chain exercises, with arm support, PNF diagonals against elastic band resistance. Typically, patients should regain at least 80–90% of their range of motion in this phase, unless they have had a large or massive tear. When the patient performs his daily activities without pain and tolerates any resistance exercise without pain, phase IV can proceed.

Phase IV - The advanced strengthening phase (16–26 weeks)
At this stage, the remodeling phase should be complete and the tissue relatively mature enough to withstand greater forces. To progressively strengthen the infraspinatus and teres minor, external rotation of the shoulder is performed at 45° of abduction using elastic resistance, and to activate the supraspinatus muscle, external rotation is performed at 90° of abduction. Additionally, exercise that strengthens the serratus anterior is the push-up with a plus progression, first starting with a push against a wall, then progressing to the edge of a table and finally to the floor. Exercises like the upper extremity plyometric exercises for improvement of the neuromuscular control, strength and proprioception are included, as well as different sports activities.

2 Material and Methods

The study was conducted with 6 patients (4 males and 2 females) after arthroscopic rotator cuff repair. The mean age was 59,5 years, ranging from 51 to 67 years. The dominant arm was injured in 4 patients.

Postoperative Rehabilitation
First four weeks, the arm was supported with an abduction sling. Patients started home therapy four weeks after surgery as prescribed by the surgeon and they performing pendulum shoulder exercises, passive forward elevation, active elbow extension and flexion and progressed towards active-assisted ROM to tolerance. The patients started their intensive rehabilitation program after approximately 2.5 months post operation. The focus of our therapy was first to overcome the disturbed scapulohumeral rhythm and restore the range of motion in the shoulder, and then to strengthen the muscle pairs and restore normal muscle control. To this end, we included muscle-inhibitory techniques and prolonged low-intensity stretching of the shortened muscles to pain level 3–4 according to VAS. For the latissimus, pectoralis, subscapularis and external rotators, with less intensity for the affected structures. From the beginning, we include PNF diagonal-spiral patterns - passively-assisted and then against light manual resistance to restore the dynamic stabilization and control of the scapula and External rotation by rotating a ball placed on a table with the elbow joint extended.We also apply the following technique introduced by us - a series of isometric contractions for the shortened muscles at different points of the arc of movement, followed by their stretching.As well as a series of isometric contractions for the weak muscles at different points of the arc of movement, followed by stretching of the shortened muscles - in the direction of the limited movement. We also used pulley stretching and exercises to strengthen the scapular stabilizers, and the rotator cuff with concentric and eccentric exercise strengthening, first against manual

resistance and then with an elastic band. We include joint-mobilization techniques grade 1–2 and mobilizations with movement techniques according to B. Mulligan to restore the arthrokinematics of the shoulder joint.

3 Results and Discussion

The measurements of the passive and active range of motion revealed a significant improvement in all planes, overcoming scapulo-humeral rhythm disorders. We performed with patients an average 12,2 procedures and they improved significantly range of shoulder motion compared to preoperative condition. Postoperative passive anterior elevation on 3 – 3,5 postoperative month was averaged 167,5°, external rotation averaged 42,5° and internal rotation was 66,7°. These measurements show that despite the later initiation of the rehabilitation program, there is a significant improvement in pain with movement, stiffness in the shoulder joint and a significant increase in range of motion in all planes.

Retear and stiffness are the main concerns of postoperative rehabilitation after rotator cuff repair. The risk of retear after surgical repair of small to intermediate tears are relatively frequent and range from 7% to 41% [20–22, 28, 29] and for massive tears range from 20 to 94% [4, 9, 10]. The postoperative period between six and twenty-six weeks is a risky period where the possibility of failure of tendon healing or retear is more frequent [1, 14, 41]. From a biomechanical point of view, normal levels of elasticity or strength of the repaired tendon are not restored until at least six months after surgery [13, 26]. This is the reason why prevent excessive loading, and judicious use of ROM is generally suggested.

Whereas, stiffness should be suspected especially in the first three months after arthroscopic rotator cuff repair [34, 35]. To decrease the possibility of developing potential adhesion and stiffness, several authors recommended early passive range of motion after rotator cuff repair [5, 36, 41]. Furthermore, limited early passive rehabilitation did not result in long-term stiffness after arthroscopic rotator cuff repair and may improve tendon healing rate [23].

To date, there is no consensus about which is the best rehabilitation protocol of surgically repaired rotator cuff, and available studies are heterogeneous [2]. Cuff et al. compared early and delayed passive motion demonstrating that there are no significant differences in terms of patient satisfaction, rotator cuff healing, or range of motion between the two groups. Furthermore, the study showed a slightly higher rotator cuff healing rate in favour of patients who received a delayed passive movement protocol. Riboh and Garrigues compared early and delayed rehabilitation protocol and concluded that there was no difference in cuff re-tears rates at one year for patients with tears less than 3 cm [31]. Kluczynski, showed an increased risk of cuff tear recurrence in patients with large tears between 3 cm and 5 cm who underwent an early rehabilitation protocol [20].

4 Conclusions

Arthroscopic repair of rotator cuff tear offers excellent functional outcome, and improvement in pain, function and strength of cuff tendons. Starting the rehabilitation program on time will necessarily lead to faster results. Nevertheless, the delayed postoperative physical therapy program that we presented in this study was associated with improvements in shoulder range of motion and muscle control, without the risk of compromising the recovery of the healing tendons. Moreover, the proposed rehabilitation approach will be used for the development of mathematical models and algorithms of the new imporved rehabilitation protocols.

Acknowledgements. The author is grateful for the support provided by the University "Prof. Dr. Asen Zlatarov"- Burgas, Scientific research sector "Scientific and artistic activity" approved by № NIH - 475 /2022 "Generalized nets models in orthopedic and traumatology rehabilitation".

References

1. Ahmad, S., Haber, M., Bokor, D.J.: The influence of intraoperative factors and postoperative rehabilitation compliance on the integrity of the rotator cuff after arthroscopic repair. J. Shoulder Elb. Surg. **24**, 229–235 (2015)
2. Bakti, N.T., Antonios, A., Phadke, B.: Singh early versus delayed mobilization following rotator cuff repair. J. Clin. Orthop. Trauma **10**, 257–260 (2019)
3. Bartolozzi, A., Andreychik, D., Ahmad, S.: Determinants of outcome in the treatment of rotator cuff disease. Clin. Orthop. Relat. Res. **308**, 90–97 (1994)
4. Bedi, A., Dines, J., Warren, R.F., Dines, D.M.: Massive tears of the rotator cuff. J. Bone Jt. Surg. Am. **92**, 1894–1908 (2010)
5. Brislin, K.J., Field, L.D., Savoie, F.H.: Complications after arthroscopic rotator cuff repair. Arthrosc. J. Arthrosc. Relat. Surg. **23**, 124–128 (2007)
6. Chang, K.-V., Hung, C.-Y., Han, D.-S., Chen, W.-S., Wang, T.-G., Chien, K.-L.: Early versus delayed passive range of motion exercise for arthroscopic rotator cuff repair: a meta-analysis of randomized controlled trials. Am. J. Sports Med. **43**, 1265–1273 (2015)
7. Cu, D.J., Pupello, D.R.: Prospective randomized study of arthroscopic rotator cuff repair using an early versus delayed postoperative physical therapy protocol. J. Shoulder Elb. Surg. **21**, 1450–1455 (2012)
8. Edwards, P.K., Ebert, J.R., Littlewood, C., Ackland, T., Wang, A.: A systematic review of electromyography studies in normal shoulders to inform postoperative rehabilitation following rotator cuff repair. J. Orthop. Sports Phys. Ther. **47**, 931–944 (2017)
9. Galatz, L.M., Ball, C.M., Teefey, S.A., Middleton, W.D., Yamaguchi, K.: The outcome and repair integrity of completely arthroscopically repaired large and massive rotator cuff tears. J. Bone Jt. Surg. **86**, 219–224 (2004)
10. Henry, P., et al.: Arthroscopic repair for chronic massive rotator cuff tears: a systematic review. Arthrosc. J. Arthrosc. Relat. Surg. **31**, 2472–2480 (2015)
11. Hersche, O., Gerber, C.: Passive tension in the supraspinatus musculotendinous unit after long-standing rupture of its tendon: a preliminary report. J Shoulder Elb. Surg. **7**, 393–396 (1998)
12. Huberty, D.P., Schoolfield, J.D., Brady, P.C., et al.: Incidence and treatment of postoperative stiffness following arthroscopic rotator cuff repair. Arthroscopy **25**, 880–890 (2009)
13. Huegel, J., Williams, A.A., Soslowsky, L.J.: Rotator cuff biology and biomechanics: a review of normal and pathological conditions. Curr. Rheumatol. Rep. **17**, 476 (2015)
14. Hristov, S.: Improved functional outcome in surgical treatment of proximal humeral fractures with Angular stable plate (LISS) and PMMA. J. Orthop. Trauma. **57**, 4 (2020). [in Bulgarian]

15. Hristov, S.: Necessary augmentation of proximal humerus fractures, when and how? J. Emerg. Med. **23**(1), 15–20 (2020). [in Bulgarian]
16. Hristov, S., Baltov, A., Sotirov, S.: Functional outcome prediction of operated proximal humerus fractures by means of artificial neural networks. In: Sotirov, S.S., Pencheva, T., Kacprzyk, J., Atanassov, K.T., Sotirova, E., Staneva, G. (eds.) Contemporary Methods in Bioinformatics and Biomedicine and Their Applications. BioInfoMed 2020. Lecture Notes in Networks and Systems, vol. 374. Springer, Cham (2022). https://doi.org/10.1007/978-3-030-96638-6_23
17. Hristov, S., Baltov, A., Sotirova, E., Bozov, H.: Intuitionistic fuzzy evaluations for analysis of the proximal humerus fractures. In: Sotirov, S.S., Pencheva, T., Kacprzyk, J., Atanassov, K.T., Sotirova, E., Staneva, G. (eds.) Contemporary Methods in Bioinformatics and Biomedicine and Their Applications. BioInfoMed 2020. Lecture Notes in Networks and Systems, vol. 374. Springer, Cham (2022). https://doi.org/10.1007/978-3-030-96638-6_30
18. Iannotti, J.P., et al.: Time to failure after rotator cuff repair: a prospective imaging study. J. Bone Jt. Surg. **95**, 965–971 (2013)
19. Kawano, Y., et al.: Evaluation of the translation distance of the glenohumeral joint and the function of the rotator cuff on its translation: a cadaveric study. Arthrosc. J. Arthrosc. Relat. Surg. **34**, 1776–1784 (2018)
20. Kluczynski, M.A., Nayyar, S., Marzo, J.M., Bisson, L.J.: Early versus delayed passive range of motion after rotator cuff repair: a systematic review and meta-analysis. Am. J. Sports Med. **43**, 2057–2063 (2015)
21. Kluger, R., Bock, P., Mittlböck, M., Krampla, W., Engel, A.: Long-term survivorship of rotator cuff repairs using ultrasound and magnetic resonance imaging analysis. Am. J. Sports Med. **39**, 2071–2081 (2011)
22. Lafosse, L., Brzoska, R., Toussaint, B., Gobezie, R.: The outcome and structural integrity of arthroscopic rotator cuff repair with use of the double-row suture anchor technique. J. Bone Jt. Surg. Am. **90**, 275–286 (2008)
23. Lee, B.G., Cho, N.S., Rhee, Y.G.: Effect of two rehabilitation protocols on range of motion and healing rates after arthroscopic rotator cuff repair: aggressive versus limited early passive exercises. Arthrosc. J. Arthrosc. Relat. Surg. **28**, 34–42 (2012)
24. Lee, Y.S., Jeong, J.Y., Park, C.-D., Kang, S.G., Yoo, J.C.: Evaluation of the risk factors for a rotator cuff retear after repair surgery. Am. J. Sports Med. **45**, 1755–1761 (2017)
25. Lin, J.C., Weintraub, N., Aragaki, D.R.: Nonsurgical treatment for rotator cuff injury in the elderly. J. Am. Med. Dir. Assoc. **9**, 626–632 (2008)
26. Longo, U.G., Petrillo, S., Rizzello, G., Candela, V., Denaro, V.: Deltoid muscle tropism does not influence the outcome of arthroscopic rotator cuff repair. Musculoskelet. Surg. **100**(3), 193–198 (2016). https://doi.org/10.1007/s12306-016-0412-5
27. Long, J.L., Ruberte Thiele, R.A., Skendzel, J.G., Jeon, J., Hughes, R.E., et al.: Activation of the shoulder musculature during pendulum exercises and light activities. J. Orthop. Sports Phys. Ther. **40**, 230–237 (2010)
28. McElvany, M.D., McGoldrick, E., Gee, A.O., Neradilek, M.B., Matsen, F.A.: Rotator cuff repair: published evidence on factors associated with repair integrity and clinical outcome. Am. J. Sports Med. **43**, 491–500 (2015)
29. Miller, B.S., et al.: When do rotator cuff repairs fail? Serial ultrasound examination after arthroscopic repair of large and massive rotator cuff tears. Am. J. Sports Med. **39**, 2064–2070 (2011)
30. Norberg, F.B., Field, L.D., Savoie, F.H., III.: Repair of the rotator cuff: mini-open and arthroscopic repairs. Clin. Sports Med. **19**(1), 77–99 (2000)
31. Riboh, J.C.; Garrigues, G.E.: Early passive motion versus immobilization after arthroscopic rotator cuff repair. Arthrosc. J. Arthrosc. Relat. Surg. **30**, 997–1005 (2014). Osteology (2021). 1 38

32. Sher, J.S., Uribe, J.W., Posada, A., Murphy, B.J., Zlatkin, M.B.: Abnormal findings on magnetic resonance images of asymptomatic shoulders. J. Bone Joint Surg. Am. **77**(1), 10–15 (1995)
33. Sonnabend, D.H., Howlett, C.R., Young, A.A.: Histological evaluation of repair of the rotator cuff in a primate model. J. Bone Joint Surg. Br. **92**–B, 586–594 (2010)
34. Tauro, J.C., Paulson, M.: Shoulder stiffness. Arthrosc. J. Arthrosc. Relat. Surg. **24**, 949–955 (2008)
35. Tauro, J.C.: Stiffness and rotator cuff tears: incidence, arthroscopic findings, and treatment results. Arthrosc. J. Arthrosc. Relat. Surg. **22**, 581–586 (2006)
36. Thomopoulos, S., Williams, G.R., Soslowsky, L.J.: Tendon to bone healing: differences in biomechanical, structural, and compositional properties due to a range of activity levels. J. Biomech. Eng. **125**, 106–113 (2003)
37. Padmanabh, V., Shah, V., Patel, N., Chudasama, C., Shah, M., Prabhakar, M.: Results of arthroscopic rotator cuff repair. Int. J. Orthop. Sci. **4**(1), 652–655 (2018). https://doi.org/10.22271/ortho.2018.v4.i1j.93
38. Yamaguchi, K., Levine, W.N., Marra, G., Galatz, L.M., Klepps, S., Flatow, E.L.: Transitioning to arthroscopic rotator cuff repair: the pros and cons. Instr. Course Lect. **52**, 81–92 (2003). [PMID: 12690842]
39. Yamamoto, A., Takagishi, K., Osawa, T., Yanagawa, T., Nakajima, D., et al.: Prevalence and risk factors of a rotator cuff tear in the general population. J. Shoulder Elb. Surg. **19**(116–120), 2 (2010)
40. Van der Meijden, O.A., et al.: Rehabilitation after arthroscopic rotator cuff repair: current concepts review and evidence-based guidelines. Int. J. Sports. Phys. Ther. **7**(2), 197–218 (2012)
41. Zumstein, M.A., Jost, B., Hempel, J., Hodler, J., Gerber, C.: The clinical and structural long-term results of open repair of massive tears of the rotator cuff. J. Bone Jt. Surg. Am. **90**, 2423–2431 (2008)

3D Technologies in Urological Practice. Application of Software for 3D Processing in Urological Practice

Zlatka Cholakova(✉) and Nikolay Mirinchev(✉)

Department of Urology, University Hospital for Active Treatment, 73 Stefan Stambolov, 8000 Burgas, Bulgaria
z_cholakova@yahoo.com, drmirinchev@abv.bg

Abstract. The aim of our research is to use free software for 3D processing of DICOM files to predict the duration of laser lithotripsy in a group of stones with different 3D configuration and to show the application of 3D reconstructions in real urological practice. To show the advantages of preoperative counseling with the help of this software and to reflect this in the individual approach in each case when planning the operative intervention - the stages and selection of the most appropriate operative approach for the patient, choice of tools and energy source for disintegration. The stones. We have the opportunity to evaluate the advantages of the above method as a suitable tool in our daily practice.

Keywords: 3D reconstruction · preoperative evaluation · software · urolithiasis

Abbreviations

RK	Right kidney
LC	Lower calyx
RMU	Right middle ureter
LMU	Left middle ureter
RDU	Right distal ureter
LDU	Left distal ureter
RPU	Right proximal ureter
RLC	Right lower calix
RPy	Right pyelon

1 Introduction

In the surgical practice of the urologist, a number of new technologies come to the rescue, which facilitate daily surgical work. 3D reconstructions have entered many areas of our lives. Urology is a discipline that is constantly advancing in the technology of instruments and equipment and the improvement of (imaging methods) images in urology that help to better understand each case in clinical practice.

S. Sotirov et al. (Eds.): BioInfoMed 2022, LNNS 658, pp. 205–215, 2023.
https://doi.org/10.1007/978-3-031-31069-0_20

Computed axial tomography (CAT) has become the gold standard for diagnosing urolithiasis and managing the healing process - conservative or surgical. Gives clear ideas about the location of structures, volume, location, shape, size (longitudinal size) of the concretion.

3D reconstructed models derived from CAT provide an opportunity for making an accurate diagnosis and hence making the most adequate decision about choosing the right access to surgical treatment and to estimate other important characteristics of pathological process in particular many stone characteristics – surface, 3d configuration, localization, density and etc.

The miniaturization of urological instruments has contributed to great progress in urological surgical practice. Surgical access through the natural openings of the human body and in particular the urological surgery of the natural openings has contributed to a very fast and effective management of many urological conditions.

The 3D reconstructions and software programs processing DICOM files are very important in the preoperative preparation of patients with urolithiasis and are increasingly used in the daily routine of the urologist.

3d reconstructions complement the images obtained from computed tomography and help to better understand the images obtained, the volume of the studied object, its configuration, location and its relationship with other structures. Builds in the diagnostic and therapeutic thinking of the urologist cognitive fusion of images obtained from 3D models and intraoperative endoscopic picture. The advantages are a deeper understanding of the problem and facilitate its further surgical treatment (solution).

3d reconstructions from CAT performed in patients with urolithiasis and software products for processing and manipulation of such images as 3D Slicer and Radiant allow to create applications for predicting the duration of surgery depending on the use of different types of laser energy and different controllable settings of laser energy - the type of pulse, the frequency of the pulse, the duration of the pulse and its power, calculation of the volume of the stone.

In 2020 A pilot study was published on the application of free 3D software for kidney stones and their surgical planning using the Kidney stone calculator extension was published [1].

The Department of Urology at the University Hospital in Burgas uses the same free software to determine the volume of concretions and the expected (approximate) duration of the endoscopic surgical intervention, and to show the application of 3D reconstructions in real urological practice on 9 patients with urolithiasis. We want to show the advantages of preoperative counseling with the help of this software and to reflect this in the individual approach in each case when planning the operative intervention - the stages and selection of the most appropriate operative access for the patient, choice of tools and energy source for disintegration of the stones. We have the opportunity to evaluate the advantages of the above method as a suitable tool in our daily practice.

2 Materials and Methods

Our study was prospective on 9 patients admitted to our ward for the first endoscopic treatment of urolithiasis detected by computed tomography. Age, sex, laterality, number of stones, location, X-ray measurements, previous operations, time for lithotripsy (TTL) were analyzed. Preoperative assessment of urolithiasis was performed with native axial computed tomography. DICOM image files were used, which were processed with two free software programs for 3D reconstructions - 3D Slicer and Radinat. We present an algorithm of data processing and work with 3D Slicer:

Open 3DSlicer and download "KidneyStoneCalculator" using the "Extension Manager"

Import DICOM data from a CT-scan (better with non-enhanced series)

Open "Crop Volume" module "Create a new annotation ROI" (Region Of Interest)

Size the ROI in x, y and z axes using axial (red), sagittal (yellow) and coronal (green) view in four-up disposition.

Create a new volume (name will be: "name cropped")

Click "Apply" button

Open "KidneyStoneCalculator" module:

Select volume: "name cropped"

Choose threshold (houndfields units) minimum and maximum to fit to the stone

If there are multiples stones select "split islands into segments" with a recommended "minimum size" of 40 voxels

Select a mode of treatment (laser source, core-diameter of laser fiber, laser settings and stone type)

Select "show 3D" if you want to visualize in 3D-view the segmented stone(s)

Click "Apply" button. A table will appear to show the segment's name, volume (mm^3) and time of lithotripsy (min) if the stone(s) and axial, coronal and sagittal views with CT-scan DICOMs (Fig. 1).

The stone load is estimated according to the following parameters: the maximum diameter (DM) corresponding to the diameter of the largest calculation, regardless of the considered cutting plane, the calculated area (SC) obtained after application of the RADIANT software program. All measurements in the study were performed by the same operator.

The volumes of the studied 17 stones were calculated in three different ways as follows:

1) Calculated volume of the concretion (VC1) according to the following formula: $VC1 = \pi * l * w * d * 0.167$, where.

L-length × w-width × D-depth × 3.14 × 0.167 as published [2, 3],

2) The stone volume estimated using Ackermann's formula (VC2): Volume = 0.6 × stone surface 1.27, as proposed in the recommendations of the EAU in 2009 [4].

3) The generation of three-dimensional volume (3D) of the concretions was performed by automated CT segmentation of the structures with 3D Slicer and the kidney stone calculator extension. The total 3D volume represents the cumulative volume of each stone. All measurements were performed by one operator using 3D Slicer and RAdinat.

"KidneyStoneCalculator" consists in a volumetric evaluation of kidney stones and also provides information on surgical duration. Moreover, it provides the estimated time of lithotripsy and consequently the operative time. It provides in a few seconds a 3D view.

The area of the stone was measured manually by the same operator by selecting the largest area of concretions in the coronal plane using Radiant (Fig. 5).

The largest diameter of the stone in the three planes (sagittal (AP), axial (T) and coronal (L)) was performed by Radiant (Fig. 5).

Density was measured with Radiant in Honsfield units (HU), representing mean density, minimum and maximum density and standard deviation (Fig. 5).

Chi-square independence test for a contingency table was applied to compare the variables according to the normality of their distribution (Table 2).

Pearson's correlation coefficient was used to estimate the relationship between lithotripsy time, calculus volume and density (Figs. 2 and 3).

ANOVA was performed to estimate each stone volume from three different ways (Fig. 4).

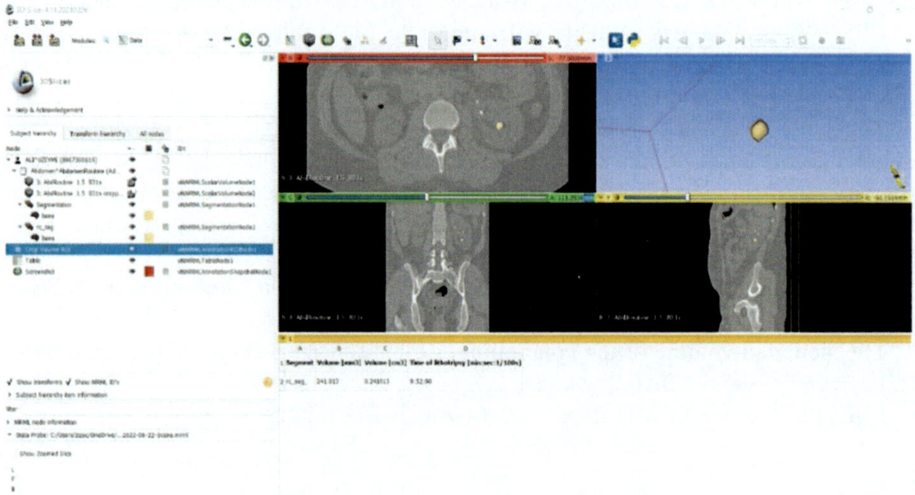

Fig. 1. Example result from 3D Slicer.

3 Results

Our study included 2 men (22%) and 7 women (78%). The average age is 45 years (min -max: 22y -71y). The total number of available stones is 17 (min-max: 1–6). All patients underwent native computed tomography peroperatively. In 5 patients a single calculus was found, and in the remaining 4 multiple lithiasis (min-max: 2–6). In 44% of patients the stone is localized in the lower calyx group of the kidney. In 22% the concretion is localized in the distal ureter. In 11% the concretion is localized in the proximal ureter. In 22% the stone is localized in the pyelonephritis of the kidney. In two of the patients, the calculus was located in the middle of the ureter. The average total 3D volume of the concretions is 421.06 mm^3 (min—max: —53.44–1845.49 mm^3) determined by a 3D Slicer - Kidney stone calculator. The average volume of concretions determined by Ackerman's formula (CV2) is 0.646 (min—max: —0.13–1.88), and with the Volume formula (CV1) the average value is 0.74 (min—max: —0.13) – 2.07). The average stone density is 1319 HU (min—max: —389.75–1169.07). The mean time for lithotripsy was 22.23 min (min—max: —4.22–80.21).

Partial coraliform stones were found in 3 patients 33%. In one of the patients an anatomical anomaly of the ureter was found - a bifurcated fused ureter with a double pyelocalyx system on the side of the calculus. One of the patients has undergone open surgery in the past - pyelithotomy on the same side of the stones. All preoperative descriptive features are presented in Table 1.

A chi-square independence test for a contingency table was performed to examine the relationship between the location of the pyelone stones, the proximal, middle, and distal parts of the ureter, and the lower calyx group with the time of lithotripsy, density and number of the stones. The ratio between these variables is significant, The chi-square statistic is 78.0923. The p-value is < 0.00001. The result is significant at $p < .05$.

Table 1. Chi-square independence test for a contingency table

Column and Row Totals						
	Pyelon	Distal ureter	middle ureter	upper ureter	lower calyx	***Row Totals***
Number of stone	2	3	3	1	8	17
Time to Lithotripsy	108	51	46	36	138	379
Stone Density	1033	686	540	743	742	3744
Column Totals	1143	740	589	780	888	**4140 (Grand Total)**

The P-Value is < .00001. The result is significant at $p < .05$.

Result Details & Calculation

X Values
$\Sigma = 381.93$
Mean = 22.466
$\Sigma(X - M_x)^2 = SS_x = 6547.072$

Y Values
$\Sigma = 7.23$
Mean = 0.425
$\Sigma(Y - M_y)^2 = SS_y = 3.249$

X and Y Combined
N = 17
$\Sigma(X - M_x)(Y - M_y) = 138.072$

R Calculation
$r = \Sigma((X - M_y)(Y - M_x)) / \sqrt{((SS_x)(SS_y))}$

$r = 138.072 / \sqrt{((6547.072)(3.249))} = 0.9467$

Meta Numerics (cross-check)
r = 0.9467

X Values	Y Values
48.56	1.09
9.52	0.24
7.5	0.16
36.24	0.44
38.25	0.47
4.22	0.04
11.33	0.14
14.29	0.18
27.50	0.62
9.35	0.21
4.21	0.20
31.28	0.71
8.46	0.20
29.21	0.36
16.38	0.20
80.21	1.84
5.42	0.13

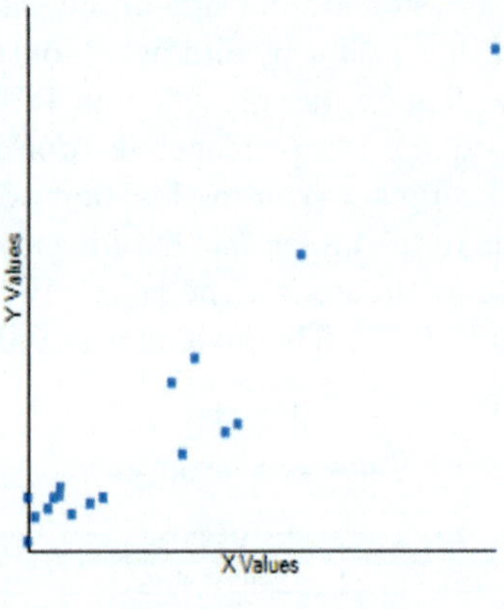

Fig. 2. Results from estimated Pearson's correlation coefficient between lithotripsy time (X Values) and volume of specific (Y Values) examined with 3D Slicer

There is a positive correlation between the time of lithotripsy and the volume of concretions and the relationship between the two quantities is statistically significant.

The P-Value is .037832. The result is significant at $p < .05$.

Summary of Data

	Treatments 1	2	3	4	5	Total
N	17	17	17			51
∑X	10.98	12.54	7.21			30.73
Mean	0.6459	0.7376	0.4241			0.603
$\sum X^2$	10.0436	16.7956	6.3171			33.1563
Std.Dev.	0.4295	0.6867	0.4513			0.5411

Result Details

Source	*SS*	*df*	*MS*	
Between-treatments	0.8834	2	0.4417	*F* = 11.13321
Within-treatments	13.7565	48	0.2866	
Error	1.2696	32	0.0397	

Fig. 4. ANOVA

A repeated measures ANOVA was performed to compare the effect of the volume measurements of 17 stones were performed according to three different formulas - CV1, CV2, 3D slicer.

The F-ratio value is 11.13321. The p-value is .000214. The result is significant at $p < .05$.

There is a positive correlation between the time for lithotripsy and the density of concretions and the relationship between the two quantities is statistically significant.

In the preoperative consultation after analyzing the results of 9 patients we proposed:

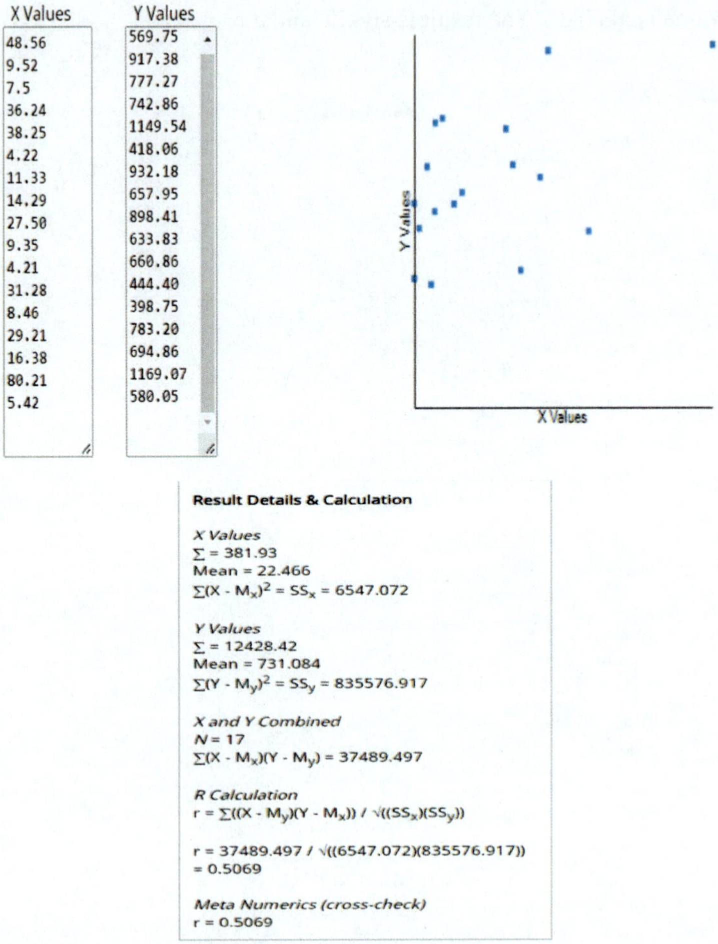

Fig. 3. Results from estimated Pearson correlation coefficient between lithotripsy time (X Values) and HU (Y Values) concrementite density tested with 3D Slicer

In patients with stone/s in the distal and middle urethra - one-act seven-rigid retrograde ureterorenoscopy with holmium lithotripsy and extraction of the resiginal fragments with a basket.

In patients with stones located in the pyelone, lower calicstan group and proximal ureter, whether multiple or single, we proposed a two-stage endoscopic approach. Presentation - unilateral or bilateral depending on the location of the calculus for 2 weeks. Second stage of retrograde intrarenal surgery with accessory sheath and single use flexible renoscopy with holmium laser lithotripsy with total disintegration of the calculus in dusting mode and without placement of a ureteral stent at the end of the operation.

4 Conclusion

Urolithiasis is a multifaceted and ancient disease. Modern treatment of urolithiasis is not unified and should be considered as the most appropriate method for the patient in accordance with its anatomical features, aesthetically pleasing and consistent with his age, comorbidities and personal beliefs and the capabilities of modern medicine - as tools and technologies and modern equipment.

Accurate estimation of the stone load, historically estimated by measuring the largest diameter of the largest stone [5], is essential in the choice of surgical technique. Because kidney stones are mostly irregular in shape and have a complex geometric structure, volume measurements made in two dimensions may be inaccurate and may not properly reflect this lithium load. If the recommendations of the scientific societies are based on the measurement of the maximum diameter of the stone [6], other measures are proposed: calculation of the volume or surface, cumulative diameters, volume measurements by three-dimensional reconstruction [6, 7].

The 3D reconstructed models of the stones facilitate preoperative consultation of the patient with urolitiaza and provide a clearer explanation of the patient's current condition, the complexity of the case and the options (alternatives) that the patient can choose as a therapeutic approach together with the urologist.

Free 3D CAT data processing software is widely used in everyday urological practice.

It can be used for all diseases of urological origin, not only for urolithiasis. The application in the preoperative consultation of patients with urolithiasis to help both the patient and the urologist for:

Better understanding of the case. Visual 3 dimensional representation of the pathological process.

Visual (through visualization) explanation of the patient about his current condition.

Alleged intervention of the operative intervention.

Choice for the most appropriate access to the calculus – endoscopic, percutaneous, conventional surgical.

Establishing congenital abnormalities or acquired conditions that lead to stone formation like (double ureters, bifurcated pyelons, confluence of the pyeloureteral joint, aberrant vessels, etc.) and helps to makes adequate decision-making for a preoperative planning.

Determination of the infundibulopelvic angle in concretions located in the lower calyx group and determination of the success of retrograde intrarenal surgery.

Determining the postoperative stone free status of patients.

Determining the stage of the operative intervention - one-act, two-act, three-act.

It can be used for training of students, postgraduates.

It can be used to perform software and to create mixed reality, virtual reality during surgical interventions and research in this field.

Allow 3D printing of the obtained 3D model.

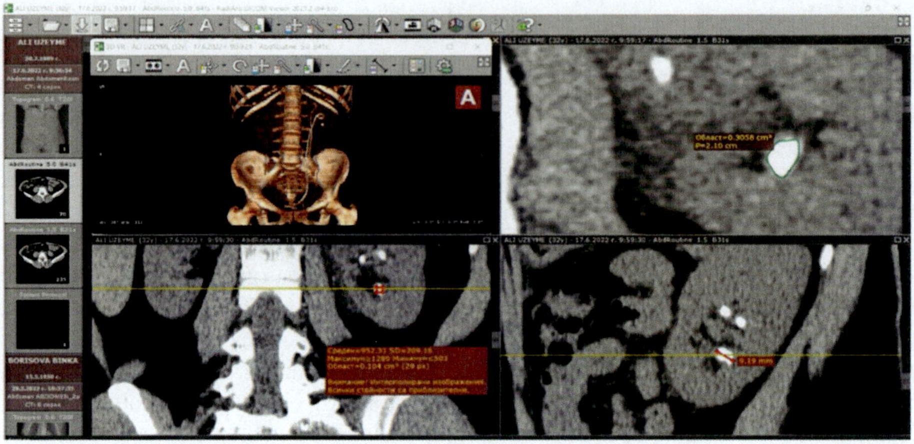

Fig. 5. Application of measurement with RADIANT

Table 2. Description of the study

Patient age Gender Stone number and localization	Stone number and localisation	Stone 3D Image From 3D Slicer	3D Slicer TTL	3D Slicer Vol (mm3)	3D Slicer VOL (sm3)	Diameter (mm) or (sm) AP	Diameter (mm) or (sm) T	Diameter (mm) or (sm) L	VC1	Area (sm2) RadiANT	VC2	Density (HU) Average	Density (HU) Max	Density (HU) Min	SD
♀/ 71y	1 RK/LC		48.56	1094.57	1.09457	2.06 sm	1.01 sm	1.90 sm	2.07	1.22	0.93	569.75	621	386	46
♀/ 32y	2 LK/LC		9.52	241.013	0.241013	7.86 mm	8.27 mm	9.79 mm	0.33	0.58	0.44	917.38	1139	650	150
♀/ 60y	3 RMU		7.5	162.806	0.162806	9.96 mm	6.03 mm	9.57 mm	0.30	0.42	0.32	777.27	889	583	88
♀/ 62y	4 RPU		36.24	444.023	0.444023	1.14 sm	1.44 sm	1.11 sm	0.95	1.15	0.88	742.86	1084	573	130
	5 RLC		38.25	468.535	0.468535	9.97 mm	1.15 sm	1.29 sm	0.77	0.92	0.70	1149.54	1415	408	272
	6 RLC		4.22	53.4435	0.04435	4.80 mm	7.74 mm	6.94 mm	0.13	0.51	0.39	418.06	469	368	28
	7RLC		11.33	141.043	0.141043	7.38 mm	8.15 mm	9.43 mm	0.29	0.41	0.31	932.18	1234	552	180
	8 RLC		14.29	176.805	0.176805	8.96 mm	1.07 sm	9.95 mm	0.50	0.44	0.33	657.95	868	511	97
♀/ 33y	9 RPy		27.50	624.676	0.624676	1.17 sm	1.09 sm	1.89 sm	1.26	1.40	1.07	898.41	1327	417	226
	10 RLC		9.35	215.263	0.215263	5.65 mm	1.21 sm	1.41 sm	0.50	0.95	0.72	633.83	780	543	59
	11 RLC		4.21	97.847	0.097847	6.67 mm	1.01 sm	7.58 mm	0.27	0.52	0.40	660.86	1045	277	194
♂/ 57y	12 RMU		31.28	706.26	0.70626	8.92 mm	8.69 mm	1.94 sm	0.79	1.32	1.00	444.40	498	378	28
	13 LMU		8.46	197.096	0.197096	4.92 mm	6.18 mm	8.20 mm	0.13	0.17	0.13	398.75	442	349	39
♀/ 37y	14 RDU		29.21	358.102	0.358102	7.72 mm	8.54 mm	2.54 sm	0.89	1.07	0.81	783.20	985	456	155
	15 LDU		16.38	202.988	0.202988	7.16 mm	6.13 mm	2.12 sm	0.49	0.51	0.39	694.86	1099	363	206
♂/36y	16 RPy		80.21	1845.49	1.84549	1.97 sm	1.15 sm	2.20 sm	2.61	2.47	1.88	1169.07	1289	1025	69
♀/ 22y	17 LDU		5.42	128.034	0.128034	6.46 mm	8.14 mm	8.56 mm	0.24	0.37	0.28	580.05	607	555	16

References

1. Panthier, F., et al.: Evaluation of a free 3D software for kidney stones' surgical planning: kidney stone calculator a pilot study. World J. Urol. **39**(9), 3607–3614 (2021). https://doi.org/10.1007/s00345-021-03671-z
2. Türk, C., Knoll, T., Petrik, A., et al.: Guidelines on Urolithiasis (EAU) (2013)
3. Breda, A., Ogunyemi, O., Leppert, J.T., et al.: Flexible ureteroscopy and laser lithotripsy for single intrarenal stones 2 cm or greater–is this the new frontier? J. Urol. **179**(3), 981–984 (2008)
4. Ackermann, D., Dunthorn, M., Newman, R.C., et al.: Calculation of stone volume and urinary stone staging with computer assistance. J. Endourol. **3**, 355–359 (1989)
5. M. Daudon, Doré, B.: Cristallographie des calculs urinaires. Encycl Méd Chir, pp. 18–104-A-25 (1999)
6. Assimos, D.: quantification of preoperative stone burden for ureteroscopy and shock, wave lithotripsy: current state and future recommendations editorial comment. J. Urol. **186**(3), 917 (2011)
7. Ben Saddik, M.A., Al-Qahtani Sejiny, S., Ndoye, M., et al.: Flexible ureteroscopy in the treatment of kidney stone between 2 and 3 cm. Prog. Urol. **21**(5), 327–332 (2011)

Emotional Intelligence of Students During Pandemic Outbreak. A Study in Higher Education

Milen Todorov[1](✉), Gergana Avramova-Todorova[1], Veselina Bureva[1], Cengiz Kahraman[2], and Guy De Tré[3]

[1] "Prof. Dr. Assen Zlatarov" University, "Prof. Yakimov" Blvd., Burgas 8010, Bulgaria
mtodorov@btu.bg, {g.avramova,vesito_ka}@abv.bg
[2] Istanbul Technical University, Management Faculty, Macka, Istanbul, Besiktas 34367, Turkey
kahramanc@itu.edu.tr
[3] Belgium Department of Telecommunications and Information Processing, Ghent University Gent, Ghent, Belgium
Guy.DeTre@UGent.be

Abstract. Emotional intelligence is a main area in educational psychology and a key factor in the academic life of students. In the current study the dimensions of emotional intelligence - self-awareness, managing emotions, motivating oneself, and social skills were examined in order to identify negatively affected aspects due to the COVID 19 outbreak.

Data was gathered from students from University "Prof. Dr. Asen Zlatarov", Burgas, Bulgaria through a self-completion electronic questionnaire which includes a total of 50 questions related to emotional intelligence. The obtained results are processed by making use of intuitionistic fuzzy set (IFS) techniques. It was found that there are is no significant influence of pandemic situation on emotional condition of the students. However, a set of elements associated to the dimensions of emotional intelligence should be taken into account in order to support the success of students during their education and further professional realization.

Keywords: emotional intelligence · COVID 19 outbreak · academic learning

1 Introduction

Nowadays, the role of emotional intelligence (EI) is widely recognized as an essential element in the process of academic students learning and their further professional realization [1]. EI could be related to student's achievements but it could be also related to psychological problems such as anxiety, depression, substance abuse, eating disorders, and other behavioral problems [2, 3]. The role of EI was shifted as key element during the COVID-19 outbreak due to the finding that college students have faced a variety of psychological and academic challenges [4, 5]. In general, the welfare of college students is considered to be affected by social isolation, uncertainty, and unforeseen transitions

S. Sotirov et al. (Eds.): BioInfoMed 2022, LNNS 658, pp. 216–222, 2023.
https://doi.org/10.1007/978-3-031-31069-0_21

compared to the daily life before pandemic changes. Thus it is expected that the ability to overcome major negative events depends on many intrinsic factors including emotional intelligence.

EI, is considered as significant element with effect for both - personal advancement and the process of organizational achievement. It becomes evident that the role of intelligence quotient (IQ) as single factor is not sufficient to predict academic performance of students. Hence, it is expected that students should not just achieve high academic results only but also to maximize individual perfection by the way of using social processes fruitfully in accordance with EI that they have already had.

Taking into account that EI is related to personal qualities toward effective interactions in daily life events, discussion continues about its actual definition and measurement. There are two widely mentioned models of EI. The first one called ability model which describes EI as "abilities that involve perceiving and reasoning abstractly with information that emerges from feelings". The second one is mixed model which defines EI as "an ability with social behaviours, traits, and competencies" [6]. By and large, different questionnaires are designed to assess individual's competencies in specific domains assessed by EI [7].

Emotional intelligence dimensions awareness, usage, understanding and controlling of emotions have positive effects on cognitive engagement [8]. Hence the evaluation of students based on questionnaire related to those dimension is expected to shed light on certain difficulties related to EI in the time of COVID 19 outbreak.

The current study aimed to explore the EI among students from University "Prof. Dr. Assen Zlatarov" by making use of intuitionistic fuzzy set (IFS) techniques. The study also aimed to identify the most negatively affected factors associated with awareness, usage, understanding and controlling of emotions.

2 Methodology for Assessment the Dimensions of EI

Electronic survey questionnaire is used to collect data. The questionnaire contains fifty statements. Each of the statements offered respondents choices on a 5-point Likert scale. The items were adapted and modified from the work of Schutte [9], and a literature review [10, 11]. Emotional intelligence, along with its domains such as self-awareness (10-items), managing emotions (10-items), motivating oneself (10-items), and social skills (10-items), was evaluated by all included items.

3 Data Processing by Intuitionistic Fuzzy Sets

The concept of intuitionistic fuzzy sets is proposed by Krassimir Atanassov in 1983. The intuitionistic fuzzy set is an extension of fuzzy set [12]. The function $\mu_A(x)$ defines the degree of membership of an element x to set A, evaluated in the interval [0; 1]. The function $\nu_A(x)$ defines the degree of non-membership of an element x to set A, evaluated in the interval [0; 1]. Thereafter the degree of uncertainty can be defined as $\pi_A(x) = 1 - \mu_A(x) - \nu_A(x)$. The intuitionistic fuzzy set has the following form $A = \{\langle x, \mu_A(x), \nu_A(x)\rangle | x \in E\}$. The Intuitionistic Fuzzy Pair (IFP) is an object with the form $\langle a, b\rangle$, where $a, b \in [0, 1]$ and the condition $a + b \leq 1$ is satisfied. Its components

(a and b) are interpreted as degrees of membership and non-membership. IFP is used as an estimation of some object or process [13]. Comparison between elements of any two IFSs is performed using pairwise comparisons between their respective elements' degrees of membership and non-membership to both sets.

The input dataset for intuitionistic fuzzy evaluation contains 50 questions containing the numerical answers of 80 respondents (students). The students answer using the numbers from 1 to 5 to assign its preferences of the employers. The numbers from the student's answers are distributed using the following scale:

- Rather positive answers (***p***) – the numbers 4 and 5;
- Rather uncertain answers (***a***) – the value 3;
- Rather negative answers (***n***) – the numbers 2, 1.
- The set of all answers (***s***) – the set of all values from 1 to 5.

The intuitionistic fuzzy evaluations are presented. The degree of membership and the degree of non-membership are calculated using the following formulas:

$$\mu = \frac{p}{s} \text{ and } \nu = \frac{n}{s}.$$

The degree of uncertainty has the following form:

$$\pi = \frac{a}{s}$$

The degree of membership and the degree of non-membership for all questions and answers are calculated in the form of intuitionistic fuzzy pairs.

Thereafter the optimistic, strongly optimistic, pessimistic and strongly pessimistic formulas are used to estimate the student emotional intelligence degree.

Optimistic formula:

$$\langle \mu_{n+1}, \nu_{n+1} \rangle = \langle \max \mu_{all}, \min \nu_{all} \rangle$$

where:

$$\mu_{all} = \{\mu_0, \mu_1, ..., \mu_n\}, \text{n} \in \{0, 1, 2, ..., m-1\},$$
$$\nu_{all} = \{\nu_0, \nu_1, ..., \nu_n\}, \text{n} \in \{0, 1, 2, ..., m-1\}$$

represent the positive estimation of the answers of the survey respondent to the selected questions.

Strongly optimistic formula:

$$\langle \mu_{n+1}, \nu_{n+1} \rangle = \langle \mu_{n+1} + \mu_n - \mu_{n+1}.\mu_n, \nu_n.\nu_{n+1}. \rangle$$

represent the strongly positive estimation of the answers of the survey respondent to the selected questions.

Pessimistic formula:

$$\langle \mu_{n+1}, \nu_{n+1} \rangle = \langle \min \mu_{all}, \max \nu_{all} \rangle$$

where:

$$\mu_{all} = \{\mu_0, \mu_1, ..., \mu_n\}, \text{n} \in \{0, 1, 2, ..., m-1\},$$
$$\nu_{all} = \{\nu_0, \nu_1, ..., \nu_n\}, \text{n} \in \{0, 1, 2, ..., m-1\}$$

represent the negative estimation of the answers of the survey respondent to the selected questions.

Strongly pessimistic formula:

$$\langle \mu_{n+1}, \nu_{n+1} \rangle = \langle \mu_{n+1}.\mu_n, \ \nu_{n+1} + \nu_n - \nu_n.\nu_{n+1} \rangle$$

represent the strongly negative estimation of the answers of the survey respondent to the selected questions.

The results of the survey estimations give us knowledge for the emotional condition of the respondents. The estimations can be interpreted as borders of the emotional intelligence of the students.

4 Evaluation of Student‘s Emotional Intelligence

The sample consisted of 80 students attending University “Prof. Dr. Asen Zlatarov” – Burgas, Bulgaria (Table 1). All participants completed electronic questionnaire based on items related to dimensions of EI. The theory of intuitionistic fuzzy sets is used to calculate the values for each statement by comparing belonging and not belonging real numbers in the interval [0, 1] and the sum of these numbers must also belongs to the interval [0, 1]. The survey was attended by 67% (54 students - Bachelors) and 33% (26 students - Masters). 77% of women and 23% of men participated in the study. The structure of the respondents does not differ from the structure of the University’s students, who are mostly female.

Table 1. Characteristics of participants in the survey

Characteristics of the analyzed population	value
Total participants	80
Male	18
Female	62
University degree Bachelor Master	54 26

Analysis of the data was focused on dimensions of EI - self-awareness, managing emotions, motivating oneself, empathy and social skills in order to identify those statements with low assessed values (below 0.7). The results are presented in Table 2.

Self-awareness is related to identification of emotion and understanding how emotions are related to one’s goal, thoughts, behaviors, and accomplishments [14, 15].

Table 2. Dimensions of EI and identified statements with low assessement

#	Competencies	Value
	Emotional self-awareness	
1	When I feel anxious I usually can account for the reason(s)	⟨0.560, 0.125⟩
2	I can let anger 'go' quickly so that it no longer affects me	⟨0.650, 0.113⟩
	Managing emotions	
4	I can consciously alter my frame of mind or mood	⟨0.538, 0.138⟩
5	Others can rarely tell what kind of mood I am in	⟨0.563, 0.138⟩
	Motivating oneself	
6	I do not prevaricate	⟨0.500, 0.188⟩
7	Delayed gratification is a virtue that I hold to	⟨0.500, 0.100⟩
	Empathy	
8	I can usually understand why people are being difficult towards me	⟨0.538, 0.088⟩
9	Reasons for disagreements are always clear to me	⟨0.60, 0.10⟩
10	I can understand why my actions sometimes offend others	⟨0.613, 0.075⟩
	Social Skills	
11	I need a variety of work colleagues to make my job interesting	⟨0.550, 0.088⟩
12	People are the most interesting thing in life for me	⟨0.563, 0.175⟩

Among the total number of statements two of them was assessed with low values (below 0.70) – "when I feel anxious I usually can account for the reason(s)" and "I can let anger 'go' quickly so that it no longer affects me".

The competence ***Managing emotions*** involves intentionally eliciting and sustaining pleasant and unpleasant emotions when considered appropriate, effectively channeling negative affect, and restraining negative emotional outbursts and impulses [14]. Here, the statements with low values are "I can consciously alter my frame of mind or mood" and "Others can rarely tell what kind of mood I am in".

Motivating oneself include components as internal strivings, attributions, and need for achievement [14] which enables the initiative to persevere in the face of obstacles and setbacks. The statements "I do not prevaricate" and "Delayed gratification is a virtue that I hold to" was assessed with low values.

The next dimension ***Empathy*** includes awareness of others' feelings, needs, and concerns, understanding and sympathising with others' emotions, and responding to others' unspoken feelings [14]. Three statements was found with low assessment – "I can usually understand why people are being difficult towards me", "Reasons for disagreements are always clear to me" and "I can understand why my actions sometimes offend others".

The last dimension ***Social skills*** is related to ability to manage, influence and inspire emotions in others. It is considered as essential foundation skills for successful teamwork and leadership. Here the statements "I need a variety of work colleagues to make my job interesting" and "People are the most interesting thing in life for me" was found with low assessment values.

Based on the total number of statements related to EI dimensions - self-awareness, managing of emotion, motivating oneself, empathy and social skills it can be concluded that cognitive engagement of students in mixed learning environment during COVID 19 outbreak is in general not negatively affected. However, a set of twelve statements related to dimensions of EI should be taken into account in developing counseling programs to support students to overcome identified difficulties.

The results of optimistic, strongly optimistic, pessimistic and strongly pessimistic estimations are presented in Table 3.

Table 3. Results of optimistic, strongly optimistic, pessimistic and strongly pessimistic estimations

№	*Competencies*	*Optimistic formula*	*Strongly optimistic formula*	*Pessimistic formula*	*Strongly pessimistic formula*
1–2	***Emotional self-awareness***	⟨0.650,0.100⟩	⟨0.846,0.014⟩	⟨0.500,0.188⟩	⟨0.364,0.224⟩
4–5	***Managing emotions***	⟨0.563,0.138⟩	⟨0.798,0.019⟩	⟨0.538,0.138⟩	⟨0.303,0.257⟩
6–7	***Motivating oneself***	⟨0.500,0.188⟩	⟨0.750,0.019⟩	⟨0.500,0.100⟩	⟨0.250,0.269⟩
8–9	***Empathy***	⟨0.600,0.088⟩	⟨0.815,0.009⟩	⟨0.538,0.100⟩	⟨0.323,0.179⟩
11–12	***Social Skills***	⟨0.563,0.088⟩	⟨0.803,0.015⟩	⟨0.550,0.175⟩	⟨0.310,0.248⟩

5 Conclusions

The psychological well-being of the students related to dimensions of emotional intelligence - self-awareness, managing emotions, motivating oneself, empathy and social skills was found slightly affected during the COVID 19 outbreak. The statements that deserves attention are highlighted and could be further investigated. This will be a suitable opportunity for intervention to improve the experience of university students in the development of academic and professional life projects. In order to increase the level of emotional intelligence of students organization of courses, seminars, and workshops, could be integrated in proper manner in the structure of educational programs.

Acknowledgments. The authors are thankful for the support provided by the European Regional Development Fund and the Operational Program "Science and Education for Smart Growth" under contract UNITe No. BG05M2OP001–1.001–0004-C01 (2018–2023).

References

1. Hasanzadeh, R., Shahmohamadi, F.: Study of emotional intelligence and learning strategies. Procedia Soc. Behav. Sci. **29**, 1824–1829 (2011)
2. Hunt, J., Eisenberg, D.: Mental health problems and helpseeking behavior among college students. J. Adolesc. Health **46**, 3–10 (2010)
3. Pajević, I., Žigić, N., Bećirović, E., Pajević, A.: Psychological disorders in childhood and adolescent age-new classifications. Psychiatr. Danub. **32**, 311–315 (2020)
4. Aslam, N., Ahmed, A.: Prospective impact of COVID-19 on adolescents: guidelines for interventions. Psychiatr. Danub. **32**, 603–604 (2020)
5. Lazzari, C., Shoka, A., Nusair, A., Mon Hein, S., Rabottini, M.: Clinical psychopathology during COVID-19 pandemic: case reports of first psychiatric presentations. Psychiatr. Danub. **32**, 597–601 (2020)
6. Goleman, D.: Emotional Intelligence. Bantam Books, New York (1995)
7. Zeidner, M., Matthews, G., Roberts, R.D.: Emotional intelligence in the workplace: a critical review. Appl. Psychol. **53**(3), 371–399 (2004)
8. Zhoc, K.C.H., King, R.B., Chung, T.S.H., Chen, J.: Emotionally intelligent students are more engaged and successful: examining the role of emotional intelligence in higher education. Eur. J. Psychol. Educ. **35**, 839–863 (2020)
9. Schutte, N.S., et al.: Development and validation of a measure of emotional intelligence. Pers. Individ. Differ. **25**(2), 167–177 (1998)
10. Ayodele, C., Adebiyi, D.: Study habits as influence of academic performance of university undergraduates in Nigeria. Res. J. Organ. Psychol. Educ. Stud. **2**, 72–75 (2013)
11. Iqbal, J., Qureshi, N., Ashraf, M.A., Rasool, S.F., Asghar, M.Z.: The effect of emotional intelligence and academic social networking sites on academic performance during the COVID-19 pandemic. Psychol. Res. Behav. Manag. **14**, 905–920 (2021)
12. Atanassov K., Generalized Nets and Intuitionistic Fuziness in Data Mining. Prof. M. Drinov Publishing House of Bulgarian Academy of Sciences (2020)
13. Atanassov, K., Szmidt, E., Kacprzyk, J.: On intuitionistic fuzzy pairs. In: 17th International Conference on IFSs, Sofia, 1–2 Nov 2013. Notes on Intuitionistic Fuzzy Sets, vol. 19, no. 3, pp. 1–13 (2013)
14. Goleman, D.: Working with Emotional Intelligence. Banta mBooks, New York (1998)
15. Weisinger, H.: Emotional Intelligence at Work: The Untapped Edge for Success. Jossey-Bass, San Francisco (1998)

Mathematical Modelling in Biomedicine and Healthcare

Generalized Net Model of Rehabilitation Algorithm for Patients with Proximal Humeral Fracture After Surgical Treatment

Simeon Ribagin[1,2](✉), Antoaneta Grozeva[2], and Stoyan Hristov[3]

[1] Department of Bioinformatics and Mathematical Modelling Institute of Biophysics and Biomedical Engineering, Bulgarian Academy of Sciences, Sofia, Bulgaria
simribagin@gmail.com
[2] Department of Health and Pharmaceutical Care, Medical College, University "Prof. D-r Asen Zlatarov", Burgas, Bulgaria
[3] University Hospital Burgas, Burgas, Bulgaria
hristov@dr.com

Abstract. The purpose of the present paper is to present an example of Generalized Nets application in orthopedics and traumatology rehabilitation. The model describes a possible algorithm protocol for rehabilitation treatment of patients with fracture of the proximal humerus and the different transitions of the model are representing respectively the different parts of the rehabilitation process. The proposed model can be implemented in the decision making support systems, telerehabilitation platforms, optimization of the physiotherapy protocols for proximal humeral fractures rehabilitation based on current 'good practices' and better rehabilitation strategies.

Keywords: Proximal humeral fractures · Rehabilitation algorithm · Generalized Nets · GN-model

1 Introduction

Proximal humeral fractures (PHFs) are the third most common upper extremity fracture across the lifespan [13] and represent approximately 5–6% of diagnosed fractures [8]. Their incidence is predicted to triple by 2030 [18] and women are affected between two and three times as often as men [8, 15]. The epidemiology of PHFs shows a clear tendency to increase the number of these fractures in the elderly population not only because of the osteoporotic bones but also because of the active way of living and the high risk of traumatic accidents and management of these fractures will continue to pose challenges on many fronts. Management of proximal humeral fractures has been highly debated in the literature in recent years [10]. The majority of these fractures is minimally displaced [7] and can be treated non-operatively [14] with good functional outcome [12]. Unstable and displaced fractures of the proximal humerus are commonly treated surgically, including closed reduction and percutaneous pinning, tension band wiring, screw osteosynthesis, intramedullary nails, proximal humerus plates and hemi

S. Sotirov et al. (Eds.): BioInfoMed 2022, LNNS 658, pp. 225–235, 2023.
https://doi.org/10.1007/978-3-031-31069-0_22

or reversed arthroplasty. Recently the most often utilized treatment method is the use of Proximal Humeral Interlocking System Plate (PHILOS) [9]. It was developed in an attempt to reduce complications associated with other fixation methods and has become the most often utilized treatment method [23]. The plate is contoured to the shape of the proximal humerus and is versatile due to being able to accept both standard and locking screws. The plate does not require any compression which reduces the risk of reduction being lost and helps to preserve blood supply to the proximal humeral head. The utilization of locking screws in conjunction with the PHILOS plate improves axial stability and further decreases the risk of reduction being lost [25] and also provides good functional outcomes with a relatively low complication rate [22, 24]. Despite that, defining appropriate treatment protocols is complicated by poor reproducibility and reliability of the commonly used classification system devised by Neer [1]. The four-part classification reported by Neer [16, 17] in 1970 represents a four-segment classification that incorporates the concepts of displacement and vascular isolation of articular segment. This system groups each fracture by the number of fracture segments and describes the fractured anatomic segment as a part. The groups are: group I, non-displaced; group II, 2-part; group III, 3-part; and group IV, 4-part. Regardless of the type, severity and the surgical management of the fracture the proper physiotherapy and rehabilitation plan plays a vital role in the post-surgical management and the overall functional outcomes. In general the modifications of the existing rehabilitation methods and specific approaches are necessary to develop new and effective rehabilitation protocols [20, 21]. The overall result of any fracture treatment should be to bring patients back as near as possible to their pre injury function and quality of life. In this manner an appropriate rehabilitation protocols are needed. The review of the literature on proximal humerus rehabilitation suggests that treatment must begin immediately if the harmful effects of immobilization are to be avoided [11]. Most of the rehabilitation protocols for patients surgically treated after fracture of the proximal humerus are based on several different stages of recovery, respectively with different short and long-term therapeutic goals. The therapeutic goals for the first phase (Phase I) of the rehabilitation treatment are: minimization of pain and inflammatory response, protection of the fracture and optimization of tissue healing, restoration of the shoulder passive range of motion (PROM), maintains of the elbow, wrist and hand function. The therapeutic goals for the second phase (Phase II) of the rehabilitation treatment are: full shoulder passive ROM, initiation of the shoulder active assisted range and active range of motion, initiation of gentle elbow isotonic strengthening and shoulder isometrics, minimization of the compensatory motions of involved upper extremity. The therapeutic goals for the third phase (Phase III) of the rehabilitation treatment are: full shoulder active ROM, initiation of the shoulder strengthening, progression of the elbow and wrist strengthening, adequate pain control. The therapeutic goals for the third phase (Phase IV) of the rehabilitation treatment are: progression of the shoulder strengthening with heavier resistance and compound movements, returning to the normal functional activities, restoration of the scapulohumeral rhythm. These goals are achieved through regular functional examinations and properly selected therapeutic interventions structured in rehabilitation protocol.

In the aim of this, we propose a mathematical model based on the apparatus of Generalized nets theory of a possible rehabilitation algorithm protocol for patients with proximal humeral fractures.

2 Material and Methods: The Apparatus of Generalized Nets

Generalized Nets (GNs, [2–4] were introduced as an extension of Petri Nets (see [19]) and the other their modifications and extensions. Similarly, to all other Petri nets type, they have places (here denoted by l), transitions (here denoted by Z), and tokens. In GNs the transitions are characterized with more complex structure, having input and output places, moment of activation, duration of the active status, a special matrix, called index matrix (IM, see [5, 6]) of transition condition predicates (here marked by r), IM of the capacities of transition arcs and type of the transition. The full definition of a GN contains a set of transitions, functions determining the priorities of the transitions and places, the capacities of the places, of the truth-values of the transition condition predicates, the moments of the transition activations and the duration of them; a set of tokens, their priorities and the moments in which they must enter the net; time-moment in which the GN will start function, the duration of its work and the elementary time-step with which the time will grow; a set of initial token characteristics, a function that give new characteristics of the tokens when they transfer from input to an output place of a some transition and the maximal number of characteristics that token can have. As it is mentioned in [3, 4], a part of the GN-components can be omitted in respect of the aims of each one concrete model and a such net is called a reduced GN. In the present paper we construct a reduced GN-model of rehabilitation algorithm for patients with proximal humeral fracture after surgical treatment. The model is based on the time period after the surgical intervention, functional outcomes of the patient and the rehabilitation goals. The proposed model gives the possibility of development of more complex and detailed model allowing implementation in decision making support systems, telerehabilitation platforms, optimization of the physiotherapy protocols for PHFs rehabilitation based on current 'good practices' and better rehabilitation strategies.

3 Results: The Generalized Net Model

The GN model (Fig. 1) has 24 places and the following set of transitions:

$$A = \{Z_1, Z_2, Z_3, Z_4, Z_5, Z_6\}$$

These transitions describe the following processes:

- Transition Z_1 represents the personal data of the patients (age, gender, symptoms, etc.),
- Transition Z_2 represents the current functional status of the patient and the current time period after the surgery intervention,
- Transition Z_3 represents the first phase (Phase I) of the rehabilitation process,
- Transition Z_4 represents the second phase (Phase II) of the rehabilitation process,

- Transition Z_5 represents the third phase (Phase III) of the rehabilitation process,
- Transition Z_6 represents the fourth phase (Phase IV) of the rehabilitation process

The net contains six types of tokens: α, β, μ, η, γ and token φ. Some of the model transitions contain a so called "special place" where a token stays and collect information about the specific parts of the screening process and serves as a database storage, which it represents as follows:

- In place l_4, token α^1 stays permanently and collects the overall information obtained from the screening in the personal record (personal data),
- In place l_9, token β stays permanently and collects information about the current status of the patient obtained from the functional evaluation and the time period after the surgery intervention (tissue healing),
- In place l_{13}, token μ stays permanently and collects information about the rehabilitation plan during the Phase I,
- In place l_{17}, token η stays permanently and collects information about the rehabilitation plan during the Phase II,
- In place l_{21}, token γ stays permanently and collects information about the rehabilitation plan during the Phase III,
- In place l_{24}, token φ stays permanently and collects information about the rehabilitation plan during the Phase IV

At the time of duration of the GN-functioning, some of these tokens can split, generating new tokens, that will transfer in the net obtaining respective characteristics, and also in some moments they will unite with some of the tokens β, μ, η, γ and φ.

Token α enters the net with additional characteristics "*patient treated surgically for proximal humeral fracture*" in place $l1$.

The transition Z_1 of the GN-model has the following form:

$$Z_1 =< \{l_1, l_4, l_8\}, \{l_2, l_3, l_4\}, r_1$$

where:

$$r_1 = \begin{array}{c|ccc} & l_2 & l_3 & l_4 \\ \hline l_1 & false & false & true \\ l_4 & W_{4,2} & W_{4,3} & true \\ l_8 & false & false & true \end{array}$$

and,

$W_{4,2}$ = "*the patient has no severe fever, no un resolving numbness/tingling, no excessive drainage from the incision, no uncontrolled pain, no mental health disorders, no high risk of cardiovascular accident*",

$W_{4,3}$ = "$\neg W_{4,2}$"

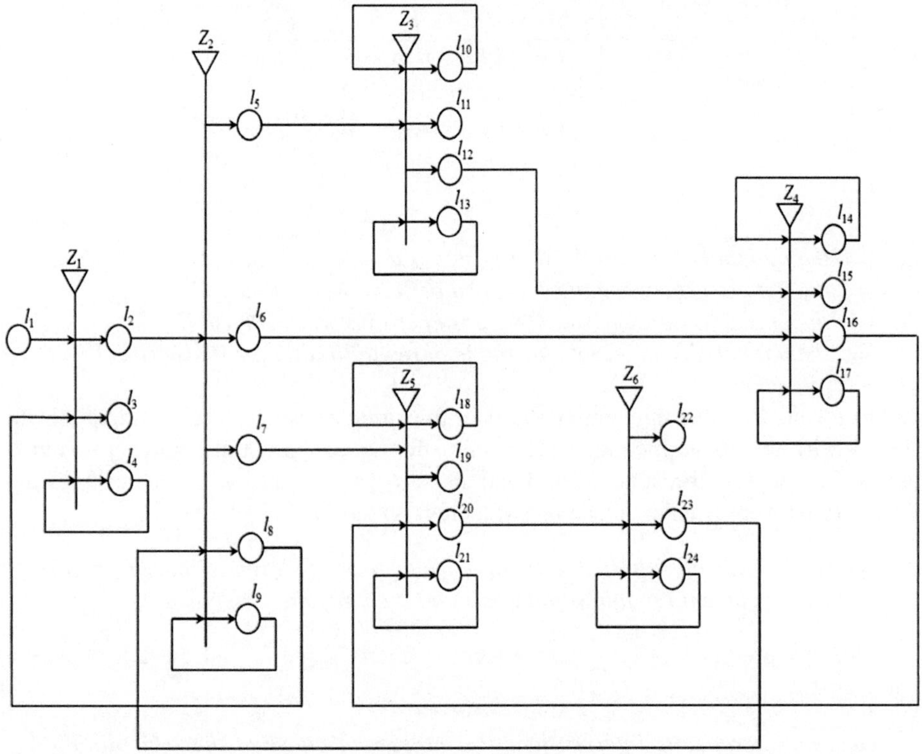

Fig. 1. Generalized Net model of rehabilitation algorithm for patients with proximal humeral fractures, after surgical treatment

The tokens from the three input places of transition Z_1 enter place l_4 and unite with token α^1 with the above mentioned characteristic. On the next time-moment, token α^1 splits to three tokens – the same token α^1 and tokens α_1 *and* α_2. When the predicate $W_{4,2}$ is true, token α_1 enters place l_2 and there it obtains a characteristic:

"perform an initial functional examination and set the rehabilitation goals for the current patient".

When the predicate $W_{4,2}$ is true, token α_2 enters place l_2 and there it obtains a characteristic:

"consider the presents of "red flags": ask for medical consultation"

The transition Z_2 has the following form:

$$Z_2 =< \{l_2, l_9, l_{23}\}, \{l_5, l_6, l_7, l_8, l_9\}, r_2$$

where:

$r_2 =$

	l_5	l_6	l_7	l_8	l_9
l_2	*false*	*false*	*false*	*false*	*true*
l_9	$W_{9,5}$	$W_{9,6}$	$W_{9,7}$	$W_{9,8}$	*true*
l_{23}	*false*	*false*	*false*	*false*	*true*

and,

$W_{9,5}$ = "*the patient is between the 0 -4 postoperative week (Phase I)*";
$W_{9,6}$ = "*the patient is between the 4 -8 postoperative week (Phase II)*";
$W_{9,7}$ = "*the patient is between the 8 -12 postoperative week (Phase III)*"
$W_{9,8}$ = "*the patient has undergone a course of rehabilitation for Phases I to IV*"

The tokens from the three input places of transition Z_2 enter place l_9 and unite with token β with the above mentioned characteristic. On the next time-moment, token β splits to five tokens – the same token β and tokens α_1, α_2, α_3 and α_4. When the predicate $W_{9,5}$ is true, token α_1 enters place l_5 and there it obtains a characteristic:

"*select the therapeutic interventions for the first phase of the rehabilitation process based on the current rehabilitation goals and the functional outcomes*".

When the predicate $W_{9,6}$ is true, token α_2 enters place l_6 and there it obtains a characteristic:

"*select the therapeutic interventions for the second phase of the rehabilitation process based on the current rehabilitation goals and the functional outcomes*".

When the predicate $W_{9,7}$ is true, token α_3 enters place l_7 and there it obtains a characteristic:

"*select the therapeutic interventions for the third phase of the rehabilitation process based on the current rehabilitation goals and the functional outcomes*"

When the predicate $W_{9,8}$ is true, token α_4 enters place l_8 and there it obtains a characteristic:

"*give patient a set of home based therapeutic exercises and sports activities*"

The transition Z_3 has the following form:

$$Z_3 =< \{l_5, l_{10}, l_{13}\}, \{l_{10}, l_{11}, l_{12}, l_{13}\}, r_3$$

where:

$r_3 =$

	l_{10}	l_{11}	l_{12}	l_{13}
l_5	*false*	*false*	*false*	*true*
l_{10}	*true*	*true*	*true*	*true*
l_{13}	$W_{13,10}$	$W_{13,11}$	$W_{13,12}$	*true*

and,

$W_{13,10}$ = *"the patient has tolerable pain, swelling, reduced active and passive ROM and wears a sling"*;
$W_{13,11}$ = *"the patient has severe pain, difficult breathing, dizziness or lightheadedness or there is abnormal hemodynamics "*;
$W_{13,12}$ = *"the patient has full active ROM in the elbow, shoulder passive flexion to 140°, external rotation of the shoulder to 40°, abduction of the shoulder to 90°"*;

The tokens from the three input places of transition Z_3 enter place l_{13} and unite with token μ with the above mentioned characteristic. On the next time-moment, token μ splits to four tokens – the same token μ and tokens α_1, α_2, and α_3. When the predicate $W_{13,10}$ is true, token α_1 enters place l_{10} and there it obtains a characteristic:

"perform: cryotherapy, reflexology and segmental massage, passive exercises of the affected shoulder in all planes, active exercises of the elbow, wrist and hand, Codman's pendulum exercises, exercises for scapular stabilization, submaximal isometric exercises".

When the predicate $W_{13,11}$ is true, token α_2 enters place l_{11} and there it obtains a characteristic:

"medical attention is needed".

When the predicate $W_{13,12}$ is true, token α_3 enters place l_{12} and there it obtains a characteristic:

"progress to the second phase of the rehabilitation process".

The transition Z_4 has the following form:

$$Z_4 =< \{l_6, l_{12}, l_{14}, l_{17}\}, \{l_{14}, l_{15}, l_{16}, l_{17}\}, r_4$$

where:

r_4 =

	l_{14}	l_{15}	l_{16}	l_{17}
l_6	*false*	*false*	*false*	*true*
l_{12}	*false*	*false*	*false*	*true*
l_{14}	*true*	*true*	*true*	*true*
l_{17}	$W_{17,14}$	$W_{17,15}$	$W_{17,16}$	*true*

and,

$W_{17,14}$ = *"the patient has undergone a course of rehabilitation for Phase I"*;
$W_{17,15}$ = *"the patient has severe pain, difficult breathing, dizziness or lightheadedness or there is abnormal hemodynamics "*;

$W_{17,16}$ = "*the patient has full shoulder passive ROM, full elbow active ROM, good pain control, good tolerance to shoulder isometrics and elbow strengthening exercises, active assisted ROM of the shoulder flexion to 90 degrees or X-ray evidence of healing*";

The tokens from the four input places of transition Z_4 enter place l_{17} and unite with token η with the above mentioned characteristic. On the next time-moment, token η splits to four tokens – the same token η and tokens α_1, α_2, and α_3. When the predicate $W_{17,14}$ is true, token α_1 enters place l_{14} and there it obtains a characteristic:

"*perform: exercises from the I phase, active exercises for the affected shoulder, strengthening exercises with resistance (teraband or weights) multi-angle isometrics and self-stretching exercises*".

When the predicate $W_{17,15}$ is true, token α_2 enters place l_{15} and there it obtains a characteristic:

"*medical attention is needed*".

When the predicate $W_{17,16}$ is true, token α_3 enters place l_{12} and there it obtains a characteristic:

"*progress to the third phase of the rehabilitation process*"

$$Z_5 =< \{l_7, l_{16}, l_{18}, l_{21}\}, \{l_{18}, l_{19}, l_{20}, l_{21}\}, r_5$$

where:

$r_5 =$

	l_{18}	l_{19}	l_{20}	l_{21}
l_7	*false*	*false*	*false*	*true*
l_{16}	*false*	*false*	*false*	*true*
l_{18}	*true*	*true*	*true*	*true*
l_{21}	$W_{21,18}$	$W_{21,19}$	$W_{21,20}$	*true*

and,

$W_{21,18}$ = "*the patient has undergone a course of rehabilitation for Phase II*";
$W_{21,19}$ = "*the patient has severe pain, difficult breathing, dizziness or lightheadedness or there is abnormal hemodynamics*";
$W_{21,20}$ = "*the patient has full shoulder active ROM with appropriate mechanics and no pain or compensatory strategies with strengthening exercises or X-ray evidence of healing*";

The tokens from the four input places of transition Z_5 enter place l_{21} and unite with token γ with the above mentioned characteristic. On the next time-moment, token γ

splits to four tokens – the same token γ and tokens α_1, α_2, and α_3. When the predicate $W_{21,18}$ is true, token α_1 enters place l_{18} and there it obtains a characteristic:

"*perform: exercises from the II phase, isotonic strengthening with weights in all directions and aggressive stretching exercises*".

When the predicate $W_{21,19}$ is true, token α_2 enters place l_{19} and there it obtains a characteristic:

"*medical attention is needed*".

When the predicate $W_{21,20}$ is true, token α_3 enters place l_{20} and there it obtains a characteristic:

"*progress to the fourth phase of the rehabilitation process*".

The transition Z_6 has the following form:

$$Z_6 = < \{l_{20}, l_{24}\}, \{l_{22}, l_{23}, l_{24}\}, r_6$$

where:

$r_6 =$

	l_{22}	l_{23}	l_{24}
l_{20}	*false*	*false*	*true*
l_{24}	$W_{24,22}$	$W_{24,23}$	*true*

and,

$W_{24,22}$ = "*the patient has severe pain, difficult breathing, dizziness or lightheadedness or there is abnormal hemodynamics*";
$W_{24,23}$ = "*the patient has 80% or > strength of involved upper extremity compared to uninvolved arm, no pain with progressive strengthening exercises, low level to no disability score on patient reported outcome measure or X-ray evidence of healing*";

The tokens from the four input places of transition Z_6 enter place l_{24} and unite with token φ with the above mentioned characteristic. On the next time-moment, token φ splits to three tokens – the same token φ and tokens α_1 and α_2. When the predicate $W_{24,22}$ is true, token α_1 enters place l_{18} and there it obtains a characteristic:

"*perform: exercises from the III phase, shoulder plyometric, interval return to sports training and activities*".

When the predicate $W_{24,23}$ is true, token α_2 enters place l_{19} and there it obtains a characteristic:

"*medical attention is needed*".

When the predicate $W_{21,20}$ is true, token α_3 enters place l_{20} and there it obtains a characteristic:

"*progress to the fourth phase of the rehabilitation process*"

4 Conclusions

The proposed model gives the possibility of development of more complex and detailed model allowing implementation in decision making support systems, telerehabilitation platforms, optimization of the physiotherapy protocols for PHFs rehabilitation based on current 'good practices' and better rehabilitation strategies. The GN-model may provide a framework that can be used by physiotherapists and other medical specialists, to guide the rehabilitation treatment process for patients with proximal humeral fracture treated surgically, enabling more accurate and efficient management of that condition and would assist in optimizing patient outcomes. This model can be complicated and detailed, which will significantly improve the reliability of the proposed algorithm.

Acknowledgements. The authors are grateful for the support provided by the University "Prof. Dr. Asen Zlatarov"- Burgas, Scientific research sector "Scientific and artistic activity" approved by № NIH - 475/2022 "Generalized nets models in orthopedic and traumatology rehabilitation".

References

1. Akel, Y., El Nahas, M., Shafei, S.R.A.: Comparative Study of Open Reduction Internal Fixation with Proximal Humerus Interlocking System and Closed Reduction and Pin-ning with K –Wire in Proximal Humeral Fracture. Egypt. J. Hosp. **76**(4), 3846–3852 (2019)
2. Atanassov, K.: Theory of generalized nets (an algebraic aspect). Adv. Modell. Simul. **1**(2), 27–33 (1984)
3. Atanassov, K.: Generalized Nets. World Scientific, Singapore (1991)
4. Atanassov, K.: On Generalized Nets Theory. Prof. M. Drinov Academic Publishing House, Sofia (2007)
5. Atanassov, K.: Generalized index matrices. C. R. Acad. Sci. **40**(11), 15–18 (1987)
6. Atanassov, K.: Index Matrices: Towards an Augmented Matrix Calculus. Springer, Cham (2014)
7. Court-brown, C.M., Garg, A., Mcqueen, M.M.: The epidemiology of proximal humeral fractures. Acta Orthop. Scand. **72**, 365–371 (2001)
8. Court-Brown, C.M., Caesar, B.: Epidemiology of adult fractures: a review. Injury **37**(8), 691–697 (2006)
9. Ethiraj, P., Venkataraman, S., Shanthappa, S.J.K., Agarawal, A.H., Does, S.: Proximal humerus inter locking system (PHILOS) plating provide a good functional outcome in proximal humerus fractures? Cureus 30, **14**(6), e26474. https://doi.org/10.7759/cureus.26474. PMID: 35919369; PMCID: PMC9339088 (2022)
10. Fisher, N.D., Driesman, A., Saleh, H., et al.: The proximal humerus out-come score at one year (POSY) predicts which patients have poor functional out-comes following operative fixation of proximal humerus fractures. Cureus **14**(7), e26631. https://doi.org/10.7759/cureus.26631 (2022)
11. Hodgson, S.: Proximal humerus fracture rehabilitation. Clin Orthop Relat Res. (2006)

12. Iyengar, J.J., Devcic, Z., Sproul, R.C., Feeley, B.T.: Nonoperative treatment of proximal hu-merus fractures: a systematic review. J. Orthop. Trauma **25**, 612–617 (2011)
13. Karl, J.W., Olson, P.R., Rosenwasser, M.P.: The epidemiology of upper extremity fractures in the UnitedStates, 2009. J. Orthop. Trauma **2015**(29), 242–244 (2009)
14. Koval, K.J., Gallagher, M.A., Marsicano, J.G., Cuomo, F., McShinawy, A., Zuckerman, J.D.: Functional outcome after minimally displaced fractures of the proximal part of the humerus. J. Bone Joint Surg Am. 1997, **79**(2), 203–207 (1997)
15. Lind, T., Kroner, K., Jensen, J.: The epidemiology of fractures of the proximal humerus. Arch Orthop Trauma Surg. **108**, 285–287 (1989). https://doi.org/10.1007/BF00932316
16. Neer, C.S. 2nd.: Displaced proximal humeral fractures: II. treatment of three-part and four-part displacement. J. Bone Joint Surg Am. **52**(6), 1090–1103 (1970)
17. Neer C.S. 2nd.: Displaced proximal humeral fractures. I. Classification and Evaluation. J. Bone Joint Surg Am. **52**(6), 1077–1089 (1970)
18. Palvanen, M., Kannus, P., Niemi, S.: Parkkari, Update in the epidemiology of proximal humeral fractures. J. Clin Orthop Relat Res. **442**, 87–92 (2006)
19. Petri, C.A.: Kommunication mit automaten, Ph. D. diss., Univ. of Bonn, 1962; Schriften des Inst. Fur Instrument. Math., no. 2, Bonn (1962)
20. Popova, G.: Effectiveness of joint mobilisation and Mulligan's mobilisation with movement techniques in patients after distal radius fracture. Int. J. Knowl. Sci. papers, vol. 13.2, Skopije, 2016, ISSN 1857–92, p. 279–285, International Scientific Conference "The teacher of the future" in Durres, R. Albania, June 17–19 (2016)
21. Popova, G.: Modified manual edema mobilization and kinesiotaping applications in reducing postimmobilizatsion edema after distal radius fracture. Int. J. Knowl. **19**(4), 447–1455 (2017)
22. Ong, C.C., Kwon, Y.W., Walsh, M., Davidovitch, R., Zuckerman, J.D., Egol, K.A.: Outcomes of open reduction and internal fixation of proximal humerus fractures managed with locking plates. Am. J. Orthop **41**, 407–412 (2012)
23. Saber, A.Y., et al.: Surgical fixation of three- and four-part proximal humeral fractures using the proximal humeral interlocking system plate. Cureus**14**(5), e25348. https://doi.org/10.7759/cureus.25348 PMID: 35774694; PMCID: PMC9236683 (2022)
24. Shulman, B.S., Ong, C.C., Lee, J.H., Karia, R., Zuckerman, J.D., Egol, K.A.: Outcomes after fixation of proximal humerus(OTA type 11) fractures in the elderly patients using modern techniques. Geriatr. Orthop. Surg. Rehabil. **4**, 21–25 (2013). https://doi.org/10.1177/2151458513498597(2013)
25. Südkamp, N., Bayer, J., Hepp, P., et al.: Open reduction and internal fixation of proximal humeral fractures with use of the locking proximal humerus plate. Results Prospective Multicenter Observational Study J. Bone Joint Surg. Am. **2009**(91), 1320–1328 (2009)

Generalized Net Model of the Malignant Melanoma Treatment

Evdokia Sotirova[1(✉)], Hristo Bozov[1,2], Yaroslava Petrova[1,2], Greta Bozova[3], and Krassimir Atanassov[1,4]

[1] Prof. Assen Zlatarov University, 1 Prof. Yakimov Blvd., 8010 Burgas, Bulgaria
esotirova@btu.bg, yaroslava_ap@abv.bg
[2] Oncology Complex Center-Burgas, 86 Demokratsiya Blvd, 8000 Burgas, Bulgaria
[3] Department of Nephrology, Military Medical Academy, 3 St. George Sofiiski, Sofia, Bulgaria
greti@abv.bg
[4] Department of Bioinformatics and Mathematical Modelling, Institute of Biophysics and Biomedical Engineering, Bulgarian Academy of Sciences, Acad. G. Bonchev Street Bl. 105, 1113 Sofia, Bulgaria

Abstract. In the article a Generalized Net model of the malignant melanoma treatment is proposed. Generalized nets are extension and generalization of the concept of Petri nets, as well as of other Petri nets extensions and modifications. For constructing of the model patients with malignant melanoma registered in Oncology Complex Center in Burgas town were analyzed. The model can be used as a uniform algorithm of behavior with patients with malignant melanoma from the initial examination to the treatment (if it is necessary).

Keywords: Generalized Net · Malignant melanoma · Modelling

1 Introduction

Melanoma is a malignant cancer with a very aggressive progression [10]. Melanoma is the most common skin cancer in which cells called melanocytes begin to divide more rapidly and spread. Melanocytes contain the melanin or pigment that gives the skin, eyes and hair the specific color [5]. The untimely diagnosis and treatment of the melanoma can quickly lead to a fatal outcome [11]. To detect early malignant melanomas, the ABCDE criteria is used (A - asymmetry of the lesion, B - border irregularity, C - color change, D - diameter, and E - evolution over time) [14]. To describe the stage of melanoma, it is necessary to specify where it is located, whether and where it has spread, whether it has affected other parts of the body [9]. This process involves surgically removing the melanoma, the lesion, and some surrounding healthy tissue and analyzing it under a microscope. A significant meaning has the thickness of the melanoma in millimeters and the status of the sentinel nodes. The sentinel node has been successfully localized using gamma-probe or lymphoscintigraphy [1]. Melanoma can spread to other parts of the body than the one where it started, i.e. to investigate metastases. To establish this, imaging studies are performed (Positron emission tomography (PET), computer

S. Sotirov et al. (Eds.): BioInfoMed 2022, LNNS 658, pp. 236–245, 2023.
https://doi.org/10.1007/978-3-031-31069-0_23

tomography (CT), ect.). Depth, mitotic rate, ulceration, and microscopic satellitosis are analyzed to determine the pathology of melanoma. After analyzing the obtained results, the stage of the melanoma is determined. Staging is necessary to develop a melanoma treatment plan and to describe the patient's prognosis.

2 Materials and Methods

In the present research a Generalized net (GN, [2, 3]) model of the process of the malignant melanoma treatment is constructed. GNs represent a significant extension and generalization of the concept of Petri nets, as well as of other Petri nets extensions and modifications. The GN model consists of objects called transitions, which have input and output places and index matrix (see [4]) with predicates that determine the directions of the tokens transfer from the input to the output transition places. Two transitions can share a place, but every place can be an input of at most one transition and can be an output of at most one transition. Through the transition from one input to another output place pass tokens, which act as information carriers and can occupy a single place at every moment of the GN execution. The information carried by a token is contained in its characteristics.

For constructing of the GN-model patients with malignant melanoma registered in Oncology Complex Center in Burgas town were analyzed. 161 patients with malignant melanoma were registered for period 2014–2020. In 2021, 28 cases of malignant melanoma out of a total of 329 cases of skin carcinoma were registered. For 2022, until the month of July, 7 cases out of 155 patients with skin carcinoma were registered. The model describes the behavior algorithm for patients from the initial examination to the treatment of malignant melanoma [6]. The transitions in the model reflect the main activities in the overall procedure. The GN-model can be extent in order to describe more detail the modelling process. The next GN-model for the behavior algorithm oh the patients with malignant melanoma in ovarian and primary cervical carcinoma registered in Oncology Complex Center in Burgas town will be constructed [7, 8].

The presented GN-model of the process of the malignant melanoma treatment expands the investigation of the authors from the [12] and [13]. In [12] the correlations between parameters, characterized 100 patients (57 man and 43 woman) with malignant melanoma and were observed. In [13] the patients registered during an 8-year period (2014–2021) were analyzed and a method for reducing the input data coming to the input of a neural network of the Multilayer Perceptron.

3 A Generalized Net Model of the Process of the Malignant Melanoma Treatment

The proposed GN-model (Fig. 1) has 10 transitions and 45 places. The places and tokens are three types: l-places and α-tokens for patients; d-places and β-tokens for medical staff; c-places and γ-tokens for criteria. For brevity, we shall use the notation α-, β-, and γ- tokens instead of α_i-, β_j-, and γ_k- tokens, where i, j, k are numerations of the respective tokens, and $i = 1, \ldots, n$; $j = 1, \ldots, m$; $k = 1, \ldots, s$.

The ten transitions have the following meaning:

- Z_1 represents the patient and their activities;
- Z_2 represents the medical staff and its activities;
- Z_3 represents the process of the history and physical examination of melanoma;
- Z_4 represents the process of diagnosing the melanoma;
- Z_5 represents the process of wide local excision (WLE);
- Z_6 represents the sentinel node biopsy;
- Z_7 represents the imaging staging of melanoma;
- Z_8 represents the process of completion lymphadenectomy;
- Z_9 represents the follow-up treatment;
- Z_{10} represents the chemotherapy or clinical treatment.

Initially the α- and β-tokens remain, respectively, in places l_3 and d_9 with initial characteristics:

"list of patients: name of patient, current status, physical examination, family history, medical test results" in place l_3,

and "list of doctors and medical specialists: name of doctor / medical specialist, specialty" in place d_9.

These tokens will stay permanently in their places and collect information about current patients with malalignment melanoma, their status (α-token) and for medical staff (β-token). At the time of the GN-model functioning, these tokens can split, generating new tokens, that will go through Z_1 and Z_2 transitions respectively obtaining respective characteristics.

The forms of the transitions are the following.

Token α_1 enters the net from place l_1 with characteristic:

"new patient".

$$Z_1 = \langle \{l_1, l_3, l_{22}, l_{25}\}, \{l_2, l_3, l_{22}, l_{25}\}, \begin{array}{c|cc} & l_2 & l_3 \\ \hline l_1 & false & true \\ l_3 & w_{3,2} & true \\ l_{22} & false & true \\ l_{25} & false & true \end{array} \rangle,$$

where:

$w_{3,2}$ = "The patient is chosen".

The α_1 token enters place l_3 where it is merges with token α and token α extends its previous characteristic with a new patient.

Token α_2 enters place l_2 with characteristic.

"name of patient i, current status, physical examination, family history, medical test results, biopsy (if available)", $i = 1, \ldots, n$.

Token β_1 enters the net from place d_1 with characteristic:

"new doctor or medical specialist".

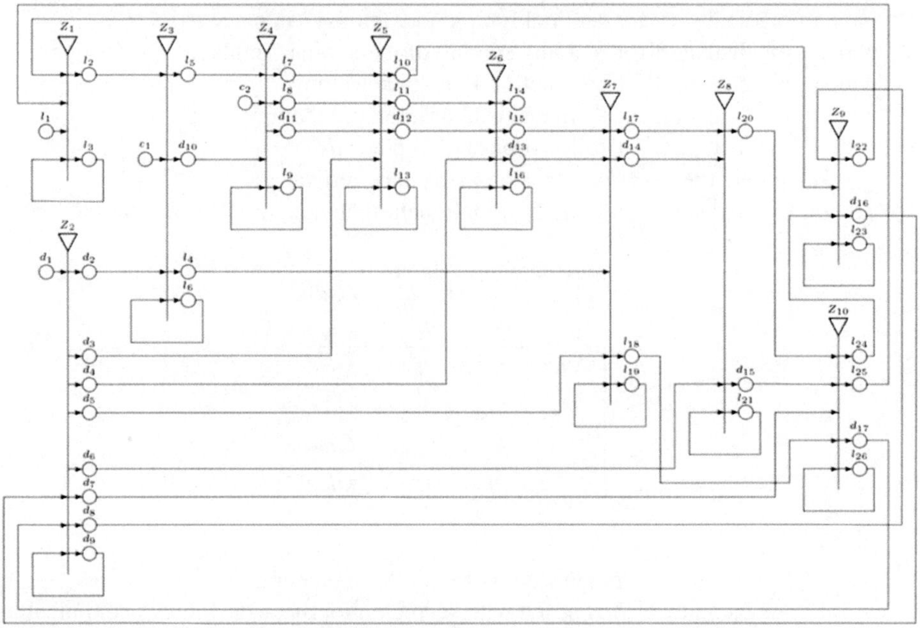

Fig. 1. Generalized net model of the treatment of malignant melanoma

$Z_2 = \langle \{d_1, d_8, d_{15}, d_{16}\}, \{d_2, d_3, d_4, d_5, d_6, d_7, d_8, d_9\},$

	d_2	d_3	d_4	d_5	d_6	d_7	d_8	d_9
d_1	*false*	*false*	*false*	*false*	*false*	*false*	*false*	*true*
d_9	$w^d_{9,2}$	$w^d_{9,3}$	$w^d_{9,4}$	$w^d_{9,5}$	$w^d_{9,6}$	$w^d_{9,7}$	$w^d_{9,8}$	*true* $\rangle$,
d_{15}	*false*	*false*	*false*	*false*	*false*	*false*	*false*	*true*
d_{16}	*false*	*false*	*false*	*false*	*false*	*false*	*false*	*true*

where:

$w^d_{9,2}$ = "The medical staff for physical examination of the patient is chosen",

$w^d_{9,3}$ = "The medical staff for WLE and for clinical study of melanoma of the patient is chosen";

$w^d_{9,4}$ = "The medical staff for sentinel lymph node biopsy of the patient is chosen";

$w^d_{9,5}$ = "The medical staff for imaging staging of melanoma of the patient is chosen";

$w^d_{9,6}$ = "The medical staff for completion lymphadenectomy of the patient is chosen";

$w^d_{9,7}$ = "The medical staff for a clinical trial or systemic therapy of the patient is chosen";

$w^d_{9,8}$ = "The medical staff for follow-up care of the patient is chosen".

The β_1 token enters place d_9 where they merge into β token. β token generate new tokens that enter places d_2, d_3, d_4, d_5, d_6, d_7, d_8 with characteristics respectively:

"patient, medical staff for physical examination" in place d_2;

"patient, medical staff for WLE and for clinical study of melanoma" in place d_3;

"patient, medical staff for sentinel lymph node biopsy" in place d_4;
"patient, medical staff for imaging staging of melanoma" in place d_5;
"patient, medical staff for completion lymphadenectomy" in place d_6;
"patient, medical staff for a clinical trial or systemic therapy" in place d_7;
"patient, medical staff for follow-up care" in place d_8.
γ_1-token enters the net from place c_1 with characteristic:
"criteria for physical examination of malignant melanoma, ABCDE criteria".

$$Z_3 = \langle \{l_2, l_6, c_1, d_2\}, \{l_4, l_5, l_6, d_{10}\}, \begin{array}{c|cccc} & l_4 & l_5 & l_6 & d_{10} \\ \hline l_2 & false & false & true & false \\ l_6 & w_{6,4} & w_{6,5} & w_{6,6} & w^d_{6,10} \\ c_1 & false & false & true & false \\ d_2 & false & false & true & false \end{array} \rangle,$$

where:
$w_{6,4}$ = "The patient was diagnosed with primary melanoma";
$w_{6,5}$ = "The patient was diagnosed with suspicion for metastatic disease or palpable lymph nodes";

$$w_{6,6} = w^d_{6,10} = \neg w_{6,4}.$$

The α-tokens (from place l_2), β-tokens (from place d_2), and γ_1-token (from place c_1) enter place l_6 without new characteristics. The β-tokens that enter place d_{10} do not obtain new characteristics.

The α-tokens enter places l_4 and l_5 with characteristics respectively:

"patient with primary melanoma" in place l_4;

and "patient with suspicion for metastatic disease or palpable lymph nodes"

in place l_5

γ_2-token enters the net from place c_2 with characteristic:
"Criteria for determining the pathology of melanoma (depth, mitotic rate, ulceration, and microscopic satellitosis)".

$$Z_4 = \langle \{l_4, l_9, c_2, d_{10}\}, \{l_7, l_8, l_9, d_{11}\}, \begin{array}{c|cccc} & l_7 & l_8 & l_9 & d_{11} \\ \hline l_4 & false & false & true & false \\ l_9 & w_{9,7} & w_{9,8} & w_{9,9} & w^d_{9,11} \\ c_2 & false & false & true & false \\ d_{10} & false & false & true & false \end{array} \rangle,$$

where:

$w_{9,7} = w_{9,8} =$ "The risk of melanoma for a patient is determined";

$$w_{9,9} = w^{d}_{9,11} = \neg w_{9,7}.$$

The α-tokens (from place l_4), β-tokens (from place d_{10}), and γ_2-token (from place c_2) enter place l_9 without new characteristics. The β-tokens that enter place d_{11} do not obtain new characteristics.

The α-tokens enter places l_7 and l_8 with characteristics respectively:

"patient with low-risk lesions (lesions ≤ 0.8 mm deep without ulceration or mitoses" in place l_7; and.

"patient with high-risk lesions" in place l_8.

$$Z_5 = \langle \{l_7, l_8, l_{13}, d_3, d_{11}\}, \{l_{10}, l_{11}, l_{13}, d_{12}\}, \begin{array}{c|cccc} & l_{10} & l_{11} & l_{13} & d_{12} \\ \hline l_7 & false & false & true & false \\ l_8 & false & false & true & false \\ l_{13} & w_{13,10} & w_{13,11} & w_{13,13} & w^{d}_{13,12} \\ d_3 & false & false & true & false \\ d_{11} & false & false & false & true \end{array} \rangle,$$

where:

$w_{13,10} = w_{13,11} =$ "The wide local excision for a patient was performed";

$$w_{13,13} = w^{d}_{13,12} = \neg w_{13,10}.$$

The α-tokens (from places l_7 and l_8) and β-tokens (from place d_3) enter place l_{13} without new characteristics. The β-tokens that enter place d_{12} do not obtain new characteristics.

The α-tokens enter places l_{10} and l_{11} with characteristics respectively:

"patient with WLE with a 1 cm margin" in place l_{10};

and "patient with WLE with a 1–2 cm margin" in place l_{11}.

$$Z_6 = \langle \{l_{11}, l_{16}, d_4, d_{12}\}, \{l_{14}, l_{15}, l_{16}, d_{13}\}, \begin{array}{c|cccc} & l_{14} & l_{15} & l_{16} & d_{13} \\ \hline l_{11} & false & false & true & false \\ l_{16} & w_{16,14} & w_{16,15} & w_{16,16} & w^{d}_{16,13} \\ d_4 & false & false & true & false \\ d_{12} & false & false & false & true \end{array} \rangle,$$

where:

$w_{16,14}$ = "The sentinel node biopsy is performed with negative result for malignancy";

$w_{16,15}$ = "The sentinel node biopsy is performed with positive result for malignancy";

$w_{16,16} = w^d_{16,13}$ = "The sentinel node biopsy is not performed".

The α-tokens (from place l_{11}) and β-tokens (from place d_4) enter place l_{16} without new characteristics. The β-tokens that enter place d_{13} do not obtain new characteristics.

The α-tokens enter places l_{14} and l_{15} with characteristics respectively:

"patient, negative for malignancy sentinel node" in place l_{14};

and "patient, positive for malignancy sentinel node" in place l_{15}.

$$Z_7 = \langle \{l_5, l_{15}, l_{19}, d_5, d_{13}\}, \{l_{17}, l_{18}, l_{19}, d_{14}\},
\begin{array}{c|cccc}
 & l_{17} & l_{18} & l_{19} & d_{14} \\
\hline
l_5 & false & false & true & false \\
l_{15} & false & false & true & false \\
l_{19} & w_{19,17} & w_{19,18} & w_{19,19} & w^d_{19,14} \\
d_5 & false & false & true & false \\
d_{13} & false & false & false & true
\end{array} \rangle,$$

where:

$w_{19,17}$ = "The imaging staging of melanoma was performed and there is no metastatic disease";

$w_{19,18}$ = "The imaging staging of melanoma was performed and there is a metastatic disease";

$w_{19,19} = w^d_{19,14}$ = "The imaging staging of melanoma is not performed".

The α-tokens (from place l_{15}) and β-tokens (from place d_5) enter place l_{19} without new characteristics. The β-tokens that enter place d_{14} do not obtain new characteristics.

The α-tokens enter places l_{17} and l_{18} with characteristics respectively:

"patient, result from imaging staging of melanoma (CT, PET/CT, MRI)" in place l_{17};

and "patient, need from clinical treatment or chemotherapy" in place l_{18}.

$$Z_8 = \langle \{l_{17}, l_{21}, d_6, d_{14}\}, \{l_{20}, l_{21}, d_{15}\},
\begin{array}{c|ccc}
 & l_{20} & l_{21} & d_{15} \\
\hline
l_{17} & false & true & false \\
l_{21} & w_{21,20} & w_{21,21} & w^d_{21,15} \\
d_6 & false & true & false \\
d_{14} & false & false & true
\end{array} \rangle,$$

where:
$w_{21,20}$ = "The completion lymphadenectomy of the patient was performed";

$$w_{21,21} = w^{d}_{21,15} = \neg w_{21,20}.$$

The α-tokens (from place l_{17}) and β-tokens (from place d_6) enter place l_{21} without new characteristics. The β-tokens that enter place d_{15} do not obtain new characteristics.
The α-token enters place l_{20} with characteristic.
"patient, result from completion lymphadenectomy".

$$Z_9 = \langle \{l_{10}, l_{14}, l_{23}, l_{24}, d_8\}, \{l_{22}, l_{23}, d_{16}\}, \begin{array}{c|ccc} & l_{22} & l_{23} & d_{16} \\ \hline l_{10} & false & true & false \\ l_{14} & false & true & false \\ l_{23} & w_{23,22} & w_{23,23} & w^{d}_{23,16} \\ l_{24} & false & true & false \\ d_8 & false & true & false \end{array} \rangle,$$

where:
$w_{23,22}$ = "The follow-up care for a patient was determined";

$$w_{23,23} = w^{d}_{23,16} = \neg w_{23,22}.$$

The α-tokens (from places l_{10}, places l_{14} and places l_{24}) and β-tokens (from place d_8) enter place l_{23} without new characteristics. The β-tokens that enter place d_{16} do not obtain new characteristics.
The α-token enters place l_{22} with characteristic.
"patient, program for follow-up care".

$$Z_{10} = \langle \{l_{18}, l_{20}, l_{26}, d_7, d_{15}\}, \{l_{24}, l_{25}, l_{26}, d_{17}\}, \begin{array}{c|cccc} & l_{24} & l_{25} & l_{26} & d_{17} \\ \hline l_{18} & false & false & true & false \\ l_{20} & false & false & true & false \\ l_{26} & w_{26,24} & w_{26,25} & w_{26,26} & w^{d}_{26,17} \\ d_7 & false & false & true & false \\ d_{15} & false & false & false & true \end{array} \rangle,$$

where:
$w_{26,24}$ = "The patient needs from follow-up care";
$w_{26,25}$ = "The patient needs from clinical trial or chemotherapy";
$w_{26,26} = w^{d}_{26,17}$ = "The program for clinical trial or chemotherapy for a patient is not ready".

The α-tokens (from places l_{18} and l_{20}) and β-tokens (from place d_7) enter place l_{26} without new characteristics. The β-tokens that enter place d_{17} do not obtain new characteristics.

The α-tokens enter places l_{24} and l_{25} with characteristic respectively:

"patient, needs from follow-up care" in place l_{24};

and "patient, need from clinical trial or chemotherapy" in place l_{25}.

4 Conclusion

The GN-model suggested here could be used as a general algorithm of behavior from the initial examination to the treatment of the patients with malignant melanoma. It can be detailed in order to trace the status of each patient - risk factors for melanoma and family history, the appropriate treatment, specific diagnostic technique, appropriatesystemic therapy or a clinical treatment, etc. Since the modeled process is very complex, the GN presented here has not been described in excessive detail. Nevertheless, it has been provided with sufficient information to apply the model in practice.

Acknowledgment. This research was funded in part by the European Regional Development Fund through the Operational 268 Programme "Science and Education for Smart Growth" under contract UNITe BG05M2OP001–1.001–0004 269 (2018–2023)."

References

1. Alex, J.C., Weaver, D.L., Fairbank, J.T., Rankin, B.S., Krag, D.N.: Gamma-probe-guided lymph node localization in malignant melanoma. Surg Oncol. **2**(5), 303–308 (1993). https://doi.org/10.1016/s0960-7404(06)80006-x. PMID: 8305972
2. Atanassov, K.: On Generalized Nets Theory. In: Prof. M. Drinov Academic Publishing House, Sofia (2007)
3. Atanassov, K.: Generalized Nets. World Scientific, Singapore (1991)
4. Atanassov, K.: Index Matrices: Towards an Augmented Matrix Calculus. Springer, Cham (2014). https://doi.org/10.1007/978-3-319-10945-9
5. Garbe, C., Leiter, U.: Melanoma epidemiology and trends. Clin. Dermatol. **27**(1), 3–9 (2009). https://doi.org/10.1016/j.clindermatol.2008.09.001
6. Khanna, A.K., Sundaran, P., Khanna, S.: necrotizing soft tissue infection. In: Khanna, A.K., Tiwary, S.K. (eds.) Ulcers of the Lower Extremity, pp. 275–288. Springer, New Delhi (2016). https://doi.org/10.1007/978-81-322-2635-2_17
7. Dobrev, P., Strashilov, S., Yordanov, A.: Metastasis of malignant melanoma in ovarian simulating primary ovarian cancer: a case report. Gazetta Medica Italiana – Archivio per le Scienze Medicine, **180**, 867–869 (2021)
8. Dobrev, P., Yordanov, A., Strashilov, S.: Synchronous primary cervical carcinoma and ovarian fibroma: challenge in surgery. Gazzetta Medica Italiana-Archivio per le Scienze Mediche, **179**(5), 375–377 (2020)
9. Holzmann, B., et al.: Tumor progression in human malignant melanoma: five stages defined by their antigenic phenotypes. Int. J. Cancer **39**(4), 466–471 (1987)
10. Karimkhani, C., et al.: The global burden of melanoma: results from the global burden of disease Study 2015. Br. J. Dermatol. **177**(1), 134–140 (2017)

11. Markovic, S., et al.: Malignant melanoma in the 21st century, part 1: epidemiology, risk factors, screening, prevention, and diagnosis. In: Mayo Clinic Proceedings, pp. 364–380 Elsevier (2007)
12. Sotirov, S., Petrova, Y., Bozov, H., Sotirova, E.: A hybrid algorithm for multilayer perceptron design with intuitionistic fuzzy logic using malignant melanoma disease data. In: Kahraman, C., Tolga, A.C., Cevik Onar, S., Cebi, S., Oztaysi, B., Sari, I.U. (eds.) Intelligent and Fuzzy Systems. INFUS 2022. Lecture Notes in Networks and Systems, vol. 504, pp. 665–672. Springer, Cham (2022). https://doi.org/10.1007/978-3-031-09173-5_77
13. Sotirova, E., Bozova, G., Bozov, H., Sotirov, S., Vasilev, V.: application of the intercriteria analysis method to a data of malignant melanoma disease for the burgas region for 2014–2018. In: Atanassov, K.T., et al. (eds.) IWIFSGN 2019 2019. AISC, vol. 1308, pp. 166–174. Springer, Cham (2021). https://doi.org/10.1007/978-3-030-77716-6_15
14. Tsao, H., et al.: Early detection of melanoma: reviewing the ABCDEs. J. Am. Acad. Dermatol. **72**(4), 717–723 (2015). https://doi.org/10.1016/j.jaad.2015.01.025

Generalized Net Model of the Prostate Cancer Early Stages of Development

Elenko Popov[1,2(✉)], Radostina Georgieva[1,2], Dmitrii Dmitrenko[1,2], Borislav Bojkov[1,2], Chavdar Slavov[1,2], Martin Lubich[3], Peter Vassilev[4], Vassia Atanassova[4], Lyudmila Todorova[4], and Krassimir T. Atanassov[4]

[1] Faculty of Medicine, Medical University-Sofia, 15 Acad. Ivan Geshov Blvd., 1431 Sofia, Bulgaria
shennyp@yahoo.com

[2] Department of Urology and Andrology, University Hospital Tsaritsa Yoanna-ISUL, 8 Byalo More Str., 1527 Sofia, Bulgaria

[3] Department of Nephrology, University Hospital Sofiamed, 16 G. M. Dimitrov Blvd., 1797 Sofia, Bulgaria
martinlub11@abv.bg

[4] Department of Bioinformatics and Mathematical Modelling, Institute of Biophysics and Biomedical Engineering, Bulgarian Academy of Sciences, Acad. G. Bonchev Str., Bl. 105, 1113 Sofia, Bulgaria
lpt@biomed.bas.bg, krat@bas.bg

Abstract. A generalized net model of the prostate cancer early stages of development is described. The relationship of the prostate with the human body respiratory, blood and circulatory, and endocrine systems are represented.

Keywords: Generalized net · Model · Prostate cancer

1 Introduction: Stages and Early Events in the Development of Prostate Cancer

Prostate cancer is the most frequent non-skin cancer diagnosed in males and is the second most common reason for cancer-related death in men, with 1,600,000 cases and 366,000 deaths annually - see the report of the Global Cancer Observatory (GCO) that is an interactive web-based platform presenting global cancer statistics to inform cancer control and research [1]. In the last decades major advances had been achieved in understanding on prostate cancer, as well as its diagnosis and treatment, but it still represent a formidable challenge in contemporary medicine. The two main obstacles are the limited efficacy in cases of metastatic disease and the observed significant overtreatment for clinically indolent cases, that never have clinical consequences during the life-time of the patient [2]. Because of that, additional research on prostate cancer nature and development are critically needed.

Three different stages of development of prostate cancer have been identified on the basis on molecular, pathological, clinical and imaging characteristics:

S. Sotirov et al. (Eds.): BioInfoMed 2022, LNNS 658, pp. 246–252, 2023.
https://doi.org/10.1007/978-3-031-31069-0_24

(1) precancerous states such as atypical small acinar proliferation (ASAP) and high-grade intraepitelial neoplasia (HGPIN), defined by the hyperplasia of acinar luminal cells and progressive disruption and loss of basal cells layer;
(2) androgen-dependent prostate cancer, characterized by the complete loss of basal cells layer and the overt malignant microspopic phenotype; because at this stage prostate cancer is still androgen-dependent, it is sensitive to androgen deprivation therapy (ADT); and
(3) androgen-independent (castration resistant) prostate cancer, which is an inevitable step in natural progression of prostate cancer and is insensitive to ADT.

ASAP and HGPIN are believed to be precursor lesions of prostate cancer owing to two facts,based on significant evidence: 1) epidemiological data that connects their existence with following finding of invasive carcinoma during follow-up, and 2) microscopic structural resemblance of epithelial cells of pre-cancerous lesions and prostate cancer; and their colocalization in the prostate tissue, as well as shared genetic changes, vastly investigated in the last decade [3].

In the following sections we propose a GN model of the early stages of prostate cancer development through the different stages of precursor lesions, incorporating the influence of endocrine, respiratory, gastro-intestinal and circulatory system on this process.

In the next section, a Generalized Net (GN) model of the prostate cancer early stages of development is described. All notation is described in [4–6].

2 The Generalized Net Model

The GN model, illustrated in Fig. 1, contains 8 transitions, 24 places and 12 types of tokens that have the following sense:

α - Respiratory System (RS),
α' - the air that enters RS,
α'' - the carcinogen agents from the air that enter the Blood Stream (BS),
β - Blood and Circulatory System (BCS),
$\beta_1, \beta_2, \beta_3, \beta_4$ - the blood for the Prostate Gland (PG) and in particular - for the Prostate Cell (PC), in its different phases,
γ - Gastrointestinal System (GS),
γ' - the food and water that enter GS,
γ'' - the carcinogen agents from the food and water that enter the BS,
ε - Endocrine System (ES),
ε' - the hormones that enter the BS,
π - a PC,
$\pi_1, \pi_2, \pi_3, \pi_4$ - the blood from the PC that enter the BS.

The GN transitions have the following forms.

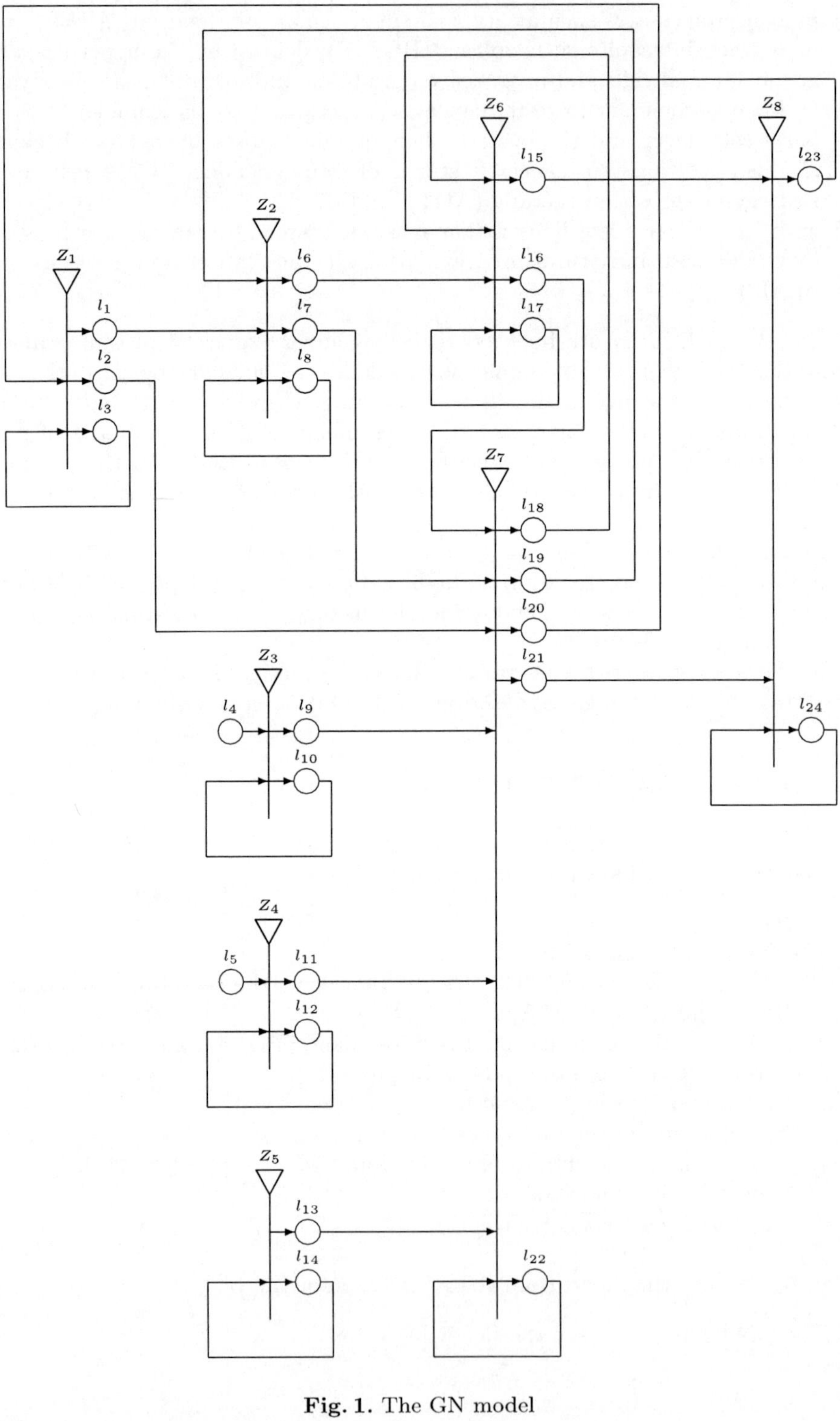

Fig. 1. The GN model

$$Z_1 = \langle \{l_3, l_{20}\}, \{l_1, l_2, l_3\}, \begin{array}{c|ccc} & l_1 & l_2 & l_3 \\ \hline l_3 & W_{3,1} & true & true \\ l_{20} & false & false & true \end{array} \rangle,$$

where
$W_{3,1}$ = "there are genetic mutations that lead the Atypical Small Acinar Protiferation (ASAP).

An each time step, token β_1 from place l_{20} enters place l_3 and unites with token π that obtains a characteristic

"current status of the (normal) PC".

At each time-step token π splits to two tokens - the same token π that continue to stay in place l_3 and the token π_1 that enters place l_2 with a characteristic

"the quantity of the blood from the PC".

When predicate $W_{3,1}$ is true, token π enters place l_1 with the characteristic

"the level of the genetic mutations that lead to ASAP".

$$Z_2 = \langle \{l_1, l_8, l_{19}\}, \{l_6, l_7, l_8\}, \begin{array}{c|ccc} & l_6 & l_7 & l_8 \\ \hline l_1 & false & false & true \\ l_8 & W_{8,6} & true & true \\ l_{19} & false & false & true \end{array} \rangle,$$

where
$W_{8,6}$ = "there are genetic mutations that lead to Prostate Intraepiteliar Neoplasma (PIN)".

An each time step after the moment in which token π had entered place l_8, token β_2 from place l_{19} enters place l_8 and unites with token π that obtains a characteristic

"current status of the ASAP in the PC".

At each time-step token π splits to two tokens - the same token π that continue to stay in place l_8 and the token π_2 that enters place l_7 with a characteristic

"the quantity of the blood from the PC".

When predicate $W_{8,6}$ is true, token π enters place l_6 with the characteristic

"the level of the genetic mutations that lead to PIN".

$$Z_3 = \langle \{l_4, l_{10}\}, \{l_9, l_{10}\}, \begin{array}{c|cc} & l_9 & l_{10} \\ \hline l_4 & false & true \\ l_{10} & true & true \end{array} \rangle.$$

In some moments, tokens denoted for brevity without their current indices from place l_4, enter place l_{10} and unite with the token γ that permanently stays in place l_{10} with a current characteristic

"current status of the GS".

At each time-step token γ splits to two tokens - the same token γ that continues to stay in place l_{10} and the token γ'' that enters place l_9 with a characteristic

"the quantity of the carcinogen agents from the food and water".

$$Z_4 = \langle \{l_5, l_{12}\}, \{l_{11}, l_{12}\}, \begin{array}{c|cc} & l_{11} & l_{12} \\ \hline l_5 & false & true \\ l_{12} & true & true \end{array} \rangle.$$

At each time-step, tokens denoted for brevity without their current indices from place l_5, enter place l_{12} and unite with the token α that permanently stays in place l_{12} with a current characteristic

"current status of the RS".

At each time-step token α splits to two tokens - the same token α that continues to stay in place l_{12} and the token α'' that enters place l_{11} with a characteristic

"the quantity of the carcinogen agents from the air".

$$Z_5 = \langle \{l_{14}\}, \{l_{13}, l_{14}\}, \begin{array}{c|cc} & l_{13} & l_{14} \\ \hline l_{14} & true & true \end{array} \rangle.$$

At each time-step token ε splits to two tokens - the same token ε that permanently stays in place l_{14} with a current characteristic

"current status of the ES"

and the token ε'' that enters place l_{13} with a characteristic

"the quantity of the sex-steroid hormones".

$$Z_6 = \langle \{l_6, l_{17}, l_{20}\}, \{l_{15}, l_{16}, l_{17}\}, \begin{array}{c|ccc} & l_{15} & l_{16} & l_{17} \\ \hline l_6 & false & false & true \\ l_{17} & W_{17,15} & true & true \\ l_{20} & false & false & true \end{array} \rangle,$$

where
$W_{17,15}$ = "there are genetic mutations that lead to cancer".

An each time step after the moment in which token π had entered place l_{17}, token β_3 from place l_{18} enters place l_{17} and unites with token π that obtains a characteristic

"current status of the PIN in the PC".

At each time-step token π splits to two tokens - the same token π that continues to stay in place l_{17} and the token π_3 that enters place l_{16} with a characteristic

"the quantity of the blood from the PC".

When predicate $W_{17,15}$ is true, token π enters place l_{15} with the characteristic

"the level of the genetic mutations that lead to cancer".

$$Z_7 = \langle \{l_2, l_7, l_9, l_{11}, l_{13}, l_{16}, l_{22}, l_{24}\}, \{l_{18}, l_{19}, l_{20}, l_{21}, l_{22}\},$$

	l_{18}	l_{19}	l_{20}	l_{21}	l_{22}
l_2	*false*	*false*	*false*	*false*	*true*
l_7	*false*	*false*	*false*	*false*	*true*
l_9	*false*	*false*	*false*	*false*	*true*
l_{11}	*false*	*false*	*false*	*false*	*true*
l_{13}	*false*	*false*	*false*	*false*	*true*
l_{16}	*false*	*false*	*false*	*false*	*true*
l_{22}	$W_{22,18}$	$W_{22,19}$	$W_{22,20}$	$W_{22,21}$	*true*
l_{24}	*false*	*false*	*false*	*false*	*true*

$, \rangle.$

where
$W_{22,18}$ = "token π is in place l_3",
$W_{22,19}$ = "token π is in place l_8",
$W_{22,20}$ = "token π is in place l_{17}",
$W_{22,21}$ = "token π is in place l_{24}".

Each one of the tokens $\alpha'', \beta_1, \beta_2, \beta_3, \beta_4, \gamma'', \varepsilon''$ enters place l_{22} and unites with token β that stays permanently in this place and that obtains a characteristic

"current status of the BCS".

When predicate $W_{22,17+i}$ is true for $i = 1, 2, 3, 4$, token β splits to two tokens - the same token β that continues to stay in place l_{22} and the token β_i that enters place $l_{20}, l_{19}, l_{18}, l_{21}$ with a characteristic

"the quantity of the blood directed to the PC in the i-th phase",

respectively.

$$Z_8 = \langle \{l_{15}, l_{21}, l_{24}\}, \{l_{23}, l_{24}\},$$

	l_{23}	l_{24}
l_{15}	*false*	*true*
l_{21}	*false*	*true*
l_{24}	*true*	*true*

$\rangle.$

An each time step after the moment in which token π had entered place l_{24}, token β_4 from place l_{21} enters place l_{24} and unites with token π that obtains a characteristic

"current status of the prostate cancer".

At each time-step token π splits to two tokens - the same token π that continues to stay in place l_{24} and the token π_4 that enters place l_{23} with a characteristic

"the quantity of the blood from the PC".

3 Conclusion

This GN model represent a broad frame in which different genetic events and the influence of the surroundings as well as regulatory systems in the human body can define their interactions in early stages of prostate cancer development. This model should be a sound basis for gaining additional insight on the subject, helping to elucidate multiple and complex connections and correlations.

Acknowledgments. This research is supported by the Bulgarian National Science Fund under Grant Ref. No. KP-06-N43/7 "Creating a prognostic model predicting life expectancy in prostate cancer patients and providing better quality of life after definitive surgical treatment" from 30.11.2020.

References

1. IARC: Global Cancer Observatory. https://gco.iarc.fr/today/home:GLOBOCAN
2. Foreman, K.J., Marquez, N., Dolgert, A., et al.: Forecasting life expectancy, years of life lost, and all-cause and cause-specific mortality for 250 causes of death: reference and alternative scenarios for 2016–40 for 195 countries and territories. Lancet **392**, 2052–2090 (2018)
3. Tolkach, Y., Kristiansen, G.: Is high-grade prostatic intraepithelial neoplasia (HGPIN) a reliable precursor for prostate carcinoma? Implications for clonal evolution and early detection strategies. J. Pathol. **244**, 389–393 (2018)
4. Atanassov, K.: Generalized Nets. World Scientific, Singapore (1991)
5. Atanassov, K.: On Generalized Nets Theory. Sofia, "Prof. Marin Drinov" Academic Publishing House (2007)
6. Atanassov, K.: Index Matrices: Towards an Augmented Matrix Calculus. Springer, Cham (2014). https://doi.org/10.1007/978-3-319-10945-9

Generalized Net Model of the Vegetative (Autonomic) Innervation of Gastrointestinal Tract

Valentina Ignatova[1] and Krassimir Atanassov[2,3](✉)

[1] Clinic of Neurology at National Cardiology Hospital,
65, Konyovitsa Str, 1309 Sofia, Bulgaria
valyaig@abv.bg

[2] Department of Bioinformatics and Mathematical Modelling,
Institute of Biophysics and Biomedical Engineering, Bulgarian Academy of Sciences,
Acad. G. Bonchev Str., Block 105, 1113 Sofia, Bulgaria
krat@bas.bg,k.t.atanassov@gmail.com

[3] Intelligent Systems Laboratory, Prof. Asen Zlatarov University,
1,"Prof. Yakimov" Blvd, 8010 Burgas, Bulgaria

Abstract. The gastrointestinal tract (GIT) is a basic part of the human body. It interacts with many organs and systems and has a complex autonomic innervation. Generalized net model of GIT innervation is described and the relations with central nervous system are shown. Some applications of the model are discussed.

Keywords: Gastrointestinal tract · Generalized net · Human body · Innervation

1 Introduction

The GastroIntestinal Tract (GIT) performs a number of complex and integrative functions which contributes not only to normal digestion, but also to homeostasis, communication with other organs and systems, and to maintaining immunity [14]. The main functions of the GIT include digestion, absorption, excretion and protection. After chewing in the mouth, the food reaches the stomach through the esophagus, where the main digestive processes begin. The stomach is the most distended part of the GIT and communicates via the cardia with the esophagus and distally with the duodenum via the pylorus [15]. Distal to the pylorus begins the small intestine. It has a length of 4–6 meters and is the main locus where early digestion takes place, which is facilitated by its huge absorption surface. Digestive contents (chyme) are pushed distally from the pylorus to the ileocecal valve at a rate of 5–20 mm/s by weak peristaltic contractions, which takes 3–5 h [12]. The breakdown of food in the stomach and small intestine results in the formation of substances that the body uses for energy, growth and tissue repair. Waste products that the body cannot use leave the organism through bowel movements [7]. The large intestine acts as a reservoir for food

S. Sotirov et al. (Eds.). BioInfoMed 2022, LNNS 658, pp. 253–263, 2023.
https://doi.org/10.1007/978-3-031-31069-0_25

waste and allows reabsorption of water from the chyme leaving the small intestine. In recent years was found the important role of the population of intestinal bacteria in the large intestine (gut microbiota) for the development and maturation of the hypothalamic-pituitary-adrenal axis, for the state of the immune system, the regulation of the blood-brain barrier, the synthesis and neurotransmitter recognition, neurogenesis, myelination formation and supporting brain function [19].

The innervation of the GIT is extremely complex [10]. On the one hand, the GIT is the only internal organ that has evolved in evolution with its own independent nervous system known as the Enteric Nervous System (ENS, see, e.g., [18]). Its main role is to control the motility, secretion, mucosal transport and blood flow of the GIT [8]. The ENS implements these functions through motor neurons located in ganglia, constituting a final common pathway to effectors cells of the GIT [9,11].

The external innervation of the GIT is carried out by splanchnic sympathetic nerves and by vagal (parasympathetic) innervation [13]. Noradrenergic fibers in the GIT wall arise from cell bodies embedded in the prevertebral sympathetic ganglia. The main sympathetic projections to the colon originate from the inferior mesenteric ganglia, and the remaining noradrenergic fibers to the rectum which originate from the pelvic ganglia. The Vagal Nerve (VN) carries parasympathetic information between the internal organs and the brainstem. The VN is composed of afferent and efferent nerve fibers and provides innervation to the small intestine and the proximal $\frac{2}{3}$ of the colon. The afferent fibers of VN terminates in the Nucleus Tractus Solitarii (NTS), situated at the level of medulla oblongata [17]. In turn, the HypoThalamus (HT) sends projections to the NTS. At the same time, the HT receives reverse projections from the NTS [16].

The caudal portion of NTS neurons provide projections to vagal efferent neurons in the nucleus Dorsalis Nervi Vagi (DVN) for control of parasympathetic gastrointestinal responses, including pancreatic insulin secretion and gastric emptying, and to InterMedioLateral (IML) columnar cells in the spinal cord along with projections from neurons in other regions in and HT and rhombencephalon [20].

Afferent sensory information from the stomach and small intestine is carried by C-type sensory afferent fibers to the posterior horns of the cervical myelon, and afferent sensory information from the large intestine is carried by C-type sensory fibers to the sacral spinal cord. On the other hand, afferent autonomic information from the stomach, duodenum, and small intestine is transmitted via the VN and spinal nerves, whose cell bodies are located in the Nodose Ganglia (NG) and the Spinal Ganglia (GS) embedded in the dorsal roots [16].

For better understanding of the complex mechanism of many biological processes and, in particular, of the anatomo-physiological interactions between the individual organs and systems in the human organism, a tendency to develop mathematical models that more clearly show the interrelationships between their constituent elements arose in recent years. One of the modern approaches that can serve as a rational tool reflecting the hierarchy and complex interrelationships at different levels in the presentation of various medical models and diseases are the Generalized Net (GN) models [1,2].

2 A Generalized Net Model

The aim of the present research is to construct a GN model of the innervation of the GIT in the form of an algorithm, schematizing the complex process of physiological interactions between the stomach, intestine and nervous system.

The GN contains 10 transitions, 34 places and 33 types of tokens that have the following sense:

HT - HypoThalamus (HT) - it stays permanently in place l_2
h - signal from the HT to the NTS
NG - Nodose Ganglia (NG) - it stays permanently in place l_4
g - signal from the NG to the DVN
NTS - Nucleus Tractus Solitarii (NTS) - it stays permanently in place l_8
n_1 - signal from the NTS to the colon
n_2 - signal from the NTS to the liver
n_3 - signal from the NTS to HT
DVN - nucleus Dorsalis Nervi Vagi (DVN) - it stays permanently in place l_{13}
d_1 - signal from the DVN to the NTS
d_2 - signal from the DVN to the stomach
d_3 - signal from the DVN to the duodenum
d_4 - signal from the DVN to the small intestine
L - liver - it stays permanently in place l_{16}
v_1 - signal from the liver to the stomach
S - stomach - it stays permanently in place l_{20}
s_1 - signal from the stomach to the NG
s_2 - food from the stomach to the duodenum
s_3 - signal from the stomach to the IML cervicothoracic spinal column
F - food that enters the stomach - place l_{14}
IML - InterMedioLateral column - it stays permanently in place l_{22}
l - signal from the IML to the HT
D - duodenum - it stays permanently in place l_{25}
u_1 - signal from the small intestine to the the IML cervicothoracic spinal column
u_2 - signal from the duodenum to the NG
SI - small intestine - it stays permanently in place l_{29}
u_1 - signal from the small intestine to the IML
u_2 - signal from the small intestine to the colon
u_3 - signal from the small intestine to the NG
C - colon - it stays permanently in place l_{33}
c_1 - signal from the colon to the IML
c_2 - excrements
c_3 - signal from the colon to the IML sacral spinal column

The GN-transitions have the following forms.

$$Z_1 = \langle \{l_2, l_5, l_{21}\}, \{l_1, l_2\}, \begin{array}{c|cc} & l_1 & l_2 \\ \hline l_2 & true & true \\ l_5 & false & true \\ l_{21} & false & true \end{array} \rangle.$$

Tokens n_3 from place l_5 and token c_1 from place l_{21} enter place l_2 and unite with token HT that stays permanently in place l_2 with characteristic

"current status of the hypothalamus".

On each time-step, token HT splits to two tokens - the same token HT that continues to stay in place l_2 and token h that enters place l_1 with a characteristic

"vegetative signal from hypothalamus to NTS".

$$Z_2 = \langle \{l_4, l_{19}, l_{24}, l_{28}, l_{32}\}, \{l_3, l_4\}, \begin{array}{c|cc} & l_3 & l_4 \\ \hline l_4 & true & true \\ l_{19} & false & true \\ l_{24} & false & true \\ l_{28} & false & true \\ l_{32} & false & true \end{array} \rangle.$$

Tokens s_3, u_2, i_3, c_3 from places $l_{19}, l_{24}, l_{28}, l_{32}$, respectively, enter place l_4 and unite with token NG that stays permanently in place l_4 with characteristic

"current status of the NG".

On each time-step, token NG splits to two tokens - the same token NG that continues to stay in place l_4 and token g that enters place l_3 with a characteristic

"parasympathetic-vegetative afferent signal from the NG to the DVN".

$$Z_3 = \langle \{l_1, l_8, l_9\}, \{l_5, l_6, l_7, l_8\}, \begin{array}{c|cccc} & l_5 & l_6 & l_7 & l_8 \\ \hline l_1 & false & false & false & true \\ l_8 & true & true & true & true \\ l_9 & false & false & false & true \end{array} \rangle.$$

Tokens h, from place l_1 and d_1 from place l_9 enter place l_8 and unite with token NTS that stays permanently in place l_8 with characteristic

"current status of the NTS".

On each time-step, token NTS splits to four tokens - the same token NTS that continues to stay in place l_8 and tokens

n_1 that enters place l_5 with a characteristic

"vegetative signal from the NTS to the hypothalamus",

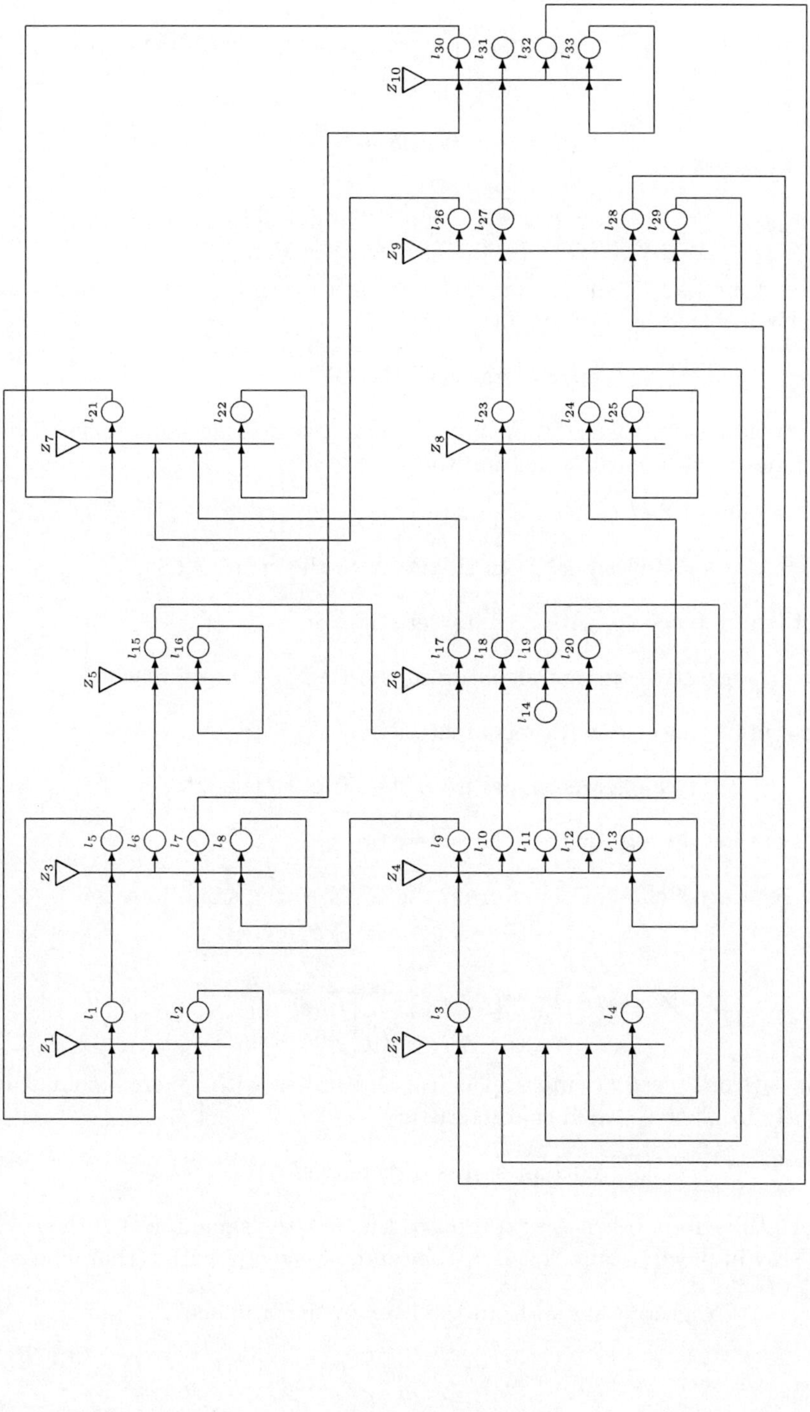

Fig. 1. A GN model

n_2 that enters place l_6 with a characteristic

"sympathetic-vegetative efferent signal from the NTS to the liver",

n_3 that enters place l_7 with a characteristic

"sympathetic-vegetative efferent signal from the NTS to the colon".

$$Z_4 = \langle \{l_3, l_{13}\}, \{l_9, l_{10}, l_{11}, l_{12}, l_{13}\}, \begin{array}{c|ccccc} & l_9 & l_{10} & l_{11} & l_{12} & l_{13} \\ \hline l_3 & false & false & false & false & true \\ l_{13} & true & true & true & true & true \end{array} \rangle.$$

Token g from place l_3 enters place l_{13} and unites with token DVN that stays permanently in place l_{13} with characteristic

"current status of the DVN".

On each time-step, token DVN splits to five tokens - the same token DVN that continues to stay in place l_{13} and tokens

d_1 that enters place l_9 with a characteristic

"efferent signal from the DVN to the small NTS",

d_2 that enters place l_{10} with a characteristic

"vegetative efferent signal from the DVN to the stomach",

d_3 that enters place l_{11} with a characteristic

"vegetative efferent signal from the DVN to the duodenum",

d_4 that enters place l_{12} with a characteristic

"vegetative efferent signal from the DVN to the small intestine".

$$Z_5 = \langle \{l_6, l_{16}\}, \{l_{15}, l_{16}\}, \begin{array}{c|cc} & l_{15} & l_{16} \\ \hline l_6 & false & true \\ l_{16} & true & true \end{array} \rangle.$$

Token n_2 from place l_6 enters place l_{16} and unites with token L that stays permanently in place l_{16} with characteristic

"current status of the liver".

On each time-step, token L splits to two tokens - the same token L that continues to stay in place l_{16} and token v_1 that enters place l_{15} with a characteristic

"digestive signal from the liver to the stomach".

$$Z_6 = \langle \{l_{10}, l_{14}, l_{15}, l_{20}\}, \{l_{17}, l_{18}, l_{19}, l_{20}\}, \begin{array}{c|cccc} & l_{17} & l_{18} & l_{19} & l_{20} \\ \hline l_{10} & false & false & false & true \\ l_{14} & false & false & false & true \\ l_{15} & false & false & false & true \\ l_{20} & false & true & true & true \end{array} \rangle.$$

Tokens d_2 from place l_{10}, F from place l_{14}, and v_2 from place l_{15}, enter place l_{20} and unite with token S that stays permanently in place l_{20} with characteristic

"current status of the stomach".

On each time-step, token S splits to four tokens - the same token S that continues to stay in place l_{20} and tokens

s_1 that enters place l_{17} with a characteristic

"afferent sympathetic signal from the stomach to the IML cervicothoracic spinal cord",

s_2 that enters place l_{18} with a characteristic

"efferent signal from the stomach to the duodenum",

s_3 that enters place l_{19} with a characteristic

"efferent parasympathetic signal from the stomach to the NG",

$$Z_7 = \langle \{l_{17}, l_{22}, l_{26}, l_{30}\}, \{l_{21}, l_{22}\}, \begin{array}{c|cc} & l_{21} & l_{22} \\ \hline l_{17} & false & true \\ l_{22} & true & true \\ l_{26} & false & true \\ l_{30} & false & true \end{array} \rangle.$$

Tokens s_1 from place l_{17}, i_1 from place l_{26}, and c_1 from place l_{30}, enter place l_{22} and unite with token IML that stays permanently in place l_{22} with characteristic

"current status of the IML".

On each time-step, token IML splits to two tokens - the same token IML that continues to stay in place l_{22} and token l that enters place l_{21} with a characteristic

"afferent signal from the IML to the HT".

$$Z_8 = \langle \{l_{11}, l_{18}, l_{25}\}, \{l_{23}, l_{24}, l_{25}\}, \begin{array}{c|ccc} & l_{23} & l_{24} & l_{25} \\ \hline l_{11} & false & false & true \\ l_{18} & false & false & true \\ l_{25} & true & true & true \end{array} \rangle.$$

Tokens d_3 from place l_{11} and s_2 from place l_{18} enter place l_{25} and unite with token D that stays permanently in place l_{25} with characteristic

"current status of the duodenum".

On each time-step, token D splits to three tokens - the same token D that continues to stay in place l_{25} and tokens

u_1 that enters place l_{23} with a characteristic

"afferent signal from the duodenum to the small intestine",

u_2 that enters place l_{24} with a characteristic

"afferent signal from the duodenum to the NG".

$$Z_9 = \langle \{l_{12}, l_{23}, l_{29}\}, \{l_{26}, l_{27}, l_{27}, l_{29}\}, \begin{array}{c|cccc} & l_{26} & l_{27} & l_{28} & l_{29} \\ \hline l_{12} & false & false & false & true \\ l_{23} & false & false & false & true \\ l_{29} & true & true & true & true \end{array} \rangle.$$

Tokens d_4 from place l_{12} and u_1 from place l_{23} enter place l_{29} and unite with token SI that stays permanently in place l_{29} with characteristic

"current status of the small intestine".

On each time-step, token SI splits to four tokens - the same token SI that continues to stay in place l_{29} and tokens

i_1 that enters place l_{26} with a characteristic

"afferent signal from the small intestine to the IML",

i_2 that enters place l_{27} with a characteristic

"afferent signal from the small intestine to the colon",

i_3 that enters place l_{28} with a characteristic

"afferent parasympathetic signal from the duodenum to the NG".

$$Z_{10} = \langle \{l_7, l_{27}, l_{33}\}, \{l_{30}, l_{31}, l_{32}, l_{33}\}, \begin{array}{c|cccc} & l_{30} & l_{31} & l_{32} & l_{33} \\ \hline l_7 & false & false & false & true \\ l_{27} & false & false & false & true \\ l_{33} & true & true & true & true \end{array} \rangle.$$

Tokens n_3 from place l_7 and i_2 from place l_{27} enter place l_{33} and unite with token C that stays permanently in place l_{33} with characteristic

"current status of the colon".

On each time-step, token C splits to four tokens - the same token C that continues to stay in place l_{33} and tokens

c_1 that enters place l_{30} with a characteristic

"afferent parasympathetic signal from the colon to the IML",

c_2 that enters place l_{31} with a characteristic

"excrements",

c_3 that enters place l_{32} with a characteristic

"afferent sympathetic signal from the colon to the NG".

In the current GN model, all signals necessarily pass from the input to the output position, without conditions, as the principles of innervation of the GIT in a healthy organism are followed, but in the future the predicates "*true*" will be replaced by specific other predicates.

The innervation of the GIT is one of the most important functions in the human body, which ensures the health not only of the gut and the stomach, but also of the entire organism. In recent years, the view on the interactions between the nervous system, the stomach and the intestine has expanded substantially. It became known, that the interconnection between the gut and the brain is based on a complex system which incorporates not only neural but also endocrine, immune, and humoral relationships [6]. For the proper functioning of the GIT, the role of the other organs of the gastrointestinal system, namely the pancreas, which secretes digestive juices into the small intestine, the liver and the biliary system, performing vital metabolic functions in addition to contributing to the digestion and assimilation of nutrients, is also indisputable [14].

The ENS was found to be an integrative system of neurons with structural complexity and functional heterogeneity close to those of the brain and spinal cord [8]. In fact, the ENS is the only system in the human body with separate innervation, independent of commands from the brain. On the other hand, in recent years, it has become known the substantive influence of GIT on the state of important brain structures. Significant grade of plasticity within CNS locations involved in coordination and regulation of gastrointestinal functions was found [6].

Research in animal models has proven as crucial in guiding the investigation on the brain-gut axis in humans. Exploring the interaction between gut microbes and the human brain will not only allow us to better understand the pathogenesis of neuropsychiatric disorders, but will also provide us with new opportunities to design new immuno- or microbe-based therapies [13]. The qualitative composition of the microbiota in the large intestine also affects the sympathetic and parasympathetic innervation not only of the GIT, but also of various departments of the CNS. An imbalance of the intestinal microbiota can lead to various gastrointestinal diseases as well as diseases of the CNS such as depression, autism, Parkinson's disease, and according to the latest data - multiple sclerosis [4]. On the other hand, the presence of independent dysfunction of the CNS, through its reciprocal relationships with the stomach and intestines, can lead to gastrointestinal dysfunction, which further worsens the condition of the body. A typical example is the so-called irritable bowel syndrome, in which no structural damage of the intestine is detected, but the patients experience frequent diarrhea and abdominal pain. Rather, in this contingent of patients, the leading factor for the GIT discomfort is increased anxiety, which through sympathetic nervous interactions at the level of the brain-brainstem-paravertebral ganglia-intestine leads to difficulties in intestinal resorption against the background of restlessness, fear, sleep disorders, often depressive states. In persons with panic attacks and anxiety attacks, aerophagia often occurs, leading to gastric discomfort when a gastroenterological problem is completely rejected by gastroenterologists [5].

There are still many unknowns about GIT-nervous system interactions, which has encouraged many researchers to work on the field.

3 Conclusion

The innervation of the GIT is one of the most complex physiological processes to understand in the human body because of the multiple pathways for delivery of nerve impulses to the stomach and intestines - autonomic and somatosensory projections, interactions at different levels of the CNS - from the spinal cord through the medulla oblongata and hypothalamus, including the fully autonomic own innervation of the GIT. The impact of the stomach and intestines on the brain further complicates the gut-brain relationship. The proposed generalized network model provides an exact and clear view of the described neurological processes and explains in an accessible way the interrelationship of the GIT - nervous system at different hierarchical levels.

Acknowledgment. The authors acknowledge the support from the project UNITe BG05M2OP001-1.001-0004 /28. 02.2018 (2018–2023).

References

1. Atanassov, K.: Generalized Nets. World Scientific, Singapore, London (1991)
2. Atanassov, K.: On Generalized Nets Theory. Prof. M. Drinov Academic Publ. House, Sofia (2007)
3. Atanassov, K., Chakarov, V., Shannon, A., Sorsich, J.: In: Generalized Net Models of the Human Body. Publishing House of the Bulgarian Academy of Sciences, Sofia (2008)
4. Bostick, J.W., Schonhoff, A.M., Mazmanian, S.K.: Gut microbiome-mediated regulation of neuroinflammation. Curr. Opin. Immunol. **76**, 102177 (2022)
5. Breit, S., Kupferberg, A., Rogler, G., Hasler, G.: Vagus nerve as modulator of the brain-gut axis in psychiatric and inflammatory disorders. Front. Psychiatry **9**, 44 (2018)
6. Browning, K.N., Travagli, R.A.: Central nervous system control of gastrointestinal motility and secretion and modulation of gastrointestinal functions. Compr. Physiol. **4**(4), 1339 (2014)

7. Cheng, L.K., O'Grady, G., Du, P., et al.: Gastrointestinal system. Wiley Interdisc. Rev. Syst. Biol. Med. **2**(1), 65–79 (2010)
8. Costa, M., Brookes, S.J.H.: The enteric nervous system. Am. J. Gastroenterol. **89**, S129–S137 (1994)
9. Fung, C., Vanden Berghe, P.: Functional circuits and signal processing in the enteric nervous system. Cell. Mol. Life Sci. **77**(22), 4505–4522 (2020). https://doi.org/10.1007/s00018-020-03543-6
10. Furness, J.B., Pustovit, R.V., Fothergill, L.J., et al.: The Innervation of the Gastrointestinal Tract. Yamada's Textbook of Gastroenterology, pp. 191–212 (2022)
11. Gillis, R.A., Quest, J.A., Pagini, F.D., Norman, W.P.: Control centers in the central nervous system for regulating gastrointestinal motility. In: Schultz, S.G., Wood, J.D., Rauner, B.B., (eds.) Handbook of Physiology. Section 6. The Gastrointestinal System. New York: Oxford University, pp. 621–683 (1989)
12. Guyton, A.C., Hall, J.E.: Propulsion and mixing of food in the alimentary tract. In: Textbook of Medical Physiology, 10 ed. Philadelphia, PA: W.B. Saunders Company, (63) (2000)
13. Ignatova, V. Influence of Gut Microbiota on Behavior and its Disturbances. In: Behavioral Neuroscience (2019)
14. Keshav, S.: The Gastrointestinal System at a Glance. Wiley, Hoboken (2009)
15. Perkin, G.D., Murray-Lyon, I.: Neurology and the gastrointestinal system. J. Neurol. Neurosurg. Psychiatry **65**(3), 291–300 (1998)
16. Riera, C.E., Dillin, A.: Emerging role of sensory perception in aging and metabolism. Trends Endocrinol. Metab. **27**(5), 294–303 (2016)
17. Sengupta, J.N., Gebhart, G.F.: Gastrointestinal afferent fibers and sensation. In: Johnson, L.J. (ed.) Physiology of the Gastrointestinal Tract, 3rd edn., pp. 483–519. Raven, NewYork (1994)
18. Spencer, N.J., Hu, H.: Enteric nervous system: sensory transduction, neural circuits and gastrointestinal motility. Nat. Rev. Gastroenterol. Hepatol. **17**(6), 338–351 (2020)
19. Taché, Y., Saavedra, J.M.: Introduction to the special issue "the brain-gut axis". Cell. Mol. Neurobiol. **42**, 311–313 (2021). https://doi.org/10.1007/s10571-021-01155-7
20. Wehrwein, E.A., Orer, H.S., Barman, S.M.: Overview of the anatomy, physiology, and pharmacology of the autonomic nervous system. Regulation **37**(69), 125 (2016)

Generalized Net Model of Density-Based Spatial Clustering of Applications with Noise (DBSCAN) Algorithm and Its Application on Diabetes Dataset

Petar Petrov[1,2], Veselina Bureva[1](✉), and Janusz Kacprzyk[3,4]

[1] Laboratory of Intelligent Systems, "Prof. Dr. Assen Zlatarov" University, "Prof. Yakimov" Blvd., Burgas 8010, Bulgaria
vbureva@btu.bg

[2] Vocational School of Electrical Engineering and Electronics "Konstantin Fotinov", Burgas, Bulgaria

[3] Systems Research Institute, Polish Academy of Sciences, Newelska 6, 01-447 Warsaw, Poland
kacprzyk@ibspan.waw.pl

[4] Warsaw School of Information Technology, Ul. Newelska 6, 01-447 Warsaw, Poland

Abstract. Density-based spatial clustering of applications with noise (DBSCAN) is a data science algorithm for density-based clustering. A Generalized net (GN) model of the DBSCAN algorithm is constructed. Intuitionistic fuzzy evaluations are defined to estimate the clustering procedure. The GN model of DBSCAN optimizes and estimates the procedure of the standard clustering algorithm. The clustering procedure is implemented using Python programming language. The method is then applied to a diabetes dataset to form clusters.

Keywords: Big Data · Data Mining · Machine Learning · Data Science · Generalized Nets · Intuitionistic Fuzzy Sets

1 Brief Introduction on Generalized Nets (GNs) and Intuitionistic Fuzzy Sets (IFS)

The theory of Generalized Nets (GNs) is introduced in [4] and extended in [1, 3, 5, 8]. GNs are an extension of the Petri nets. Similar GN models in the field of cluster analysis are already published in [10, 11]. **The concept of Intuitionistic Fuzzy Sets (IFS)** is presented in [6, 7, 9]. The theory of IFS includes the interval-valued intuitionistic fuzzy sets and intuitionistic fuzzy pairs, operators and implications. In the current research work the intuitionistic fuzzy evaluations are constructed to estimate the clustering procedure.

The paper has the following structure. Section 1 presents the notation of generalized nets and intuitionistic fuzzy sets. Section 2 introduces the constructed GN model of Density-based spatial clustering of applications with noise (DBSCAN) with Intuitionistic Fuzzy Evaluations. Section 3 presents an example of the implementation of the DBSCAN with IFE using Python language. Section 4 presents some findings about the clusters in the dataset. In Sect. 5 some conclusion remarks are discussed.

S. Sotirov et al. (Eds.): BioInfoMed 2022, LNNS 658, pp. 264–271, 2023.
https://doi.org/10.1007/978-3-031-31069-0_26

2 Generalized Net Model of Density-Based Spatial Clustering of Applications with Noise (DBSCAN)

DBSCAN clustering method is applied in the big data environment using the single machine approach [12]. Compared to other clustering algorithms DBSCAN does not require the number of clusters to be defined in the beginning. Another valuable quality of the algorithm is that it produces better distribution of the clusters when the data is odd-shaped. The DBSCAN algorithm relies on two parameters:

- *minPoints* - minimum number of points to form a cluster considering their density
- ε, *eps* - distance that will be used to consider the neighboring of points

The DBSCAN algorithm includes three steps. In the first step DBSCAN picks arbitrary point from the dataset. Then a check is performed to see if there are at least *minPoints* points in the ε radius. If this happens, then we consider these points a cluster. Afterwards we attempt to recursively expand each cluster with neighboring points. The three steps are repeated until all points are visited [14].

In the current research work a Generalized Net (GN) Model of DBSCAN with Intuitionistic Fuzzy Evaluations is presented (Fig. 1). The GNDraw software is used for GN model construction [13]. The GN model contains the following set of transitions:

$$A = \{Z_1, Z_2, Z_3, Z_4, Z_5\},$$

where the transitions describe the processes:

- Z_1 – Datasets;
- Z_2 – Preprocessing the received dataset;
- Z_3 – Scanning data for selection initial core point/s;
- Z_4 – Forming clusters with minimum points;
- Z_5 – Intuitionistic fuzzy evaluation of the received clusters;

Initially, there is one token that is located in place l_4 with initial characteristic: "*datasets*". In the next time-moments this token is split into two ones. The original token will continue to stay in place L_4, while the other tokens will move to the next transitions.

A token enters the net via place l_1 with initial characteristics: "*new data*". Transition Z_1 has the form:

$$Z_1 = \langle \{l_1, l_4\}, \{l_2, l_3, l_4\}, R_1, \vee(l_1, l_4) \rangle,$$

where

$$R_1 = \begin{array}{c|ccc} & l_2 & l_3 & l_4 \\ \hline l_1 & false & false & true \\ l_4 & W_{4,2} & W_{4,3} & W_{4,4} \end{array}$$

and

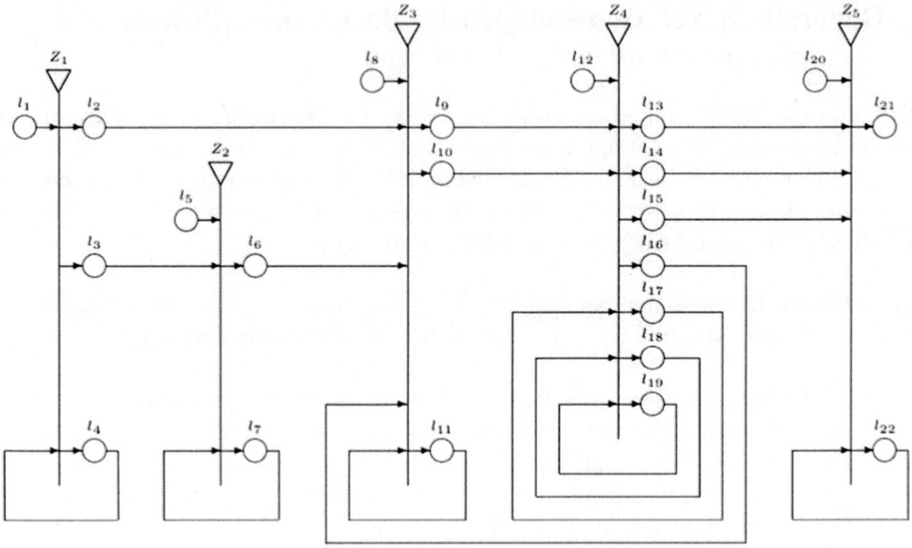

Fig. 1. Generalized net model of the process of cluster analysis using DBSCAN algorithm with intuitionistic fuzzy evaluations

- $W_{4,2}$ = "there is dataset for applying DBSCAN algorithm";
- $W_{4,3}$ = "there is dataset for preprocessing";
- $W_{4,4} = \neg (W_{4,2} \wedge W_{4,3})$.

The token, entering from place l_1 in place l_4 don't obtain new characteristic. The tokens that enter the places l_2 and l_3 have the following characteristics:

"*dataset for applying DBSCAN algorithm*"

in place l_2

and

"*dataset for preprocessing*"

in place l_3.

A token enters the net via place l_5 with initial characteristics: "*techniques and parameters for data preprocessing*". Transition Z_2 has the form:

$$Z_2 = \langle\{l_3, l_5, l_7\}, \{l_6, l_7\}, R_2, \vee(\wedge(l_3, l_5), l_7)\rangle,$$

where

$$R_2 = \begin{array}{c|cc} & l_6 & l_7 \\ \hline l_3 & false & true \\ l_5 & false & true \\ l_7 & W_{7,6} & W_{7,7} \end{array}$$

and

- $W_{7,6}$ = "there is preprocessed dataset for applying DBSCAN algorithm";
- $W_{7,7} = \neg W_{7,6}$.

The tokens, entering from places l_3 and l_5 in place l_7 don't obtain new characteristics. The token that enters place l_6 has the following characteristics:

"*preprocessed dataset for applying DBSCAN algorithm*".

A token enters the net via place l_8 with initial characteristics: "*parameter* ***minPoints*** *and metric* ***eps*** *for DBSCAN cluster analysis*". Transition Z_3 has the form:

$$Z_3 = \langle\{l_2, l_6, l_8, l_{11}, l_{16}\}, \{l_9, l_{10}, l_{11}\}, R_3, \vee(\wedge(\vee(l_2, l_6), l_8, l_{11}), l_{16})\rangle,$$

where

$R_3 =$

	l_9	l_{10}	l_{11}
l_2	*false*	*false*	*true*
l_6	*false*	*false*	*true*
l_8	*false*	*false*	*true*
l_{11}	$W_{11,9}$	$W_{11,10}$	$W_{11,11}$
l_{16}	*false*	*false*	*true*

and

- $W_{11,9}$ = "there are initial core point/s";
- $W_{11,10}$ = "there are initial outliers (noise points)";
- $W_{11,11} = \neg(W_{11,9} \wedge W_{11,10})$.

The tokens, entering from places l_2, l_6, l_8 and l_{16} in place l_{11} don't obtain new characteristics. The tokens that enter the places l_9 and l_{10} have the following characteristics:

"*initial core point/s*"

in place l_9

and

"*initial outliers*"

in place l_{10}.

A token enters the net via place l_{12} with initial characteristics: "*metric* ***distance*** *for DBSCAN cluster analysis*". Transition Z_4 has the form:

$Z_4 = \langle\{l_9, l_{10}, l_{12}, l_{17}, l_{18}, l_{19}\}, \{l_{13}, l_{14}, l_{15}, l_{16}, l_{17}, l_{18}, l_{19}\}, R_4, \vee(\wedge(l_9, l_{10}, l_{12}), l_{17}, l_{18}, l_{19})\rangle$,

where

$$R_4 = \begin{array}{c|ccccccc} & l_{13} & l_{14} & l_{15} & l_{16} & l_{17} & l_{18} & l_{19} \\ \hline l_9 & false & false & false & false & false & false & true \\ l_{10} & false & false & false & false & false & false & true \\ l_{12} & false & false & false & false & false & false & true \\ l_{17} & W_{17,13} & W_{17,14} & W_{17,15} & W_{17,16} & W_{17,17} & false & false \\ l_{18} & false & false & false & false & W_{18,17} & W_{18,18} & false \\ l_{19} & false & false & false & false & false & W_{19,18} & W_{19,19} \end{array}$$

and

- $W_{19,18} =$ "there are groups of core points";
- $W_{19,19} = \neg W_{19,18}$;
- $W_{18,17} =$ "there are groups of core points with connected border points";
- $W_{18,18} = \neg W_{18,17}$;
- $W_{17,13} =$ "there are output core points of the clusters";
- $W_{17,14} =$ "there are output border points of the clusters";
- $W_{17,15} =$ "there are outliers after the step of cluster expansion";
- $W_{17,16} =$ "there is need of new initial core points or parameters and metrics";
- $W_{17,17} = \neg(W_{17,13} \wedge W_{17,14} \wedge W_{17,15} \wedge W_{17,16})$.

The tokens, entering from places $l_9, l_{10}, l_{12}, l_{17}$ and l_{18} in place l_{19} don't obtain new characteristics. The token that enters in place l_{18} has the following characteristic:

"*groups of core points*".

At the second activation of the transition the token from place l_{18} generates a new token that enters in places l_{17} with characteristics:

"*groups of core points with connected border points*".

At the third activation of the transition the token from place l_{17} generates four new tokens that enter in places l_{13}, l_{14}, l_{15} and l_{16} with characteristics respectively:

"*output core points of the clusters*"

in place l_{13},

"*output border points of the clusters*"

in place l_{14},

"*outliers after the step of cluster expansion*"

in place l_{15},

"*need of new initial core points or parameters and metrics*"

in place l .

A token enters the net via place l_{20} with initial characteristic: "*formulas for intuitionistic fuzzy evaluations*". Transition Z_5 has the form:

$Z_5 = \langle\{l_{13}, l_{14}, l_{15}, l_{20}, l_{22}\}, \{l_{21}, l_{22}\}, R_5, \vee(\wedge(l_{13}, l_{14}, l_{15}, l_{20}), l_{22})\rangle$,

where

$$R_5 = \begin{array}{c|cc} & l_{21} & l_{22} \\ \hline l_{13} & false & true \\ l_{14} & false & true \\ l_{15} & false & true \\ l_{20} & false & true \\ l_{22} & W_{22,21} & w_{22,22} \end{array}$$

and

- $W_{22,21}$ = "there are intuitionistic fuzzy evaluation of the clustering procedure";
- $W_{22,22} = \neg W_{22,21}$.

The tokens, entering from places l_{13}, l_{14}, l_{15} and l_{20} in place l_{22} don't obtain new characteristics. The token that enters the place l_{21} has the following characteristic:

"*intuitionistic fuzzy evaluation of the clustering procedure*".

The intuitionistic fuzzy evaluations are calculated using the following formulas:

- Degree of membership

$$\mu = \frac{c}{t}$$

- Degree of non-membership

$$\nu = \frac{n}{t}$$

- Degree of uncertainty

$$\pi = 1 - \mu - \nu$$

where c = core points, n = noise points (outliers) and t = all points.

3 Results of the DBSCAN Implementation

An implementation of the DBSCAN algorithm with IFEs is implemented using Python programming language. An example using diabetes data is presented in Fig. 2. The parameters *minPoints* (minimum number of points in cluster) and ε (distance) are set. The clusters and their respective outliers are calculated. The *DBSCAN*(), *fit_predict*() and *analyze*() are used to derive the data points related to clusters and perform extra checks on the data in each cluster.

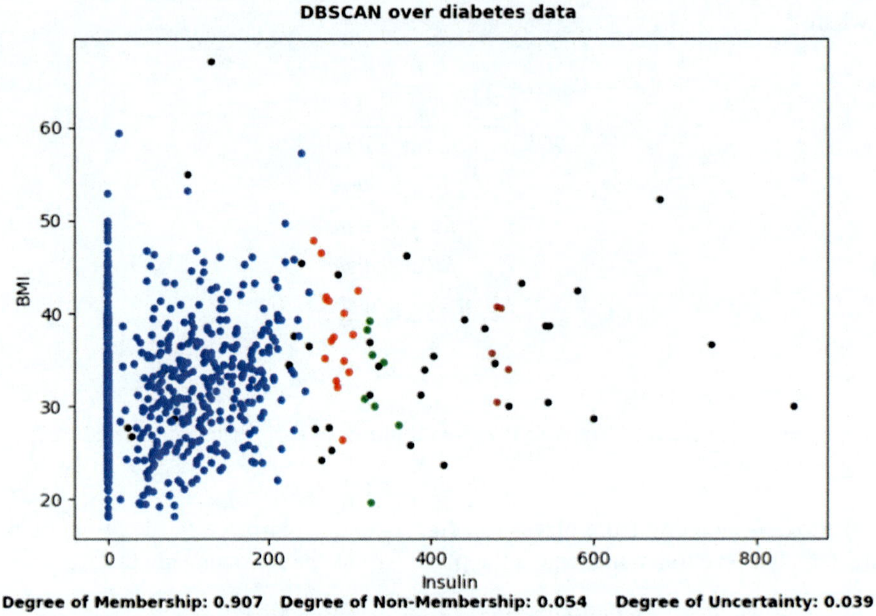

Fig. 2. DBSCAN over diabetes dataset

4 Analysis of the Clusters from DBSCAN

The DBSCAN algorithm is applied on a diabetes dataset [15] containing data about females that are at least 21 years old of Pima Indian heritage. The black points are outliers. The total number of observed records is 724. The outliers contain 22 diabetes cases out of 39, the blue cluster contains 656 records, out of them 211 are not healthy (32% of the cluster). We will consider this the "mostly-healthy" cluster. The red cluster contains 16 records with 8 diabetes cases (50% of the cluster). The green cluster contains 5 records with 3 cases (60% of cases). The brown cluster contains 5 cases out of 8 records (62.5% of the cases). Despite the fact that the red, green and brown clusters contain significantly smaller numbers of people, the model shows that these classification is significantly more correct in terms of detecting diabetes cases. It is also evident that the outliers are mostly diabetes cases (in 56% of the cases) meaning that a point with parameters which is considered an outlier by the algorithm is highly likely to be a diabetes case.

By observing the related data in the dataset in each case it was found that in 19 cases of the outliers had high levels of glucose, 18 had high levels of insulin, 22 were overweight. The red cluster contains 8 cases and all of them had high levels of glucose, insulin and were overweight. The brown cluster contains 5 cases and all of them had high levels of glucose, insulin and were overweight. The same applies for the items in the green cluster.

5 Conclusion

In the current paper a GN model of DBSCAN is presented. It allows us to better monitor the DBSCAN clustering algorithm. The presented GN model can be used for optimizing the process of cluster analysis. Practical examples on different datasets can be performed using the Python implementation. This particular paper shows application on diabetes dataset and forms multiple conclusions based on the data.

Acknowledgments. The authors are thankful for the support provided by the European Regional Development Fund and the Operational Program "Science and Education for Smart Growth" under contract UNITe No. BG05M2OP001–1.001–0004-C01 (2018–2023).

References

1. Atanassov, K.: Generalized Nets and Intuitionistic Fuzziness in Data Mining. Professor Marin Drinov Publishing House of Bulgarian Academy of Sciences, Sofia (2020)
2. Atanassov, K.: Index Matrices: Towards an Augmented Matrix Calculus. Springer, Cham (2014). https://doi.org/10.1007/978-3-319-10945-9
3. Atanassov, K., Sotirova, E.: Generalized Nets. Professor Marin Drinov Publishing House of Bulgarian Academy of Sciences, Sofia (2017). (in bulgarian)
4. Atanassov, K.: Theory of Generalized Nets (an algebraic aspect). AMSE Rev. **1**(2), 27–33 (1984)
5. Atanassov, K.: Generalized Nets. World Scientific, Singapore (1991)
6. Atanassov, K.: Intuitionistic fuzzy sets. Fuzzy Sets Syst. **20**(1), 87–96 (1986)
7. Atanassov, K.: Intuitionistic Fuzzy Sets: Theory and Applications. Physica-Verlag, Heidelberg (1999)
8. Atanassov, K.: On Generalized Nets Theory, "Prof. M. Drinov". Academic Publishing House, Sofia (2007)
9. Atanassov, K.: On Intuitionistic Fuzzy Sets Theory. Springer, Berlin (2012)
10. Bureva, V., Sotirova, E., Popov, S., Mavrov, D., Traneva, V.: Generalized net of cluster analysis process using STING: a statistical information grid approach to spatial data mining. In: Christiansen, H., Jaudoin, H., Chountas, P., Andreasen, T., Legind Larsen, H. (eds.) FQAS 2017. LNCS (LNAI), vol. 10333, pp. 239–248. Springer, Cham (2017). https://doi.org/10.1007/978-3-319-59692-1_21
11. Bureva, V., Traneva, V., Zoteva, D., Tranev, S.: Generalized net model simulation of cluster analysis using CLIQUE: clustering in quest. In: Dimov, I., Fidanova, S. (eds.) HPC 2019. SCI, vol. 902, pp. 48–60. Springer, Cham (2021). https://doi.org/10.1007/978-3-030-55347-0_5
12. Hassanien, A.E., Azar, A.T., Snasael, V., Kacprzyk, J., Abawajy, J.H. (eds.): SBD, vol. 9. Springer, Cham (2015). https://doi.org/10.1007/978-3-319-11056-1
13. Ikonomov, N.: GNDraw – software application for creating generalized nets. Issues Intuitionistic Fuzzy sets Generalized Nets **13**, 61–71 (2017)
14. Ester, M., Kriegel, H.P., Sander, J., Xu, X.: A density-based algorithm for discovering clusters in large spatial databases with noise. In: Evangelos, S., Jiawei, H., Usama, M.F. (eds.) Proceedings of the Second International Conference on Knowledge Discovery and Data Mining (KDD-96), pp. 226–231. AAAI Press, CiteSeerX (1996). 10.1.1.121.9220. ISBN: 1-57735-004-9
15. Diabetes dataset. https://www.kaggle.com/datasets/mathchi/diabetes-data-set. Accessed 16 July 2022

Check for updates

Generalized Net Model of Multimodal Biometric System for Authenticating an Individual by Keystroke Dynamics and Eye Tracking Techniques

Veselina Bureva(✉), Todor Petkov, and Stanislav Popov

Laboratory of Intelligent Systems, "Prof. Dr. Assen Zlatarov" University, "Prof. Yakimov" Blvd., 8010 Burgas, Bulgaria
{vbureva,todor_petkov}@btu.bg

Abstract. In the current research work the behavioral biometrics eye tracking and keystroke dynamics are investigated. The methods are frequently used for integration between patients and medical devices. A GN model of the multimodal biometric system is constructed. It monitors the process of behavioral biometrics eye tracking and keystroke dynamics. The optimization step is added to provide the necessity for better results.

Keywords: Biometrics · Behavioral Biometrics · Generalized Nets · Multimodal Biometric System

1 Behavioral Biometrics: Keystroke Dynamics and Eye Tracking

In the literature the biometrics are separated in two categories: physical biometrics and behavioral biometrics. Physiological biometrics contains face recognition, fingerprint recognition, hand recognition, iris/retina recognition, etc. Behavioral biometrics include traits for keystroke dynamics, mouse/coursor movements, eye/head movements, signature recognition, voice recognition, gait recognition, gesture analysis of an individual. The behavioral biometrics present the way of which a person interacts with the objects. Nowadays the behavioral biometrics are frequently used for annualizing the user behavior of its mobile device or internet of things devices. The traits are typically appropriate habits of the individual. Gait and gesture recognition analyze the type of the movements of an individual. Voice recognition and signature recognition present the unique traits of the person related to its speech manners and handwriting. Mouse movements and eye/head tracking present the interaction of the user with devices. Behavioral biometrics can be used in banking, fraud detection, healthcare, education, social media and websites [11–13].

Keystroke dynamics represent rhythm of typing characters on a keyboard or keypad. The idea is to record the unique user template of individual keyboard typing and use it for future investigations related to person authentication. Keystroke dynamics determine

S. Sotirov et al. (Eds.): BioInfoMed 2022, LNNS 658, pp. 272–280, 2023.
https://doi.org/10.1007/978-3-031-31069-0_27

the identity of the person by its way of typing on the keyboard: the latency between consecutive keystrokes, dwell time on the key press up, flight time, overall typing speed, frequency of errors and control keys. Dwell time is the time duration that a key is pressed while the flight time is the time duration in between releasing a key and pressing the next key. Different types of techniques can be used for keystroke authentication: the most of them are from the field of statistical analysis and artificial intelligence. Keystroke dynamics are used for authentication, surveillance, security [12, 14].

Eye tracking is the process of measuring the point of gaze or the motion of an eye relative to the head. It estimates the eye rotations using several of the following types of measurements: measurement of the movement of an object (lens), optical tracking without direct contact to the eye and measurement of electric potentials using electrodes placed around the eyes. The eye positions and eye movements are measured using the device called eye tracker. The eye tracking is applied in the field of the visual system, in psychology, in psycholinguistics, marketing research, in human-computer interaction, medical research, distraction detection of drivers and pilots, computers that are operated by people with severe motor impairment. In the last years the use of eye tracking in rehabilitative and assistive applications increases. It is frequently applied to control of wheelchairs, robotic arms and prostheses [11, 13].

2 Generalized Net Model of Multimodal Biometric System Authenticating the Individual by Keystroke Dynamics and Eye Tracking Techniques

The concept of generalized nets is described in series of papers [1–5]. Generalized net (GN) models of biometric identification process [6], fingerprint recognition [10], biometric authentication system based on palm geometry and palm vein matching [7, 8] and biometric multifactor authentication system [9] are already published. In the current research work a generalized net model of multimodal biometric system authenticating the individual by keystroke dynamics and eye tracking techniques is presented (Fig. 1). The GN model contains the following set of transitions:

- Z_1 – Capturing data;
- Z_2 – Preprocessing keystroke dynamics data;
- Z_3 – Preprocessing eye tracking data;
- Z_4 – Keystroke dynamics features extraction;
- Z_5 – Eye tracking features extraction;
- Z_6 – Optimization;
- Z_7 – Database with templates;
- Z_8 – Decision making;

Initially, there is one token that is located in place l_{27} with initial characteristic: "*database with templates*". In the next time-moments this token is split into two ones. The original token will continue to stay in place l_{27}, while the other tokens will move to the next transitions.

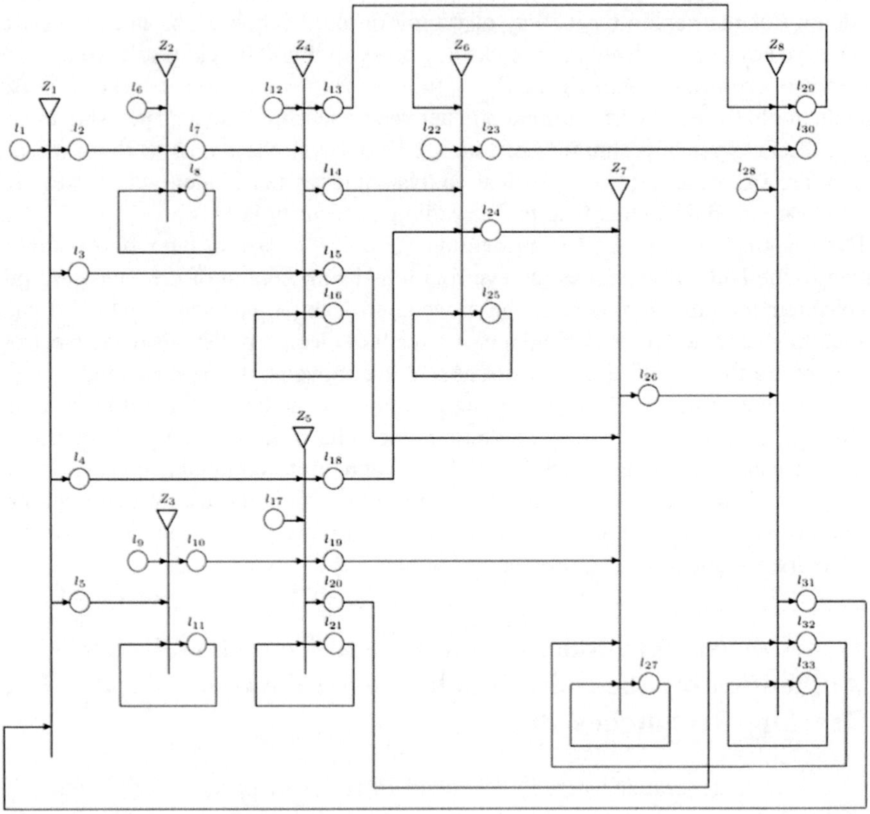

Fig. 1. Generalized net model of multimodal biometric system authenticating the individual by keystroke dynamics and eye tracking techniques

Via place l_1 a token enters the net with initial characteristics: "*an individual*". Transition Z_1 has the form:

$$Z_1 = \langle \{l_1, l_{31}\}, \{l_2, l_3, l_4, l_5\}, R_1, \vee(l_1, l_{31})\rangle$$

where

$R_1 =$

	l_2	l_3	l_4	l_5
l_1	*true*	*true*	*true*	*true*
l_{31}	$W_{31,2}$	$W_{31,3}$	$W_{31,4}$	$W_{31,5}$

and

- $W_{31,2}$ = "there is keystroke dynamics data for preprocessing";
- $W_{31,3}$ = "there is keystroke dynamics data for features extraction";
- $W_{31,4}$ = "there is eye tracking data for features extraction";

- $W_{31,5}$ = "there is eye tracking data for preprocessing".

The tokens that enter the places l_2, l_3, l_4 and l_5 have the following characteristics:

"*keystroke dynamics data for preprocessing*"
in place l_2,
"*keystroke dynamics data for features extraction*"
in place l_3,
"*eye tracking data for features extraction*"
in place l_4 and
"*eye tracking data for preprocessing*"
in place l_5.

Via place l_6 a token enters the net with initial characteristics: "*keystroke dynamics preprocessing methods and parameters*". Transition Z_2 has the form:

$$Z_2 = \langle\{l_2, l_6, l_8\}, \{l_7, l_8\}, R_2, \vee(\wedge(l_2, l_6), l_8)\rangle$$

where

$R_2 =$

	l_7	l_8
l_2	*false*	*true*
l_6	*false*	*true*
l_8	$W_{8,7}$	$W_{8,8}$

and

- $W_{8,7}$ = "there is preprocessed keystroke dynamics data";
- $W_{8,8} = \neg W_{8,7}$.

The tokens, entering from places l_2 and l_6 in place l_8 don't obtain new characteristics. The token that enters place l_7 has the following characteristics: "*preprocessed keystroke dynamics data*".

Via place l_9 a token enters the net with initial characteristics: "*eye tracking preprocessing methods and parameters*". Transition Z_3 has the form:

$$Z_3 = \langle\{l_5, l_9, l_{11}\}, \{l_{10}, l_{11}\}, R_3, \vee(\wedge(l_5, l_9), l_{11})\rangle$$

where

$R_3 =$

	l_{10}	l_{11}
l_5	*false*	*true*
l_9	*false*	*true*
l_{11}	$W_{11,10}$	$W_{11,11}$

and

- $W_{11,10}$ = "there is preprocessed eye tracking data";
- $W_{11,11} = \neg W_{11,10}$.

The tokens, entering from places l_5 and l_9 in place l_{11} don't obtain new characteristics. The token that enters place l_{10} has the following characteristics: "*preprocessed eye tracking data*".

Via place l_{12} a token enters the net with initial characteristics: "*keystroke dynamics feature extraction methods and parameters*". Transition Z_4 has the form:

$Z_4 = \langle\{l_3, l_7, l_{12}, l_{16}\}, \{l_{13}, l_{14}, l_{15}, l_{16}\}, R_4, \vee(\wedge(\vee(l_3, l_7), l_{12}), l_{16})\rangle$

where

$$R_4 = \begin{array}{c|cccc} & l_{13} & l_{14} & l_{15} & l_{16} \\ \hline l_3 & false & false & false & true \\ l_7 & false & false & false & true \\ l_{12} & false & false & false & true \\ l_{16} & W_{16,13} & W_{16,14} & W_{16,15} & W_{16,16} \end{array}$$

and

- $W_{16,13}$ = "there are extracted keystroke dynamics features for decision making";
- $W_{16,14}$ = "there are extracted keystroke dynamics features for optimization";
- $W_{16,15}$ = "there are extracted keystroke dynamics features for storing in database";
- $W_{16,16} = \neg(W_{16,13} \wedge W_{16,14} \wedge W_{16,15})$.

The tokens, entering from places l_3, l_7 and l_{12} in place l_{16} don't obtain new characteristics. The tokens that enter the places l_{13}, l_{14} and l_{15} have the following characteristics:

"*extracted keystroke dynamics features for decision making*"

in place l_{13},

"*extracted keystroke dynamics features for optimization*"

in place l_{14} and

"*extracted keystroke dynamics features for storing*"

in place l_{15}.

Via place l_{17} a token enters the net with initial characteristics: "*keystroke dynamics feature extraction methods and parameters*". Transition Z_5 has the form:

$Z_5 = \langle\{l_4, l_{10}, l_{17}, l_{21}\}, \{l_{18}, l_{19}, l_{20}, l_{21}\}, R_5, \vee(\wedge(\vee(l_4, l_{10}), l_{17}), l_{21})\rangle$

where

$R_5 =$	l_{18}	l_{19}	l_{20}	l_{21}
l_4	*false*	*false*	*false*	*true*
l_{10}	*false*	*false*	*false*	*true*
l_{17}	*false*	*false*	*false*	*true*
l_{21}	$W_{21,18}$	$W_{21,19}$	$W_{21,20}$	$W_{21,21}$

and

- $W_{21,18} =$ "there are extracted eye tracking features for optimization";
- $W_{21,19} =$ "there are extracted eye tracking features for storing";
- $W_{21,20} =$ "there are extracted eye tracking features for decision making";
- $W_{21,21} = \neg(W_{21,18} \wedge W_{21,19} \wedge W_{21,20})$.

The tokens, entering from places l_4, l_{10} and l_{17} in place l_{21} don't obtain new characteristics. The tokens that enter the places l_{18}, l_{19} and l_{20} have the following characteristics:

"*extracted eye tracking features for optimization*"

in place l_{18},

"*extracted eye tracking features for storing*"

in place l_{19} and

"*extracted eye tracking features for decision making*"

in place l_{20}.

Via place l_{22} a token enters the net with initial characteristics: "*optimization methods and parameters*". Transition Z_6 has the form:

$Z_6 = \langle\{l_{14}, l_{18}, l_{22}, l_{25}, l_{29}\}, \{l_{23}, l_{24}, l_{25}\}, R_6, \vee(\wedge(\vee(l_{14}, l_{18}, l_{29}), l_{22}), l_{25})\rangle$

where

$R_6 =$	l_{23}	l_{24}	l_{25}
l_{14}	*false*	*false*	*true*
l_{18}	*false*	*false*	*true*
l_{22}	*false*	*false*	*true*
l_{25}	$W_{25,23}$	$W_{25,24}$	$W_{25,25}$
l_{29}	*false*	*false*	*true*

and

- $W_{25,23} =$ "there are optimized eye tracking and/or keystroke dynamics features for decision making";
- $W_{25,24} =$ "there are optimized eye tracking and/or keystroke dynamics features for storing";
- $W_{21,21} = \neg(W_{25,23} \wedge W_{25,24})$.

The tokens, entering from places l_{14}, l_{18}, l_{22} and l_{29} in place l_{25} don't obtain new characteristics. The tokens that enter the places l_{23} and l_{24} have the following characteristics:

"*optimized eye tracking and/or keystroke dynamics features for decision making*" in place l_{23},

"*optimized eye tracking and/or keystroke dynamics features for storing*" in place l_{24}.

Transition Z_7 has the form:

$Z_7 = \langle\{l_{15}, l_{19}, l_{24}, l_{27}, l_{32}\}, \{l_{26}, l_{27}\}, R_7, \vee(l_{15}, l_{19}, l_{24}, l_{27}, l_{32})\rangle$

where

$$R_7 = \begin{array}{c|cc} & l_{26} & l_{27} \\ \hline l_{15} & false & true \\ l_{19} & false & true \\ l_{24} & false & true \\ l_{27} & W_{27,26} & W_{27,27} \\ l_{32} & false & true \end{array}$$

and

- $W_{27,26}$ = "there are stored eye tracking and keystroke dynamics templates for decision making";
- $W_{27,27} = \neg W_{27,26}$.

The tokens, entering from places l_{15}, l_{19}, l_{24} and l_{32} in place l_{27} don't obtain new characteristics. The token that enters the place l_{26} has the following characteristic: "*eye tracking and keystroke dynamics templates for decision making*".

Via place l_{28} a token enters the net with initial characteristics: "*decision making methods and parameters*". Transition Z_8 has the form:

$Z_8 = \langle\{l_{13}, l_{20}, l_{23}, l_{26}, l_{28}, l_{33}\}, \{l_{29}, l_{30}, l_{31}, l_{32}, l_{33}\}, R_8, \vee(\wedge(l_{13}, l_{20}, l_{26}, l_{28}, l_{23}), l_{33})\rangle$

where

$$R_8 = \begin{array}{c|ccccc} & l_{29} & l_{30} & l_{31} & l_{32} & l_{33} \\ \hline l_{13} & false & false & false & false & true \\ l_{20} & false & false & false & false & true \\ l_{23} & false & false & false & false & true \\ l_{26} & false & false & false & false & true \\ l_{28} & false & false & false & false & true \\ l_{33} & W_{33,29} & W_{33,30} & W_{33,31} & W_{33,32} & W_{33,33} \end{array}$$

- $W_{33,29}$ = "there are templates for optimizing";
- $W_{33,30}$ = "there are compared templates";
- $W_{33,31}$ = "there is a need of capturing data";
- $W_{33,32}$ = "there are templates for storing";
- $W_{33,33} = \neg(W_{33,29} \wedge W_{33,30} \wedge W_{33,31} \wedge W_{33,32})$.

The tokens, entering from places $l_{13}, l_{20}, l_{23}, l_{26}$ and l_{28} in place l_{33} don't obtain new characteristics. The tokens that enter the places l_{29}, l_{30}, l_{31} and l_{32} have the following characteristics:

"*templates for optimizing*" in place l_{29},

"*compared templates*" in place l_{30},

"*need of capturing data*" in place l_{31},

"*templates for storing*" in place l_{33}.

3 Conclusion

In the current research work a multimodal biometric system is represented. Generalized net model of multimodal biometric system based on the keystroke dynamics and eye tracking is constructed. The biometrics of keystroke dynamics and eye tracking are used in the medical field to allow interaction between patients and biometric systems on computers/mobile phones/devices, patients and wheelchairs, robotic arms and prostheses. The presented GN model monitors the process and optimized results using the transition Z_6.

Acknowledgements. The authors are thankful for the support provided by Project "Analysis and modelling of artificial intelligence algorithms and their application", №NIH – 462/2021.

References

1. Atanassov, K.: Generalized Nets and Intuitionistic Fuzziness in Data Mining. Professor Marin Drinov Publishing House of Bulgarian Academy of Sciences, Sofia (2020)
2. Atanassov, K.: Index Matrices: Towards an Augmented Matrix Calculus. Springer, Cham (2014). https://doi.org/10.1007/978-3-319-10945-9
3. Atanassov, K., Sotirova, E.: Generalized Nets. Professor Marin Drinov Publishing House of Bulgarian Academy of Sciences, Sofia (2017). (in bulgarian)
4. Atanassov, K.: Theory of generalized nets (an algebraic aspect). AMSE Review **1**(2), 27–33 (1984)
5. Atanassov, K.: Generalized Nets. World Scientific, Singapore (1991)

6. Bureva, V., Sotirova, E., Bozov, H.: Generalized net model of biometric identification process. In: 2018 20th International Symposium on Electrical Apparatus and Technologies (SIELA), pp. 1–4 (2018). https://doi.org/10.1109/SIELA.2018.8447104
7. Ivanova, Z., Bureva, V.: Generalized net model of biometric authentication system based on palm geometry and palm vein matching. In: Sotirov, S.S., Pencheva, T., Kacprzyk, J., Atanassov, K.T., Sotirova, E., Staneva, G. (eds.) Contemporary Methods in Bioinformatics and Biomedicine and Their Applications. BioInfoMed 2020. Lecture Notes in Networks and Systems, vol. 374, pp. 121–130. Springer, Cham (2022). https://doi.org/10.1007/978-3-030-96638-6_13
8. Ivanova, Z., Bureva, V.: Generalized net model of a biometric authentication system based on palm geometry and palm vein matching using intuitionistic fuzzy evaluations. Notes Intuition. Fuzzy Sets **26**(4), 71–79 (2020). ISSN: 1310–4926, Online ISSN: 2367–8283. https://doi.org/10.7546/nifs.2020.26.4.71-79
9. Ivanova, Z., Bureva, V., Sotirov, S.: Generalized net model of biometric multifactor authentication system. In: et al. Uncertainty and Imprecision in Decision Making and Decision Support: New Advances, Challenges, and Perspectives. IWIFSGN BOS/SOR 2020 2020. Lecture Notes in Networks and Systems, vol. 338, pp. 419–435. Springer, Cham (2022). https://doi.org/10.1007/978-3-030-95929-6_32
10. Bureva, V., Yovcheva, P., Sotirov, S.: Generalized net model of fingerprint recognition with intuitionistic fuzzy evaluations. In: Kacprzyk, J., Szmidt, E., Zadrożny, S., Atanassov, K.T., Krawczak, M. (eds.) IWIFSGN/EUSFLAT -2017. AISC, vol. 641, pp. 286–294. Springer, Cham (2018). https://doi.org/10.1007/978-3-319-66830-7_26
11. Cantoni, V., Dimov, D., Tistarelli, M.: Biometric Authentication. In: First International Workshop, BIOMET 2014 Sofia, Bulgaria, 23–24 June 2014. Revised Selected Papers. Springer, Cham (2014)
12. Jiang, R., Al-maadeed, S., Bouridane, A., Crookes, D., Beghdadi, A.: Biometric Security and Privacy: Opportunities & Challenges in The Big Data Era. Springer, Cham (2017). https://doi.org/10.1007/978-3-319-47301-7
13. Zhang, D., Lu, G., Zhang, L.: Advanced Biometrics. Springer, Cham (2018).https://doi.org/10.1007/978-3-319-61545-5
14. Moskovitch, R., et al.: Identity theft, computers and behavioral biometrics. In: Proceedings of the IEEE International Conference on Intelligence and Security Informatics, pp. 155–160 (2009)

Generalized Net Model of the Consequences of Earthquake

Stefka Fidanova[1(✉)], Krassimir Atanassov[2], Leoneed Kirilov[1], Vanya Slavova[3], and Veselin Ivanov[3]

[1] Institute of Information and Communication Technologies, Bulgarian Academy of Sciences, Sofia, Bulgaria
stefka.fidanova@gmail.com, l_kirilov_8@abv.bg

[2] Institute of Biophysics and Biomedical Engineering, Bulgarian Academy of Sciences, Sofia, Bulgaria
krat@bas.bg

[3] Faculty of Medicine, Trakia University, Stara Zagora, Bulgaria
veskoasenov@abv.bg

Abstract. In order to prevent victims of natural disasters and to be able to help the injured people, local medical authorities must be provided in advance. One of the worst natural disasters is strong earthquake. Therefore, it is good for decision makers to have a preliminary assessment of the possible consequences and injured people of earthquakes of various kinds. In this work we propoze a model of the consequences of eventual earthquake and as well as the number and type of injured people. The model is based on the apparatus of Generalized Nets. After assessing the probability of an earthquake with a given intensity, a decision can be made on the necessary medical supplies to provide adequate assistance to the victims.

1 Introduction

Earthquakes are natural phenomenons that can cause infrastructure damage and casualties. They are difficult to predict, occur due to accumulated stress and deformation in the earth's crust. The seismic activity of a certain area is determined by the frequency, type and size of earthquakes during a certain period. Seismic processes are caused by the release of heat in the earth's interior, the formation of tectonic faults and volcanic activity. Tectonic causes predominate, and earthquakes caused by them have the greatest scope and cause the greatest damage. Although the time of occurrence and intensity of an earthquake is difficult to predict, it is possible to make a statistical estimate of the probability of an earthquake of a certain intensity. Thus, recommendations can be made for the construction of buildings, as well as an assessment of the necessary response by local authorities and medical equipment for adequate response and assistance to victims [16].

In this paper we have proposed a model of possible steps and the corresponding reaction in the event of an earthquake of different intensity. The model is

S. Sotirov et al. (Eds.): BioInfoMed 2022, LNNS 658, pp. 281–292, 2023.
https://doi.org/10.1007/978-3-031-31069-0_28

based on the Generalized Net (GN) apparatus [10,14]. The aim is to be able to replay a variety of scenarios and make the most appropriate decision. The model is aimed at supporting local authorities and allows for preliminary preparation of institutions such as hospitals, fire departments and others.

The rest of the paper is organised as follows. In Sect. 2 a literature review is made. Section 3 gives brief description of GN. In Sect. 4 we propose a GN for modeling earthquake. In Sect. 5 are given a conclusion and directions for future work.

2 Literature Review

The devastating earthquake consequences for the environment are a major cause for the occurrence of numerous medical emergencies. There are usually large numbers of permanent (deaths) and temporary (injured) medical losses. The types of injuries are diverse in type and structure. The direct impact trauma factor is the most common in cases of earthquake, as the direct impact is usually caused by airborne debris and falling objects from demolishing buildings, thus inflicting mostly traumatic injuries - see Clark (2018) [17]. In cases of secondary disaster area occurrences, it is possible to have impact factors like radiation; toxic substance release; thermal impact, biological impact, etc., which can cause radiation damage; acute poisonings; thermal injuries; outbreaks of infectious diseases; cases of drowning. The negative effects of the psychological stress factor are inevitable and they cause acute neuropsychological conditions in vast numbers of people - see Farooqui et al. (2017) [18], Todorova et al. (2020) [19], Etova (2021) [20].

In Tirkolaee et al. (2020) [1] a robust bi-objective mixed-integer linear programming model to allocate disaster rescue units is proposed. The authors apply the model to a real case study for Mazandaran province in Iran. The same problem for resource allocation for emergency response is studied in Fiedrich et al. (2000) [2]. They develop dynamic optimization model and a method for solving it. The goal is to present a schedule for optimizing of the available technical resources. Robust Model for Logistics Management (RMLM) is proposed in Najafi et al. (2012) [3]. The model is multi-objective, multi-mode, multicommodity, and multi-period. It is proposed to manage the logistics of commodities and injured people in the case of earthquake. A three objective mixed integer stochastic model for locations of storage areas for shelters pre-earthquake and distribution of shelters is proposed in Yenice and Samanlioglu (2020) [4]. The authors consider four event scenarios according to two different earthquake scenario likelihoods. The model was applied to Kadikoy municipality of Istanbul, Turkey. Dawei et al. (2015) [5] the problem about vehicle scheduling in the medicine dispatching process is studied in order to minimize the total transportation time under several types of vehicles. The problem is solved by genetic algorithm.

Transient modelling with quadratic regression analysis for simulation of hospital operations in emergency situations is studied in Paul et al. (2006) [6]. A double exponential function is used to model the transient waiting time. De Boer and Debacker (2006) [7] determine the medical resources for disasters in the Netherlands. They study the medical rescue capacity, the medical transport capacity and the hospital treatment capacity using medical severity index model. Several cases are considered under different assumptions of severity. A stochastic Petri net is used for modelling and optimizing the emergency medical rescue (EMR) process in Sun et al. (2021) [8]. The approach is tested with the data of the 2008 Wenchuan earthquake. Ghasemi et al. (2020) [9] propose a stochastic multi-objective mixed-integer model for logistic distribution and evacuation planning during an earthquake. The model is converted into deterministic one and then solved by means of genetic algorithm NSGA-II. The model is tested for a probable earthquake in Tehran.

3 Generalised Nets

In this work we propose a model of the consequence of earthquake with Generalized Nets.

Generalized Nets are extension and generalization of Petri nets. For a first time they were proposed in 1982 (see, [10,14]). Later they were used by a lot of scientists for describing and modeling different processes and algorithms (see, e.g., [11–13]). They are powerful tool for universal description of models of complex systems with many different and in most of the cases not homogeneous components, with simultaneous activities. The static structure consists of objects called **transitions**, which have input and output **places**. Two transitions can share a place, but every place can be an input of at most one transition and can be an output of at most one transition.

The dynamic structure consists of **tokens**, which act as information carriers and can occupy a single place at every moment of the GN execution. The tokens pass through the transition from one input to another output place. The tokens' movement is governed by conditions (predicates), contained in the index matrix (see, [15]) of the transition.

The information carried by a token is contained in its **characteristics**, which is obtained by a **characteristic function**. Every place possesses at most one characteristic function, which assigns new characteristics to the incoming tokens. Tokens can split and merge in the places. A transition can contain m input and n output places where $n, m \geq 1$ (see Fig. 1).

The GN are expandable. Each place and each transition can be replaced with new GN. Thus the process described by GN can be developed and complicated on more deeper level. The GN can help us to understand the processes and to see possibilities for their optimization.

In our model every particular situation will be represented by a transition. Several tokens will entry the transition with different characteristics, representing the possibilities to continue to other particular situation.

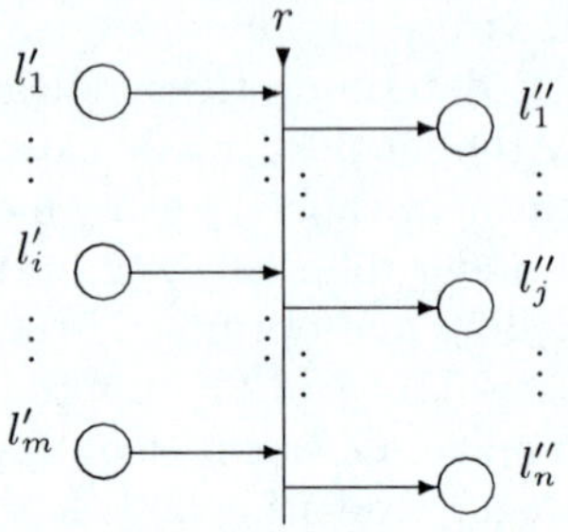

Fig. 1. The form of transition in a GN

4 Generalized Net for Modeling Earthquake Consequences

Earthquakes are one of the most destructive phenomena. Their strength is measured on a magnitude and intensity scales. This is the relative characteristic of the energy released during the earthquake. The most popular scale for estimating energy in an earthquake is the local Richter scale, and the scale of Medvedev-Sponheuer-Karnik reflects the intensity of the earthquake and its aftermath. In it, the increase of one degree of intensity corresponds to a 2-fold increase. An intensity up to 4 earthquake is barely perceptible, while a intensity 9 earthquake is the lower limit of destructive earthquakes that cover large areas [16].

The input data for our model are the strength of the earthquake, the seismological stability of each of the buildings in the area, the number of inhabitants of each building, the type of heating. The heating can be solid fuel, then there is a risk of local fire. If the building has a gas installation, in case of strong earthquakes there is a danger of damage and explosion, then the danger of fires and burns is great and more people are affected.

In case of an earthquake with a intensity less than 5 there is no danger of destruction and severe injuries. There may be cracked walls and fallen objects and people with minor injuries.

When the earthquake has a intensity of 6–7, there may be collapse in weakly stable buildings. There is a risk of local fires from solid fuel heating. People with fractures and mild to moderate burns may be present.

In earthquakes of 9 and higher intensity, damage to buildings and destruction, damage to water supply and gas installations are expected. In this case there is a high risk of serious injuries and fractures. There is a danger of explosion of gas installations and large fires. As a result, there will be people with severe burns. There is a risk of outbreaks of infections and epidemics as secondary consequences.

The consequences from earthquakes of different intensities and the corresponding decisions to be taken are described in the Generalized Net below (Fig. 2).

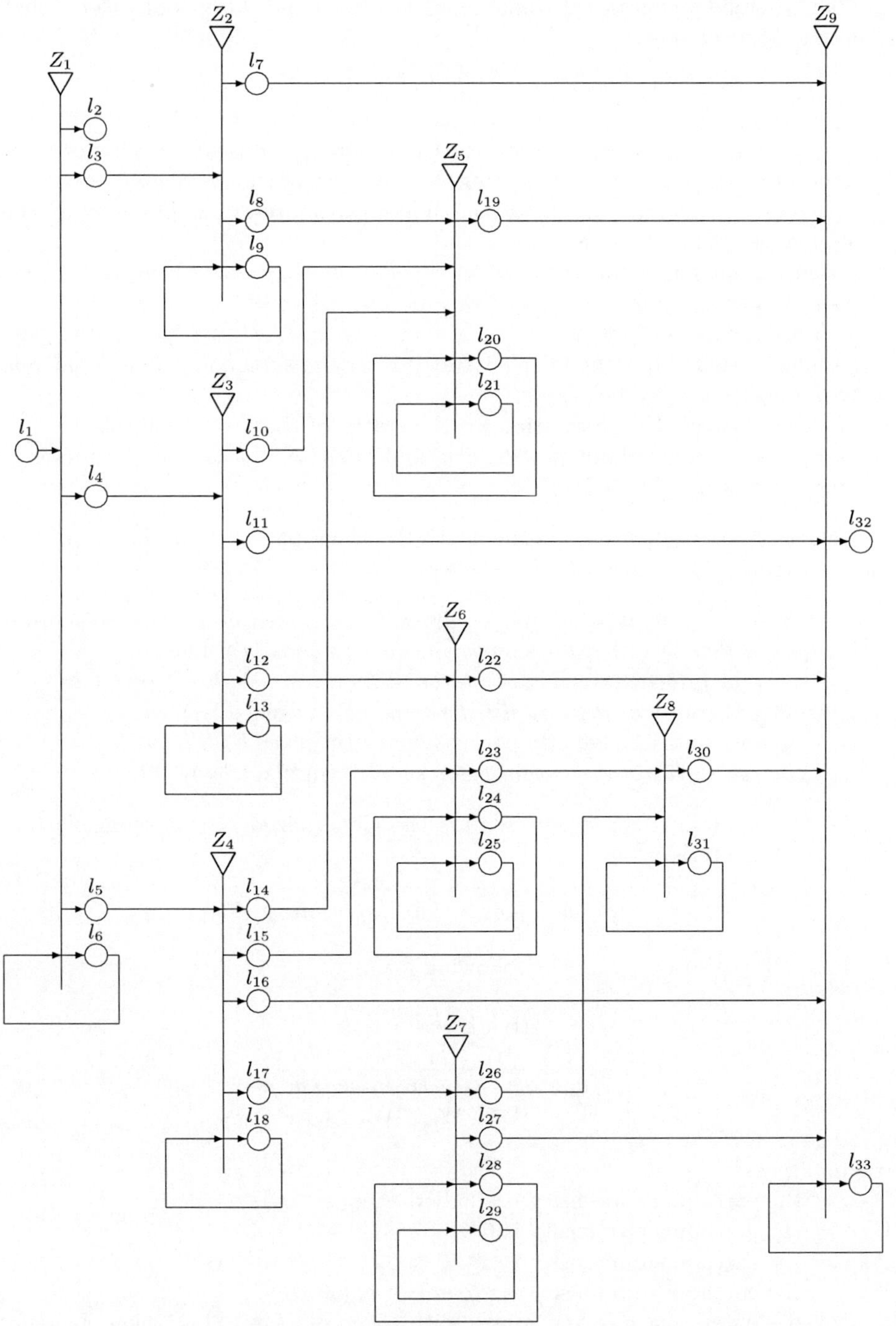

Fig. 2. Generalized Net model of earthquake consequences

The GN-model contains 9 transitions, 33 places and 9 types of tokens, that represent, respectively

ε - earthquake,
δ_1 - Data Base with information for the buildings/houses, the type of their structures, the type of their heating, the number of the residents,
$\varphi_1, \varphi_2, \varphi_3$ - information about the buildings/houses damages and of the injured people,
α - city fire department,
β - city power supply,
γ - urban water supply,
ζ - information about the people's status after the earthquake (e.g., epidemic beginning, mass psychosis, etc.),
δ_2 - Data Base with information for the status of the town hospitals,
$\sigma_1, \sigma_2, \ldots$ - signals about particular situations (type of damages, number of fires, injured people etc.)

Tokens $\delta_1, \delta_2, \alpha, \beta, \gamma$ stay permanently in their places with the respective characteristics:

δ_1 - "*current information for the buildings/houses, the type of their structures, the type of their heating, the number of the residents*" (in place l_6),
δ_2 - "*current information for the status of the town hospitals*" (in place l_{33}),
α - "*current status of the city fire department*" (in place l_{21})
β - "*current status of the city power supply*" (in place l_{25}),
γ - "*current status of the urban water supply*" (in place l_{29}),

At one point in time, token ε enters the Generalized Net through place l_1 with the characteristic

$$\text{“}earthquake\ power, date, hour, minute\text{”}.$$

The GN-transitions have the following forms.

$$Z_1 = \langle \{l_1, l_6\}, \{l_2, l_3, l_4, l_5, l_6\},
\begin{array}{c|ccccc}
 & l_2 & l_3 & l_4 & l_5 & l_6 \\
\hline
l_1 & false & false & false & false & true \\
l_6 & W_{6,2} & W_{6,3} & W_{6,4} & W_{6,5} & true
\end{array} \rangle,$$

where
$W_{6,2} =$ "the earthquake intensity is smaller or equal to 5",
$W_{6,3} =$ "the earthquake intensity is 6–7",
$W_{6,4} =$ "the earthquake intensity is 8",
$W_{6,5} =$ "the earthquake intensity is greater or equal to 9".

Token ε enters place l_6 and unites with token δ_1. After this, token δ_1 splits to two tokens - the same token δ_1 and a token σ_1 that enters one of the places

l_2, l_3, l_4 and l_5 is respect of the truth-value of predicates $W_{6,2}, ..., W_{6,5}$. Token σ obtains a characteristic

"no damaged buildings and no injured people"

in place l_2,

"number of damaged buildings, number of buildings with danger of local fire and injuries"

in place l_3,

"number of destroyed buildings, number of buildings with danger of local fire and injuries"

in place l_4,

"number of destroyed buildings, number of buildings with damage to gas and water supply installations and injuries"

in place l_5, respectively.

$$Z_2 = \langle \{l_3, l_9\}, \{l_7, l_8, l_9\}, \begin{array}{c|ccc} & l_7 & l_8 & l_9 \\ \hline l_3 & false & false & true \\ l_9 & W_{9,7} & W_{9,8} & true \end{array} \rangle,$$

where
$W_{9,7}$ ="damaged buildings",
$W_{9,8}$ ="local fire from heating with solid fuel".

If predicates $W_{6,3}$ is true, after one time-step token σ enters place l_9 and unites with token φ_1. After this, token φ_1 splits to two or three tokens - the same token φ_1 that continues to stay in place l_9 and, if predicate $W_{9,7}$ is true - a token σ_2 that enters place l_7 and/or if predicate $W_{9,8}$ is true - a token σ_3 that enters place l_8. These tokens obtain the characteristics

"number of people with fractures"

in place l_7,

"fire from heating with solid fuel"

in place l_8.

$$Z_3 = \langle \{l_4, l_{13}\}, \{l_{10}, l_{11}, l_{12}, l_{13}\}, \begin{array}{c|cccc} & l_{10} & l_{11} & l_{12} & l_{13} \\ \hline l_4 & false & false & false & true \\ l_{13} & W_{13,10} & W_{13,11} & W_{13,12} & true \end{array} \rangle,$$

where
$W_{13,10}$ ="local fires from heating with solid fuel",

$W_{13,11}$ = "damaged and distroyed buildings",
$W_{13,12}$ = "damage to the electrical installation".

If predicates $W_{6,4}$ is true, after one time-step token σ enters place l_{13} and unites with token φ_2. After this, token φ_2 splits to two, three or four tokens - the same token φ_2 that continues to stay in place l_{13} and, if predicate $W_{13,10}$ is true - a token σ_2 that enters place l_{10} and/or if predicate $W_{13,11}$ is true - a token σ_3 that enters place l_{11} and/or if predicate $W_{13,12}$ is true - a token σ_4 that enters place l_{12}. These tokens obtain the characteristics

"number of fires from heating with solid fuel"

in place l_{10},

"number of people with fractures"

in place l_{11},

"damaged electrical installations"

in place l_{12}, respectively.

$$Z_4 = \langle \{l_5, l_{18}\}, \{l_{14}, l_{15}, l_{16}, l_{17}, l_{18}\},
\begin{array}{c|ccccc}
 & l_{14} & l_{15} & l_{16} & l_{17} & l_{18} \\
\hline
l_5 & false & false & false & false & true \\
l_{18} & W_{18,14} & W_{18,15} & W_{18,16} & W_{18,17} & true
\end{array} \rangle,$$

where
$W_{18,14}$ = "fires from heating with solid fuel and damaged gas instalations",
$W_{18,15}$ = "damage to electrical installations",
$W_{18,16}$ = "distroyed buildings",
$W_{18,17}$ = "damage to water supply and flood risk".

If predicates $W_{6,5}$ is true, after one time-step token σ enters place l_{18} and unites with token φ_3. After this, token φ_3 splits to two, three, four or five tokens - the same token φ_3 that continues to stay in place l_{18} and, if predicate $W_{18,14}$ is true - a token σ_5 that enters place l_{14} and/or if predicate $W_{18,15}$ is true - a token σ_6 that enters place l_{15} and/or if predicate $W_{18,16}$ is true - a token σ_7 that enters place l_{16}, and/or if predicate $W_{18,17}$ is true - a token σ_8 that enters place l_{17}. These tokens obtain the characteristics

"number of fires from heating with solid fuel and damaged gas instalations"

in place l_{14},

"number of buildings without electricity"

in place l_{15},

"number of people with fractures"

in place l_{16},

"number of buildings and people without water supply and risk of flood"

in place l_{17}, respectively.

$$Z_5 = \langle \{l_8, l_{10}, l_{14}, l_{20}, l_{21}\}, \{l_{19}, l_{20}, l_{21}\}, \begin{array}{c|ccc} & l_{19} & l_{20} & l_{21} \\ \hline l_8 & false & false & true \\ l_{10} & false & false & true \\ l_{14} & false & false & true \\ l_{20} & false & false & true \\ l_{21} & W_{21,19} & W_{21,20} & true \end{array} \rangle,$$

where
$W_{21,19}$ = "there are people with burns",
$W_{21,20}$ = "firefighting is necessary".

All σ-tokens enter place l_{21} and unite with token α. After this, token α splits to three tokens - the same token α that continues to stay in place l_{21} and, if predicate $W_{21,19}$ is true - a token α_1 that enters place l_{19} and a token α_2 that enters place l_{20}. These tokens obtain the characteristics

"number of people with burns"

in place l_{19},

"types of firefighting actions"

in place l_{20}, respectively.

$$Z_6 = \langle \{l_{12}, l_{15}, l_{24}, l_{25}\}, \{l_{22}, l_{23}, l_{24}, l_{25}\}, \begin{array}{c|cccc} & l_{22} & l_{23} & l_{24} & l_{25} \\ \hline l_{12} & false & false & false & true \\ l_{15} & false & false & false & true \\ l_{24} & false & false & false & true \\ l_{25} & W_{25,22} & W_{25,23} & W_{25,24} & true \end{array} \rangle,$$

where
$W_{25,22}$ = "risk of electric shock",
$W_{25,23}$ = "risk of an epidemic",
$W_{25,24}$ = "damage to the electrical system".

All σ-tokens enter place l_{25} and unite with token β. After this, token β splits to four tokens - the same token β that continues to stay in place l_{25} and, if predicate $W_{25,22}$ is true - a token β_1 that enters place l_{22}, and if predicate $W_{25,23}$ is true - a token β_2 that enters place l_{23}, and if predicate $W_{25,24}$ is true - a token β_3 that enters place l_{24}. These tokens obtain the characteristics

"number of people with electric shock"

in place l_{22},

"number of people at high epidemic risk"

in place l_{23},

"electrical system restoration activities"

in place l_{24}, respectively.

$$Z_7 = \langle \{l_{17}, l_{28}, l_{29}\}, \{l_{26}, l_{27}, l_{28}, l_{29}\}, \begin{array}{c|cccc} & l_{26} & l_{27} & l_{28} & l_{29} \\ \hline l_{17} & false & false & false & true \\ l_{28} & false & false & false & true \\ l_{29} & W_{29,26} & W_{29,27} & W_{29,28} & true \end{array} \rangle,$$

where
$W_{29,26}$ = "risk of an epidemic",
$W_{29,27}$ = "flood risk",
$W_{29,28}$ = "damage to the water supply".

The token σ_8 enters place l_{29} and unite with token γ. After this, token γ splits to four tokens - the same token γ that continues to stay in place l_{29} and, if predicate $W_{29,26}$ is true - a token γ_1 that enters place l_{26}, and if predicate $W_{29,27}$ is true - a token γ_2 that enters place l_{27}, and if predicate $W_{29,28}$ is true - a token γ_3 that enters place l_{28}. These tokens obtain the characteristics

"number of people with high epidemic risk"

in place l_{26},

"number of people affected by the flood"

in place l_{27},

"water supply restoration activities"

in place l_{28}, respectively.

$$Z_8 = \langle \{l_{23}, l_{26}, l_{31}\}, \{l_{30}, l_{31}\}, \begin{array}{c|cc} & l_{30} & l_{31} \\ \hline l_{23} & false & true \\ l_{26} & false & true \\ l_{31} & true & false \end{array} \rangle.$$

The tokens β_2 and γ_2 (but only one of them) enter place l_{31} and it obtains the characteristic

"number of injured people"

In the next time-step, the token from place l_{31} enters place l_{30} without a new characteristic.

$$Z_9 = \langle \{l_7, l_{11}, l_{14}, l_{19}, l_{22}, l_{27}, l_{30}, l_{33}\}, \{l_{32}, l_{33}\}, \begin{array}{c|cc} & l_{32} & l_{33} \\ \hline l_7 & false & true \\ l_{11} & false & true \\ l_{14} & false & true \\ l_{19} & false & true \\ l_{22} & false & true \\ l_{27} & false & true \\ l_{30} & false & true \\ l_{33} & W_{33,32} & true \end{array} \rangle,$$

where
$W_{33,32}$ = "medical help is needed".

All tokens from the input places of transition Z_9 enter place l_{33} and unite with the token δ_2 that stays permanently there. When all GN-processes are completed, token δ_2 enters place l_{32} with the characteristic

"necessary medicines and materials"

5 Conclusion

Earthquakes are difficult to predict events with significant consequences for people and infrastructure. It is important that local governments and medical authorities are trained and equipped in advance to make the right decisions and prevent casualties. We propose a basic model for estimation of the earthquake consequences and the necessary reactions and medical equipment. The model is based on the aparatus of Generalized Nets. The aim is to make the right decisions and to provide the most effective medical care to the victims. Input parameters of the GN are the intensity of earthquake, and the signals received in the time from Civil Protection department about different types of damages and risk situationsr. The output of the GN is the number of injured people with classification of inhures. As a future research, the model will be expanded by adding additional features and details. Another direction of research is to simulate an earthquake at given intensity in the region of Stara Zagora in order to study the state of readiness of the medical authorities. Development of a software product implementing the model is envisaged.

Acknowledgement. The authors are grateful to the Municipality of Stara Zagora and in particular to Eng. Veselin Kuzmanov for their advices. The work of Stefka Fodanova is supported by the grant No BG05M2OP011-1.001-0003, financed by the Science and Education for Smart Growth Operational Program and co-financed by European Union through the European structural and Investment funds and by National Scientific Fund of Bulgaria grant DFNI KP-06-N52/5. The work of Leoneed Kirilov is supported by the grant No BG05M2OP011-1.001-0003, financed by the Science and Education for Smart Growth Operational Program and co-financed by European Union through the European structural and Investment funds and by National Scientific Fund of Bulgaria grant DFNI KP-06-N52/7. The work of Veselin Ivanov and Vanya Slavova is supported by project 9/2019 of Medical Faculty, Trakia University, Stara Zagora 6000, Bulgaria.

References

1. Tirkolaee, E.B., Aydın, N.S., Ranjbar-Bourani, M., Weber, G.-W.: A robust bi-objective mathematical model for disaster rescue units allocation and scheduling with learning effect. Comput. Ind. Eng. (2020). https://doi.org/10.1016/j.cie.2020.106790
2. Fiedrich, F., Gehbauer, F., Rickers, U.: Optimized resource allocation for emergency response after earthquake disasters. Saf. Sci. **35**, 41–57 (2000)

3. Najafi, M., Eshghi, K., Dullaert, W.: A multi-objective robust optimization model for logistics planning in the earthquake response phase. Transp. Res. Part E **49**, 217–249 (2013). https://doi.org/10.1016/j.tre.2012.09.001
4. Yenice, Z. D., Samanlioglu, F.: A multi-objective stochastic model for an earthquake relief network. J. Adv. Transp. **2020**, Article ID 1910632 (2020). https://doi.org/10.1155/2020/1910632
5. Dawei, L., Yingying, X., Li, W.: Vehicle scheduling of the emergency medicines in the early period after earthquake disaster. Open Cybern. Syst. J. **9**, 1329–1333 (2015)
6. Paul, J.A., George, S.K., Yi, P., Lin, L.: Transient modeling in simulation of hospital operations for emergency response. Prehospital Disaster Med. **21**(4), 223–236 (2006)
7. de Boer, J., Debacker, M.: Quantifying medical disaster management. Int. J. Disaster Med. 1–5 (2006). https://doi.org/10.1080/15031430600975569
8. Sun, H., Liu, J., Han, Z., Jiang, J.: Stochastic petri net based modeling of emergency medical rescue processes during earthquakes. J. Syst. Sci. Complex. **34**(3), 1063–1086 (2020). https://doi.org/10.1007/s11424-020-9139-3
9. Ghasemi, P., Khalili-Damghani, K., Hafezalkotob, A., Raissi, S.: Stochastic optimization model for distribution and evacuation planning (A case study of Tehran earthquake). Socioecon. Plann. Sci. **71**, 100745 (2020). https://doi.org/10.1016/j.seps.2019.100745
10. Atanassov, K.: Generalized Nets. World Scientific. Singapore, London (1991)
11. Krawczak, M.: Multilayer Neural Systems and Generalized Net Models. Akademicka Oficyna Wydawnicza EXIT, Warszawa (2003)
12. Fidanova, S., Atanassov, K., Marinov, P.: Generalized Nets in Artificial Intelligence, vol. 5: Generalized nets and Ant Colony Optimization. Prof. M. Drinov Academic Publishing House, Sofia (2011)
13. Atanassov, K.: Generalized Nets and Intuitionistic Fuzziness in Data Mining. Prof. Marin Drinov Academic Publishing House, Sofia (2020)
14. Atanassov, K.: On Generalized Nets Theory. Prof. M. Drinov Academic Publishing House, Sofia (2007)
15. Atanassov, K.: Index Matrices: Towards an Augmented Matrix Calculus. Springer, Cham (2014). https://doi.org/10.1007/978-3-319-10945-9
16. Shearer, P. M.: Introduction to Seismology. Cambridge University Press, Cambridge (2009). ISBN 978-0-521-88210-1
17. Clark, K.R.: Imaging earthquake-related injuries. Radiol. Technol. **89**(4), 351–367 (2018)
18. Farooqui, M., et al.: Posttraumatic stress disorder: a serious post-earthquake complication. Trends Psychiatry Psychother. **39**(2), 135–143 (2017). PMID: 28700042. https://doi.org/10.1590/2237-6089-2016-0029
19. Todorova, D., Mihaylova, T.S.V., Etova, R.: Social psychological support for disasters. Health Policy and Management, vol. 20. Extraordinary issue, pp. 169–172 (2020). ISSN 1313–4981
20. Etova, R.: Disasters and stress. Scientific works of the union of scientists in Bulgaria Plovdiv. series G. Med. Pharm. Dent. Med. **25**, 190–193 (2021). ISSN 1311–9427 (Print), ISSN 2534–9392 (On-line)

A Generalized Net Model of Time-Delay Recurrent Neural Networks with the Stochastic Gradient Descent and Dropout Algorithm

Plamena Yovcheva[1,2(✉)], Sotir Sotirov[1,2], Vanya Georgieva[1,2], Radovesta Stewart[1,2], and Maciej Krawczak[1,2]

[1] University of Prof. d-r Assen Zlatarov – Bourgas, Bourgas, Bulgaria
plamena.iovcheva@abv.bg, ssotirov@btu.bg, krawczak@ibspan.waw.pl
[2] Intelligent Systems Laboratory, Warsaw School of Information Technology, Warsaw, Poland

Abstract. Tapped delay lines and recurrent connections are two different components that are used along to design a time-delay recurrent neural network with a stochastic gradient descent algorithm in combination with a dropout method.

Keywords: dropout algorithm · generalized net · neural network · stochastic gradient descent algorithm · time-delay recurrent neural network

1 Introduction

Neural networks are an inspired programming paradigm which enables a computer to learn from observational data. Many types of Neural networks exist [9]. Most of them are represented by generalized nets [1–3].

Typically, in static neural networks such as MLP, the input information is given in the form of a vector or an array. Figure 1 shows a neural network with one hidden layer.

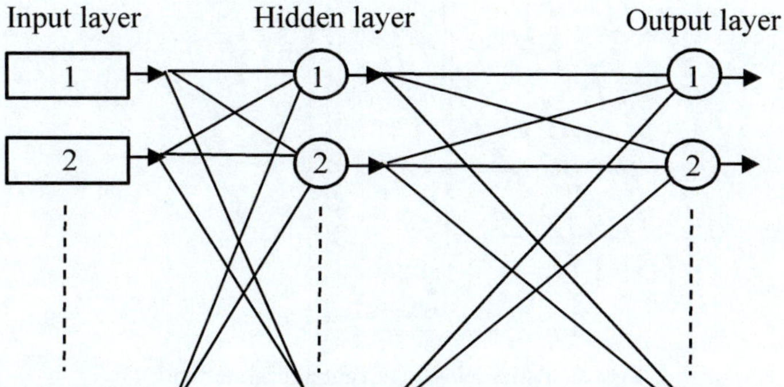

Fig. 1. MLP network with one hidden layer

S. Sotirov et al. (Eds.): BioInfoMed 2022, LNNS 658, pp. 293–302, 2023.
https://doi.org/10.1007/978-3-031-31069-0_29

There are many problems such as speech recognition where the input comprises one or more temporal signals. TDLs and recurrent connections are two different components that could be attached to static networks. TDRNN [19] using both components in the architecture of a neural network. A major feature of this architecture is that the non-linear hidden layer receives the contents of both the input time delays and the context unit, which makes it suitable for complex sequential input learning (Fig. 2).

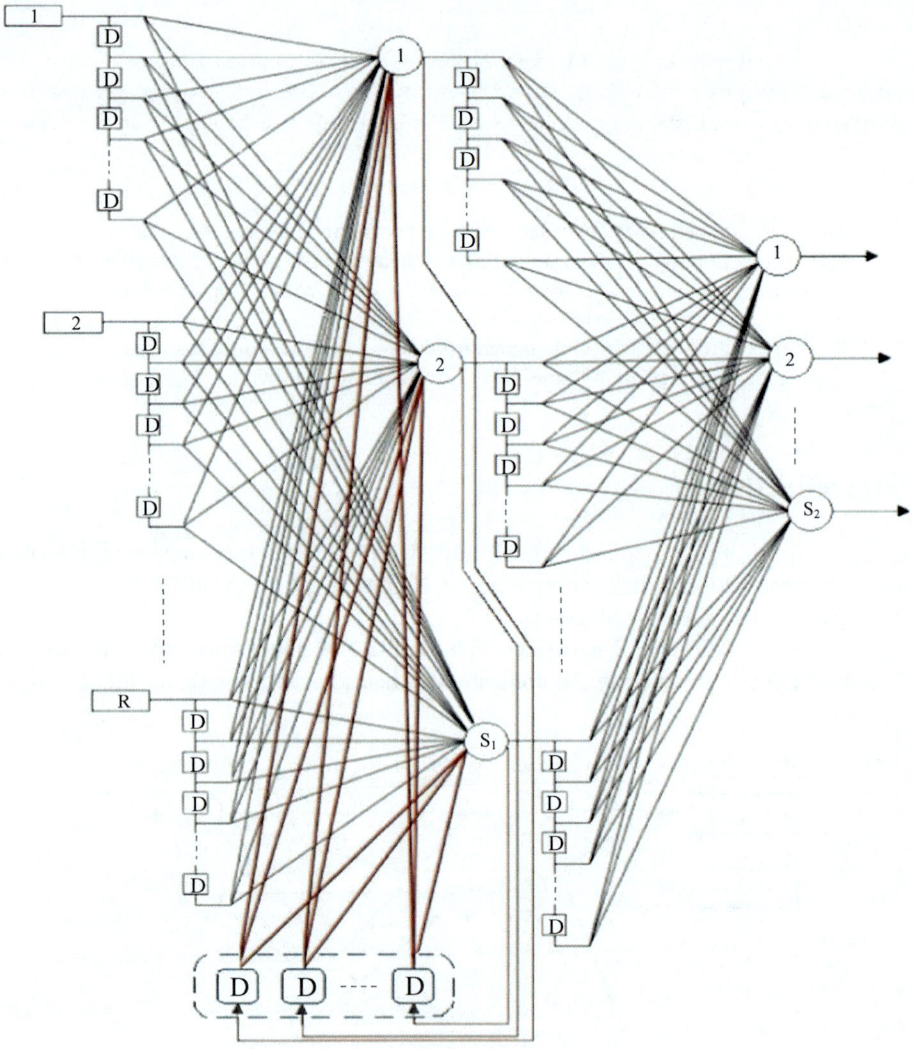

Fig. 2. Time-delay recurrent neural network

Formulation of TDRNN is expressed as follows:

$$a_{j,0}^{1}(t) = F\left(\sum_{d_1=0}^{D_1}\sum_{i=1}^{R} w_{j,i,d_1}^{1} p_{i,d_1} + 1(t) + \sum_{c=1}^{S_1} w_{j,c}^{c} a_c^{1}(t-1) + b_j^{1}\right) \quad 1 \le j \le S_1 \quad (1)$$

$$a_{k,0}^{2}(t) = G\left(\sum_{d_2=0}^{D_2}\sum_{j=1}^{S_1} w_{k,j,d_2}^{2} a_{j,d_2}^{1}(t) + b_k^{2}\right) \quad 1 \le k \le S_2 \quad (2)$$

where t denotes a discrete time, R is the number of input signals, S_1 and S_2 are the numbers of hidden and output neurons respectively, w_1 and w_2 are the weight matrices of the hidden and output layers respectively, w_C is the weight matrix of the hidden layer for the context unit, b_1 and b_2 are the bias vectors of the hidden and output layers respectively, F and G are the activation functions of the hidden and output layers respectively, D_1 and D_2 are the memory lengths of the input and hidden layers respectively, and p is the input matrix.

Gradient descent is a method of finding a local extremum (minimum or maximum) of a function by moving along the gradient. Dropout [22] works by switching off neurons in a network during training to force the remaining neurons to take on the load of the missing neurons. This is typically done randomly with a certain percentage of neurons per layer being switched off. In this paper a generalized net model of TDRNN with the Stochastic gradient descent and Dropout Algorithm is presented.

1.1 Remarks on Generalized Nets

Generalized nets (GNs) [1–3] are defined in a way that is principally different from the ways of defining the other types of Petri nets. During the time GN have become a tool for modelling parallel operating systems. Models for neural networks [4, 5, 12–15] and data mining methods [6–9] have been developed.

The first basic difference between GNs and ordinary Petri nets is the "place – transition" relation. Here the transitions are objects of a more complex nature. A transition may contain m input places and n output places where m, n $\geq$ 1.

Formally, every transition is described by a seven-tuple (Fig. 3):

$$Z = \langle L', L'', t_1, t_2, r, M, \square \rangle,$$

where:

(a) L' and L'' are finite, non-empty sets of places (the transition's input and output places, respectively). For the transition in Fig. 3 these are

$$L' = \{l'_1, l'_2, \ldots, l'_m\}$$

and

$$L'' = \{l''_1, l''_2, \ldots, l''_n\};$$

(b) t1 is the current time-moment of the transition's firing (Fig. 3);

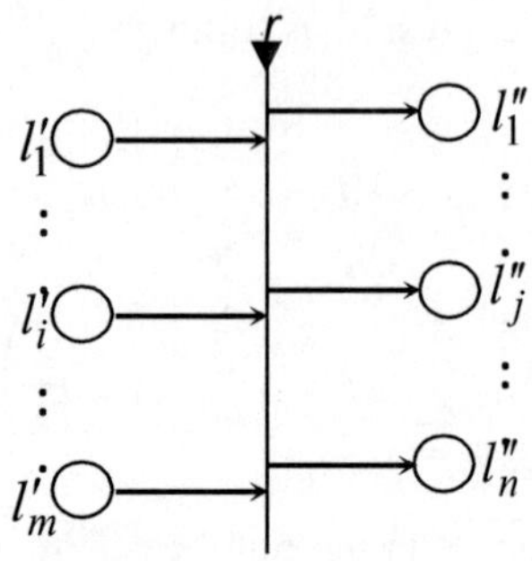

Fig. 3. A GN-transition

(c) t2 is the current value of the duration of its active state;

(d) r is the condition of the transition to determine which tokens will pass (or transfer) from the inputs to the outputs of the transition; it has the form of an Index Matrix:

$$r = \begin{array}{c|ccccc} & l''_1 & \cdots & l''_j & \cdots & l''_n \\ \hline l'_1 & & & & & \\ \vdots & \text{ri,j} & & & & \\ l'_i & \text{(ri,j – predicate)} & & & & \\ \vdots & (1 \le i \le m,\ 1 \le j \le n) & & & & \\ l'_m & & & & & \end{array},$$

ri,j is the predicate that corresponds to the i-th input and j-th output place. When its truth value is "true", a token from the i-th input place transfers to the j-th output place; otherwise, this is not possible;

(e) M is an IM of capacities of transition's arcs:

$$M = \begin{array}{c|ccccc} & l''_1 & \cdots & l''_j & \cdots & l''_n \\ \hline l'_1 & & & & & \\ \vdots & \text{mi,j} & & & & \\ l'_i & \text{(mi,j} \ge 0 \text{ – natural number)} & & & & \\ \vdots & (1 \le i \le m,\ 1 \le j \le n) & & & & \\ l'_m & & & & & \end{array};$$

(f) □ is an object of a form similar to a Boolean expression. It may contain as variables the symbols that serve as labels for a transition's input places, and □ is an expression